Materials Science and Engineering Serving Society

The 3rd Okinaga Symposium on
Materials Science and Engineering Serving Society
Employment, Environment, Health, etc.

Materials Science and Engineering Serving Society

Proceedings of the Third Okinaga Symposium on Materials Science and Engineering Serving Society, Chiba, Japan, 3-5 September, 1997

Edited by

S. Sōmiya
Teikyo University of Science and Technology
Tokyo, Japan

R.P.H.Chang
Northwestern University
Il, U.S.A.

M. Doyama
Teikyo University of Science and Technology
Tokyo, Japan

R. Roy
Pennsylvania State University
University Park, PA, U.S.A.

1998
ELSEVIER
Amsterdam - Lausanne - New York - Oxford - Shannon - Singapore - Tokyo

ELSEVIER SCIENCE B.V.
Sara Burgerhartstraat 25
P.O. Box 211, 1000 AE Amsterdam, The Netherlands

First edition 1998

Library of Congress Cataloging in Publication Data
A catalog record from the Library of Congress has been applied for.

ISBN: 0 444 82793 5

♾ The paper used in this publication meets the requirements of ANSI/NISO Z39.48-1992 (Permanence of Paper).
Printed in The Netherlands.

Preface

The goal of this symposium is to encourage collaboration in international science and engineering communities for the benefit of human kind.

The first Okinaga Symposium was held professors between the Pennsylvania State University and the Teikyo University group in Tokyo in 1993. The title was "Intelligent Engineering of Materials, What Have We Learned About Innovation".

The second one was held in Washington D.C., U.S.A. and at the Pennsylvania State University in 1994 on "Real Advances in Materials and Materials and the Engineering".

The third symposium was held in Education Center Hostel, Nippon Steel Corp. Makuhari, Chiba, Japan in September 1997 on "Materials Science and Engineering Serving Society, Employment (creating real new materials and processing), Environment (including waste management), Health (biomaterials), etc."

The symposium consisted of invited talks by experts on materials and poster presentation papers. Participants of this symposium were about 140. We hope this proceedings promotes science and engineering in materials and also human relationship among each others.

Finally we would like to express our appreciation to Professor Yoshiaki Kogure and Mr. Miyoshi Inoue, Teikyo University of Science and Technology. Without their efforts, we did not make this symposium.

S. Sōmiya (Teikyo University of Science and Technology, Editor in Chief)
R.P.H. Chang (Northwestern University)
M. Doyama (Teikyo University of Science and Technology)
R. Roy (The Pennsylvania State University)

Symposium Organizers

Prof. R.P.H. Chang (Northwestern University)
Prof. M. Doyama (Teikyo University of Science and Technology)
Prof. R. Roy (Pennsylvania State University)
Prof. S. Sōmiya (Teikyo University of Science and Technology)

International Advisory Committee Members

Prof. R.P.H. Chang
Prof. L.J. Chen
Dr. M.C. Chon
Prof. M. Doyama
Prof. O.C.C. Lin
Prof. C.N.R. Rao
Prof. R. Roy
Prof. S. Sōmiya
Prof. R. Yamamoto
Prof. D.S. Yan

Screening Committee Members : Awards for Young Scientists and Engineers

Prof. K.H. Auh (Korea)
Prof. R.P.H. Chang (U.S.A.)
Prof. L.J. Chen (Taiwan)
Dr. M.C. Chon (Korea)
Dr. Kee Hyong Kim (Korea, Chairman)
Prof. Y. Kogure (Japan, Secretary)
Prof. O.C.C. Lin (Singapore)
Prof. C.N.R. Rao (India)
Prof. D.M. Roy (U.S.A.)
Prof. D.S. Yan (China)

Acknowledgment

The Organizing Committee expresses our appreciation to the followings:

Dr. Shoichi Okinaga
Okinaga Foundation (U.S.A.)
Okinaga Foundation (Japan)
The U.S. Navy Asian Office
The U.S. Army Asian Office
Chon International Ltd.
Dr. M.C. Chon
Nippon Sheet Glass Foundation for Material Science and Engineering
Inoue Foundation for science
Nippon Steel Corp.
Nittetsu Human Development
JEOL, Inc. (Japan)
JEOL Datum Ltd.
Shimadzu Rika K.K.
Dr. Koichi Kugimiya of Matsushita Electric Industrial Co. Ltd.
Dr. Junpei Asahara and Dr. K. Nagai of Tokyo Comunication Co. Ltd.
Dr. Nobuo Kohda of K.Y. Corp.
Dr. Takeo Yokobori, Teikyo University
Dr. Yoshinosuke Yasojima, Teikyo Heisei Univresity
Dr. Yasuo Kobayashi, Teikyo University of Science and Technology
Prof. Tadashi Miyazaki, Mr. Jun Arai, Mr. Teruyuki Tanaka, Mr. Yuji Yoshida, Teikyo University
Dr. Victor Rehn, Dr. Gulian Wu, Mr. Frank H. Nagashima, ONR Asian Office
Dr. Minoru Tanaka, Mr. Takaaki Ishii, Nippon Steel Corp.
Mr. Yoshinori Yamada, Nittetsu Human Development Ltd.
Teikyo University
Teikyo Heisei University
Teikyo University of Science and Technology

Winners of Okinaga Award

Maria Cristina D'Arrigo (The Pennsylvania State University, U.S.A.)
Microwave-Hydrothermal Synthesis of Nanophase Ferrite

Shiuh-Ke Jang Jean (National Cheng kung University, Taiwan)
Characteristics of High μ_i Mn-Zn Ferrite Powders Prepared by Hydrothermal Method

Hiroshi Masuyama (University od Science & Technology, Japan)
Molecular Dynamics Simulations of Fine Particles

Yasuhiro Sugaya (Matsushita Electric Ind. Co.,Ltd.)
Insulating Metal; Al/Al_2O_3 Nano-Structure-Controlled Materials

Yoshimasa Kobayashi (Tohoku University, Japan)
Reaction Synthesis of SiC-AlN Ceramic Alloys

Hui Gu (Japan Science and Technology Corporation, Japan)
Structure, Composition and Chemical Bonding of Interfaces in Si_3N_4 Ceramics: A new Approach

Naoki Ohashi (Tokyo Institute of Technology, Japan)
Growth of Zinc Oxide Crystals and Its Ultraviolet Light Emitting Properties

Nobuaki Kitazawa (National Defense Academy, Japan)
Optical Properties of the Natural Quantum-Well System $(C_6H_5C_2H_4NH_3)_2Pb(Cl,Br,I)_4$

Jianrong Qiu (ERATO, JST-C, Japan)
Three-dimensional Optical Storage Inside Various Glasses by Using a Focused Femtosecond Pulsed Laser

Yurika Ishibashi (Tokyo Institute of Technology, Japan)
Low-Temperature Grouth of Epitaxial Barium Titanate Thin Films by Using Low-Energy Positive Oxygen Ions

Li Fei (North-West Institute of Light Industry, China)
Solidification of Colorded CaCO3 Compact by Hydrothermal Hot-pressing Technique

Kimiaki Kageyama (Kochi University, Japan)
Single Crystal Growth of Calcite in Carboxylate Solutions under Hydrothermal Conditions - Influence of pH on Growth Rate and Quality of Grown Crystal

Hisashi Hasegawa (Kochi University, Japan)
Sinterability of Stabilized Zirconia Compacts Prepared by Hydrothermal Hot-pressing

Hontei Lin (Kochi University, Japan)
Characteristics of HHP-glass Compact

Masaki Sumida (University of Tokyo, Japan)
Interface Morphological Stability of Re 123 Superconductor via Peritectic Reaction

Masayoshi Hiramoto (Matsushita Electric Industrial Co.Ltd., Japan)
Magnetic Thin Films with Anisotropic Grain-growth in the 1 or 3-dimensional Direction

Makoto Kambara (University of Tokyo, Japan)
A New Process to Obtain Re 123 Bulk Superconductors with Enhanced Superconducting Properties in Air

Junichi Tatami (Japan Society for the Promotion of Science, Japan)
New Theory on Fracture Toughness of Polycrystalline Alumina in Cosideration of Fracture Toughness of Grain and Grain Boundary

Hiromi Kitahara (Tokyo Institute of Technology, Japan)
Distorted Chalcopyrite-type Structure of $AgGaS_2$ under High Pressure

Kyeong Seok Lee (Pohang University of Science and Technology, Korea)
Domain Structure in Epitaxial PLZT Thin Film

Shunji Kikuhara (Teikyo University of Science & Technology, Japan)
Mechanical Properties and Oxidation Behavior of TiB_2 Particle Dispersion Strengthened TiAl Intermetallic Compounds

Hidetoshi Saitoh (Nagaoka University of Technology, Japan)
Initial Growth of Epitaxially Chemical-vapor-deposited Titanium Dioxide Crystallites

Minoru Takemoto (Kanagawa Institute of Technology, Japan)
Electronic Properties in $La_{n-nx}Sr_{1+nx}Mn_nO_{3n+1}$ ($x=0.4$: $n=1,2,\infty$) with Low Dimensinal Crystal Structure

Seong Ho Shin (Hanyang University , Korea)
A Study on Plasma Etching of Ta-Al Alloy Thin Films using CF_4 Gas

Jeong Ho Ryu (Hanyang University , Korea)
Growth and Characterization of LiTaO3 Single Crystals by the Floating Zone Method

CONTENTS

Materials Science and Engineering Serving Society
S. Sōmiya, R.P.H. Chang, M. Doyama and R. Roy (Editors)
1998 Elsevier Science B.V.

The 3rd Okinaga Symposium Opening Remarks

It is my great pleasure to open the 3rd Okinaga Symposium here today. The First was held in 1993 in Tokyo and the Second was in the U.S.A. in 1994.

First of all, I'd like to welcome you all to this symposium, particularly those of you who have traveled a long distance. About 40 scientists and engineers from Germany, Slovenia, China, India, Korea, Singapore, Taiwan and the U.S.A. have gathered at this symposium. In total, there are more than 100 scientists and engineers here today.

Due to the rapidly changing environment, today's topics will focus on employment, creating real new materials and processes, and the environment, including waste management and health biomaterials. It is imperative we solve these problems by creating new ways of thinking in science and technology.

One strategy to solve these problems, is to develop and promote various international opportunities. The Teikyo University Group and the Okinaga Foundation were founded with the spirit of encouraging one's best efforts for the purpose of cultivating fresh talent, nurturing people who are free from any prejudice, and who can make decisions from an international point of view. By providing global education, we strive to develop unique individuals who are creative thinkers with global perspectives. Through practial education, students are given unique education which is beneficial to society. And finally, in supplying boundless education, we reach across the boundaries of language, culture, and physical borders while striving to cultivate individuals with vision and sound characters.

Another important strategy in the quest to achieve new thinking in the area of science and technology is through collaboration and cooperation with others. Through various cooperative agreements with affiliated universities in the United States, Europe and Asia, the Teikyo University Group and the Okinaga Foundation strive to pave the way for new innovations which will benefit society. We continue to nurture strong materials research groups and take pleasure in having opportunities to support events which recognize and reward research groups whose ingenuity is ultimately advantageous to society.

On one final note, I'd like to thank the following people, without whose support, it would not have been possible to hold this symposium. The U.S. Navy, the U.S. Army, Dr. Chon of Chon International Co., Ltd., Nippon Sheet Glass Foundation of Materials Science and Engineering, Inoue Scientific Foundation, Dr. Asahara of Toyo Communication Co. Ltd., Dr. Kugimiya of Matsushita Electric Industries, Co. Ltd., Mr. Kohda of K Y Corporation, Mr. Kuraji of Shimadzu Rika Co. Ltd., and Mr. Shimizu of JEOL Inc. I wish you much success for this symposium.

Shoichi Okinaga
Chairman of the Board of Trustees
and President of Teikyo University

Materials Science and Engineering Serving Society
S. Sōmiya, R.P.H. Chang, M. Doyama and R. Roy (Editors)
1998 Elsevier Science B.V.

Keys to Successful Materials Synthesis: Goldschmidt Crystallochemical Theory and Edisonian Experimentation. No Cray Needed.

Rustum Roy

Materials Research Laboratory, The Pennsylvania State University, University Park, PA 16802 USA; Fax: 814 863 7039; e-mail: rroy@psu.edu

Abstract

The real first principles of materials synthesis, based on the work of V.M. Goldschmidt (and Linus Pauling), have proved to be very successful. By combining such principles with Edison's methodology of careful observation, precise experimentation and rapid feedback, thousands of new (inorganic) phases have been, and continue to be made (several hundred from our own laboratory), producing dozens of useful, widely commercialized materials.

In contrast, in spite of seventy years of trying, the so-called "first principles" calculations (from the quantum mechanics of cores and electrons) have utterly failed to actually predict and synthesize a single new material or add anything to the deep and reliable insights of well known chemical theories. It is clear that what students and researchers in contemporary materials synthesis need most is a thorough grounding in crystal chemistry and phase equilibria (including metastable equilibria).

1. INTRODUCTION

There are two radically different approaches to making new materials. Synthesis is the domain of the chemist. For nearly one hundred years, chemical science has succeeded admirably in creating tens of thousands of new organic (and thousands of new inorganic) materials. Some of these materials are produced in millions of lbs/year (various polymers including Teflon, superalloys, ferrites, $BaTiO_3$, etc., etc.), some have exquisitely tailored properties, as in any semiconductor circuit or GaN diode. The other approach is a more recent attempt by theoretical physicists to "calculate" the structure of a phase of a given composition — usually extremely simple compositions and hope or guess that it has certain desirable properties. We compare these approaches below.

2. THE CHEMISTS' APPROACH

This approach, different in every case, is a combination of the approaches by Edison, Sherlock Holmes and Prince Serendip. It is based on reliable data, inductive reasoning (highly perceptive observations from very different sources) and a large number of simple, cheap experiments which allow repeatability. For inorganic or ceramic materials, the 1926 textbook by V.M. Goldschmidt, entitled "Geochemische Verteilungsgesetze der Elemente"(1), is an absolutely essential and nearly sufficient guide. Recently, Ian Ross, a former President of Bell Labs, described (2), in detail, this very process as it was applied to the transistor: applications pulled basic research with new ideas translated into simple, albeit crude, experiments. The synthesis of diamond by: (a) High Pressure, High Temp; (b) CVD; (c) Flame; and, (d) Liquid Metal Low Pressure Processes, is another outstanding example. The author has relied on such approaches and probably synthesized perhaps several hundred totally new phases including dozens in each of the following areas:

- Perovskites and related structures
- ABX_4, ABX_3 and A_2BX_4 phases (see O. Muller and R. Roy (3) for a summary)
- Synthetic clays
- Synthetic zeolites
- Quenchable, high pressure phases
- New complex phosphates such as NZP and apatites.

Spectacular examples of the scientific success of Goldschmidt principles are the ability to predict and make, for example, BeF_2 in the coesite structure; or Na_2BeF_4 in the larnite structure. Ironically, the inability to predict the appearance of coesite in p-t space for the SiO_2 composition itself, illustrates the limitations of inductive reasoning (of course, no theoretical approach could predict that such a structure would appear).

The reason why subatomic theory can add so little is very simple. The typical material in a very complex assemblage of cores and electrons, the number of potential wells of more or less stable configuration is essentially infinite and many are separated by only a few kcals in the G function. Bigger computers cannot help select among these, even if all the interactions could be calculated, because the kT noise is so large.

3. THE "FIRST PRINCIPLES" OR PHYSICISTS' APPROACH

Already in 1929, Dirac, one of the fathers of quantum mechanics, wrote that "The underlying physical laws necessary...for a large part of physics and the whole of chemistry are completely known, and the difficulty is only that the exact applications of laws lead to equations much too complicated to be solvable. Yet, the total failure of the so-called "first principles" approach to predicting the structure of even the simplest composition, under given thermodynamic constraints, continues (even though computational power has increased by many orders of magnitude). Rigorous calculations can reach the H_2 molecule, but consider the quandary for such calculators with the very simplest larger "molecules" or crystalline 'solids.' Did theory predict graphite, diamond, buckyballs or glassy carbon as the structure of carbon? Another monatomic solid such as sulfur has a dozen crystalline and several noncrystalline structures with radically different properties as a function of modest p-t changes. To which of these structures will the calculations point? Water, H_2O, is a simple diatomic solid also with a half dozen crystalline and some non-crystalline forms. Yet, we have seen no papers purporting to predict or explain all this variety from physical theories.

One may take as the paradigmatic example of the current state-of-the-art of "first principles" calculations the work of Cohen (4), which claims to predict new materials and illustrates its application by the example of "C_3N_4." To start with, undergraduate crystal chemistry principles allow one to "predict" that it is conceivable that such a phase may be formed at high pressures (approx. 100kb range) and if so, one of the candidate structures would be that of Si_3N_4. The principle of parsimony demands that new theories not be constructed unless they can do better than existing theory. Si_3N_4 is modestly hard, C_3N_4 is expected to be harder. However, the concept of "harder than diamond" was the rhetorical flourish without any basis in data that attracted the attention. In the six years since the first prediction of this obvious guess, a most extraordinary phenomenon has occurred. First, it is fairly certain that no one, so far, has prepared even a few milligrams of a crystalline phase with the composition of C_3N_4. Certainly, no two lab reports agree on reproducing the composition and structure of the claimed phase. The structure of any phase(s) is obviously even less certain. The claim by Niu et al. (5), to have prepared a "harder than diamond" phase in the predicted composition and structure, has been essentially discredited and not confirmed by the original author's own later work. Yet, probably misled by the repeated allusions to "harder than diamond" in the original papers — based on no hardness data whatsoever — a worldwide flurry of research on this non-existent C_3N_4 phase began. DeVries (6) has catalogued over 400 papers on "C_3N_4," into a half dozen categories, from claimed synthesis to outright discrediting (see Table 1). Sociological analysis of this phenomenon will reveal a great deal about the status of contemporary materials synthesis. DeVries notes that a very high percentage of the authors are Chinese: why? If the authors were interested in materials that are putatively harder than diamond, why would some not have studied "$B_{22}O$," where clear evidence of scratching 111 diamond faces have been shown (7-8). DeVries points out that this futile research has employed possibly some 1200 scientists, who presumably could find no more significant problems to study. Clearly, some theorists were unaware that bulk modulus' bond length and bond strength are not, by themselves, indicators of bulk hardness, since it is the shear modulus that is critical to hardness, as Gilman has pointed out (9). It is also very well known to experimentalists that bond length, by itself, is meaningless since the C-C bond in graphite (in the plane) is shorter than in diamond.

4. POLICY ANALYSIS

Of course, sociologists and policy makers, observing such data, might infer that there is too much money chasing too few problems and that materials scientists and their supporting government agencies choose problems largely when they become scientific fads, such as superconductors, buckyballs, and diamonds. But, at least, the latter were all based on real discoveries, albeit they were soon over subscribed, possibly by an order of magnitude. "C_3N_4" (and metallic hydrogen) are unique. They are "hypothetical" goals of materials with no particular use or significance, and hence, are examples of extremely poor management of personal and national resources and strategic directions of research. Consider: one thousand papers will be written on C_3N_4 before even the existence of a material, of "no particular value," is confirmed or denied.

5. A FRUITFUL APPROACH TO CERAMIC MATERIAL SYNTHESIS

At the turn of the millennium, it is now certain that the fundamental assumption or paradigm on which American post W.W.II research is based was completely incorrect. This is the so-called "linear model."

Basic science -> applied science -> technology ->products -> prosperity

In our 1985 book, Shapley and Roy (10) first sharply pointed out the absurdity of this model. In 1997, Stokes (11), writing from the Brookings Institution (where our work was also done), completed the demolition of the linear model. Worldwide high-tech industry has completely vindicated our analysis. They have acted to abolish their immense ivory tower laboratories, built on that absurd proposition. All "real world" research in major industry has returned to the successful, time tested applications-driven, discovery-following model. Unfortunately, the governments of the U.S., now followed by Japan and W. Europe, have not yet been able to change their policies, partly because of the outcries from many of their constituents who have known no other reality.

However, very clear and simple guidelines exist for successful contemporary materials synthesis. The following algorithm has been followed in the author's lab for 50 years with some success in making new materials and new processes (sol-gel, hydrothermal, microwave) used for synthesis:

(1) The synthesis must have a specific goal or application. (A new ferroelectric, or ferrimagnet, or hard, or photoconducting, material).

(2) The classical "textbook" literature, starting with Goldschmidt, Pauling, Eitel, etc. should be thoroughly assimilated before turning to contemporary literature.

(3) The whole range of synthesis tools in p-t space, including hydrothermal, high pressure, plasma, ultrasound, microwave and laser assisted, should be available.
(4) XRD, SEM, and chemical composition characterization of all samples is essential.
(5) Repeatability must be assured, before any publication.
(6) Overclaim or exaggeration of the significance of any discovery, specifically to the public, must be absolutely abjured.

6. ACKNOWLEDGMENTS

Research in new materials synthesis is currently supported by the Office of Naval Research, Materials Science program.

7. REFERENCES

1. V.M. Goldschmidt, "Geochemische Verteilungsgesetze der Elemente," Dybwad, Oslo, Vols. 5,7,8 (1925).
2. Ian Ross, "Physics Today," p. 34 (December, 1997).
3. O. Muller and R. Roy, "The Major Ternary Structural Families," Springer-Verlag, Heidelberg (1974).
4. M.L. Cohen, et al., *Proc. Roy Soc.*, London, 334A,501 (1991), see also *Phys. Rev.*, 38B, 9488 (1988).
5. C. Niu, Y.Z. Liu and C.M. Lieber, *Science*, 261, 334 (1993).
6. R.C. DeVries, *Materials Res. Innovations*, Vol. 1:3 (1997)(in press).
7. A. Badzian, *Appl. Phys. Lett.*, 53, 2495 (1988)
8. V. Srikanth, et al., *J. Am. Cer. Soc.*, 74, 3145 (1991).
9. J.J. Gilman, *Mat. Sci. Engg.*, A209, 74 (1996).
10. D. Shapley and R. Roy, "Lost at the Frontier," ISI Press, Philadelphia (1985).
11. D.E. Stokes, "Pasteur's Quadrant Basic Science and Technological Innovation," Brookings Press, Washington D.C. (1997).

Table I (Ref. 6)
Literature Analysis of Papers on C_3N_4

1. Reviews	30
2. Theory (synthesis, structure, properties)	30
3. Claimed synthesis	65
4. Highly skeptical or negative on synthesis	15
5. C-N (other phases than C_3N_4)	28
6. Synthesis, properties of CN_x *not* C_3N_4	194
7. C-N-H	27
8. CN x in a composite (e.g., metal nitrides)	13
Total	402

Since this paper was submitted to the Journal; DeVries (personal communication) reports that another 98 papers have appeared (in less than six months) with no more success, and asymptotically diminishing significance.

Materials Science and Engineering Serving Society
S. Sōmiya, R.P.H. Chang, M. Doyama and R. Roy (Editors)
1998 Elsevier Science B.V.

Materials science education in Japan

Masao Doyama
Teikyo University of Science and Technology, Uenohara, Yamanashi 409-0193 Japan

The educational system in Japan is changing fifty years after World War II, emphasizing the upgrading graduate schools and matching the diverse demands from the public. Almost all the metallurgy departments have been changed to Materials Science with a variation of names.

1. INTRODUCTION

Japanese educational system was changed greatly after the end of World War II. There is a growing call again for the change in educational systems matched with recent social and cultural changes since the end of World War II. In this respect, the National Council on Educational Reform presented in the four reports to the Prime Minister. (1) More emphasis on individual; (2) lifelong learning system; and (3) educational systems practices matching with internationalization and information society.

2. EDUCATIONAL SYSTEM IN JAPAN

Compulsory education starts at the age of six years old in Japan. Six years in elementary schools and three years in lower secondary schools are compulsory. The enrolment rates for the elementary and lower secondary schools are both 99.99%. 64% of the children go to kindergartens for two to three years before going to elementary schools. After the compulsory school training of nine years, about 97% go to upper secondary school for three years. 46.2% of the graduates of the upper secondary schools go to universities and junior colleges. (Fig. 1)

There are 576 universities, and 598 junior colleges in Japan. The numbers of enrolment are 2.60 million students in universities. and 0.67 million students in junior colleges. The enrolment of the graduate schools in master's course and doctor's course are 56,202 and 14,345 in 1995, respectively.

Hihger education in Japan has made a rapid quantitative expansion since the middle of the 1960's. The surrounding circumsstances have been changing, such as the advancement in scientific research, internationalization and information society, decrease in the 18 year old population and increasing demands for a

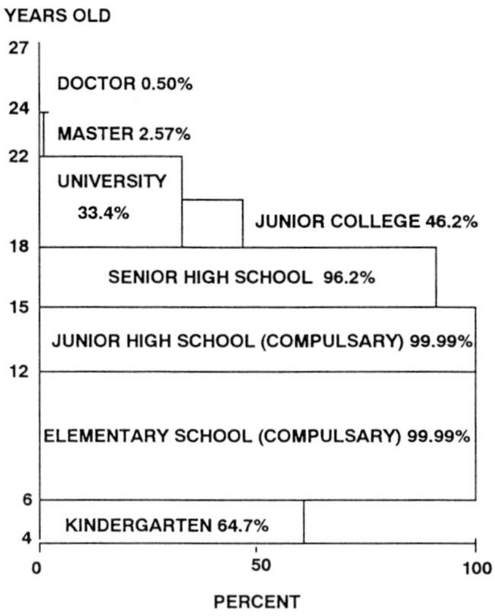

Fig. 1. Percentage of attending schools in Japan

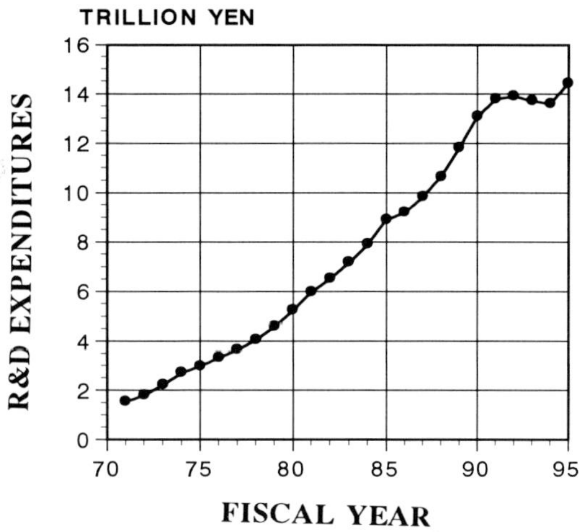

Fig. 2. Trends in R&D expenditures in in Japan

lifelong learning.

Many companies spend much time and effort to re-educate and also refresh employees in their fields. Almost all companies have educational periods right after the employment. Ten or fifteen years after graduation, employees have opportunity of refreshing their knowledge and learning newly developed materials. Recently researchers at companies enter graduate schools and can write their doctors thesis while they are working at compaanies.

3. EDUCATIONAL POLICY IN JAPAN

In the Report entitled "Planned Improvement of Higher Education for 1993 and Thereafter", the following points are emphasized. (1) strengthening the educational function; (2) promoting a world class standard of education and research; (3) the demand for lifelong learning. In order to meet world class standards of education and research, a particularly urgent task is to upgrade graduate schools both quantitatively and qualitatively. The report entitled "Quantative Improvement of Graduate school" issued by the University Council in November 1991, suggests that the total number of graduate students should be doubled by the end of the century.

The educational system of higher education in Japan changed in 1949, graduate schools in Japan emphasized only research and no masters degree was offered, nor Ph. D., but submitting and passing the theses, Doctor of Science or Doctor of Engineering can be received. This system still remains as well as attending graduate schools.

Matching with the changing and diverse demands from the public, the higher education should develop freely and in diverse ways. Standards of the framework should be as flexible as possible to universities to demonstrate their inovative and creative measure. Accepting the flexibility, self-monitoring and self evaluation system are introduced. The Ministry is encouraging the training of personnel in accordance with the need for education and research in frontier technology field giving priority to graduate school. The Ministry is promoting important basic research; cosmology, accerelators, space science, nuclear fusion, life science, prediction of earthquakes and volcanic explosion, ocean science global circumstances, observation in the south pole, and new programs. The new program includes artificial controlled functional molecular system, development of nano-scale functional materials, artificial intelligent materials, etc.

In "Stratgies for Comprehensive Promotion of Scientific Research with Prospect of the 21 Century", the Ministry is emphasizing the following points:

1) improving the scientific research environment by increasing research funding and prioritization of allocation, promoting the diversification of sources of research fundings and upgrading facilities and equipment. Total research fund spent in Japan during 1996 was 1.36×10^{13} Japanese Yen (1 U.S.Dollar is about 130 Japanese Yen as of Jan. 1, 1998), among these, 1.69×10^{12} was spend at universities, U.S.\$$1.5 \times 10^{10}$ at national laboratories and U.S.\$$1.2 \times 10^{11}$ at industries. The percentage spent by the government is lower than U.S., Germany, France and other countries. The trends in R&D expenditures in Japan is shown in Fig. 2.

2) promoting private educational institutions. The government endevors to promote univer-sities through the Japan Private School Promotion Foundation. There are two grants, one is the General Grants which depend on the numbers of teachers, clerks and students, the other is the Special Grants which emphasis in graduate shools, educational international-ization, refreshing education and local educa-tion. In 1993, the total grants amount U.S.\$$2.5 \times 10^{9}$ and 15% is granted by the Special Grants. Special Grants support the undergraduate and graduate research facilities above U.S.\$$3.7 \times 10^{5}$.

3) promoting scientific research through grants-in-aid system. The grants-in-aids was U.S. \$$7.6 \times 10^{8}$.

4) promoting universities-industry research cooperation as well as "contract research".

5) promoting international scientific exchange.

In November, 1995 the Science and Technology Basic Law, which prescribes the basic frameworks of future science and technology policy in Japan, was enacted. It calls for intensive promotion of the development of science and technology aiming at nation-building on the basis of creation of science and technology toward 21st century. In Fig. 2. the R&D expenditures increased. The financial situation in Japan is not good recently, and the government has to compress the expenditures.

4. DEPARTMENTS OF METALLURGY TO MATERIALS

The movement of the change from individual materials; metals, inorganic materials and organic materials to materials general has started during World War II, according Professor Seitz. Real change in the United States started in late 1950's. Almost all the departments of metallurgy in the United States changed to materials science and engineering with minor variations. That time in Japan the the movement to materials science followed that in the U.S.A., but no departments of metallurgy changed their names to departments of materials that time. The Japanese metallurgical industries were expanding very rapidly at that time, and the departments of metallurgy were expanding to meet the demand for graduates by industry. The production of iron was expanding very rapidly and the engineering schools in Japan were busy doubling their size in 1960's to meet the demand of the graduates. Therefore we did not pay attention to the reformation of the departments of metallurgy at that time. Only about fifteen years ago departments of metallurgy stated to change their names to departments of materials in Japan. The main reason was that Metallurgy Departments could not attract good students.

DEPARTMENT OF MATERIALS

	BE	ME	DE
NATIONAL	31	35	34
PUBLIC	3	3	3
PRIVATE	9	8	8
TOTAL	43	46	45

NUMBER OF STUDENTS RECEIVED DEGREE PER YEAR

BE	ME	DE
2,877	856	237
0.5%	2.5%	OF TOTAL
4%	10%	OF ENG.

Figure 3. Location of materials departments in Japan

Special feature of the education in Japan is the field of material (no s) science and engineering. In the chemistry field the molecular properties often reflects the properties of products. They use material science instead of materials science.

The materials Department Council consists of the Department Heads of the Materials Departments in Japan, 21 national universities (35 departments), 3 local universities (Prefec-tural or city universities, 3 departments) and 9 private universities (9 departments). The materials department group produces 2,877 bachelors, 856 masters and 237 doctors graduates per year. They correspond about 0.5% of total bachelors, 2.5% of masters and 3% of doctors in Japan. Also these are 4% of bachelors in engineering, 10% of masters in engineering in Japan. The location of materials departments in Japan is shown in Fig. 3.

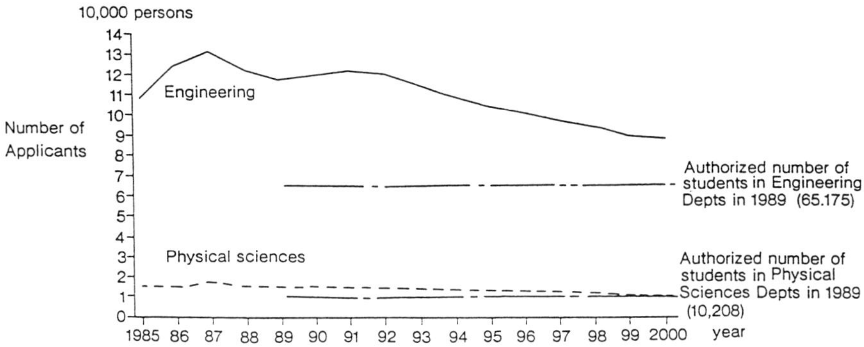

Figure 4. Estimated number of applicants to departments of engineering and physical sciences. (White paper on Science and Technology)

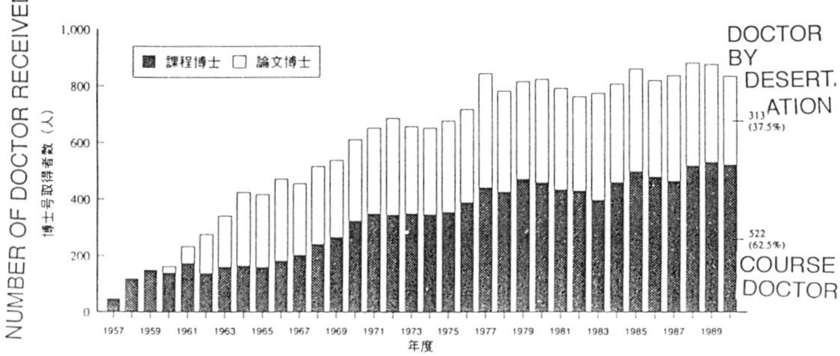

Figure 5. Trends in numbers of doctors in engineering and physical sciences by desertation and by courses. (White papers on Science and technology)

5. UNDERGRADUATE AND GRADUATE SCHOOLS IN MATERIALS

The 18 years old population in Japan had a peak a few years ago and now the 18 years old population is decreasing. Higher rate of 18 years old people can go to universities.

The graduate schools in Japan emphasize more in research than classes. They try to expand their knowledge through their research field. Professors do not need to support students. Government agency loans money to students, and they have to return with or without paying interest depending upon their grade. (If they work at government gencies or universities they do not return their loans.) Students who go to work at companies can write their thesis and take examinations and they can obtain doctors degree without paying tuitions and, instead they get salaries.(Fig. This discourage students to go into graduate school. The numbers of doctors in science by desertation and by course are shown in Fig. 4.

Unfortunately less people apply to enter science courses recently (Fig. 5). A fewer students go into materials field. Materials departments cannot attract many excellent students.

5. CONCLUSION

Education system in Japan has been much changed after the World War II and fifty years passed after the chang. Now the reformation is starting in emphasizing the garduate school and matching with the diverse demands of the society. Many departments have changed to Materials Departments but the change in real contents takes time.

Acknowledgement

"Japanese Government Policies in Education, Science and Culture 1992" published by the Ministry of Eduction, Science and Culture and "White Paper on Education" (1993) written in Japanese published by the Ministry of Education, Science, Cultue and Sports, "White Paper on Science and Technology" edited by Science and Technology Agency were quite useful for preparing the paper.

Materials Science and Engineering Serving Society
S. Sōmiya, R.P.H. Chang, M. Doyama and R. Roy (Editors)

New Processes as Drivers for Revolutions in Materials Research: Microwave Sintering and Multiple Pulsed Laser Forming

Rustum Roy

Materials Research Laboratory, The Pennsylvania State University, University Park, PA 16802 USA; Fax: 814 863 7039; e-mail rroy@psu.edu

Abstract

The authors' research has previously been associated with three major **processing** developments: glass ceramics, hydrothermal, and sol-gel, all originating 50 ±1 years ago. The history of the transfer of the science from technology will be described to illustrate the inherent intellectual conservatism of basic scientists and the opportunities that it creates for **applications-driven** research.

The first example will describe the development of microwave sintering of white ceramics and its subsequent very successful application to the WC industry. Spectacular savings of at least one order of magnitude in cycle time and energy cost, coupled to capital savings and improved properties will likely accelerate this process's entry into the marketplace. It is probable that little understanding research will have even been started before most applications have been developed beyond the patent stage, and many products are out.

The second case is the use of multiple pulsed laser fluxes for radically new materials creation. In 1994, Mistry and Turchan showed that high quality diamonds could be grown on WC substrates at a rate of 1μ/sec in an open system at 1 atm. pressure, when exposed to a programmed sequence of separate, simultaneous laser **pulses** from excimer, YAG and CO_2 lasers. They subsequently have shown how the same system could cause major changes in the mechanical properties of commercial tools made of the whole range of steels and super alloys. Phase changes resulting in optimum combination of hardness and toughness were engineered down to depths of hundreds of microns to mms.

1. INTRODUCTION

For fifty years from the July 1945 test of nuclear fission in Alamogordo, New Mexico, to the defeat of the Supercollider in October 1993, the world experienced the *Golden Age of Science*. Based on a gross overestimate that the H-bomb had won the War (the Japanese were negotiating for months before) the importance of the role of science, especially physics and nuclear physics in national life remained exaggerated and distorted for fifty years. Along with *this* distortion came a much more serious and egregious error—the linear model of S and T. Basically it is the absurd concept that science leads to technology. In several earlier papers I provide the precise quotes, references, etc. to demonstrate how nonsensical this claim is (1-3). This theory completely dominated and distorted national science policy in the U.S. as developed in my book *Lost at the Frontier* (4). It appears that in Japan the same error is being repeated, whereas in England and Germany, the changeover to the accurate theory is moving ahead. This is: that technology leads to science, or that applications guide and drive basic research.

That is the empirical fact in materials science. Serendipitous discovery of new materials (teflon, polyethylene, W-filaments, glass-ceramics, high Tc superconductors, etc.) often focus, create and shape entire areas of science. This paper will deal with the role of two new *processes* which have since 1995 led to some extraordinary advances in materials research: microwave absorption/ scattering and pulsed lasers.

Learnings from sol-gel, hydrothermal and glass-ceramics processes:

The author was one of the key players in the original science of three processes, during the period 1947-1949. This history of their subsequent development will make the point that science rarely if ever drives technology: the opposite is the case.

1. Glass ceramics. From early 1947, I worked closely with my colleague F.A. Hummel who discovered and patented the negative thermal expansion of eucryptite, $Li_2O.Al_2O_3.2SiO_2$. In my Ph.D. thesis I worked out the phase diagram for the system Li_2O-Al_2O_3-$2SiO_2$ including eucryptite, spodumene and petalite (5). Corning eventually acquired the Penn State composition of matter patent, and D. Stookey remarkably productive inventor based on a double screndipitous discovery, developed the glass-ceramic process. By the late 1950's Corningware the first major commercial glass-ceramic product line had become a major worldwide success.

Our work on both the crystal chemistry, synthesis and phase diagrams of the many relevant systems containing Li_2O, $Al_2O_3SiO_2$ and especially with TiO_2 was the only relevant science. The claims by Stookey in his patents and papers that crystals of TiO_2 were the key nuclei on which the small silicate phases crystallized were shown to be wrong. The metastable immiscibility of glasses containing several percent of TiO_2 was experimentally shown by us to be the fundamental mechanism underlying all glass ceramic nucleation and crystallization [see Mazurin and Porai-koshits (6, p. 11), for role of paper by Roy (7) in this discovery]. Long after the applications (Corningware) had reached the market, the intense academic interest in glass ceramics began, illustrating our thesis that **technology leads to science.**

The sol-gel and hydrothermal processes prove the thesis from the converse viewpoint. The basic science of the sol-gel process was essentially completely discovered and developed between 1948 and 1952. Hundreds to thousands of ceramic compositions were made into chemically ultrahomogeneous, nanoscale powders, and melted to glasses or crystallized to ceramics. The 1956 paper (8) reviewing our extensive sol-gel research, to that date, later became the first Citation Classic ever is the field of ceramics. Yet the ceramic science community had shown essentially zero interest in the field for over ten years. Not until the late sixties, indeed only when, first the technologies of nuclear fuel pellets, then abrasive grains, etc. appeared on the horizon was the interest in the **science of** the sol-gel process advanced. In both the cases above, it was the pull of a realized technology that (re)-created the interest in the science although in both cases the technology no doubt utilized our early science to some extent.

I have presented this rather long introduction in this symposium address to make the audience aware that it is up to them to make every society, every agency and every leader or legislator aware of the real successes of materials research, and **how most are applications driven,** and why/how it is neglected in most policy making. We turn now to describing two current examples again from our own laboratories, of new processes which will have as major an effect on the entire field of ceramics as the processes described above.

2. MICROWAVE PROCESSING: STELLAR EXAMPLE OF APPLICATIONS-DRIVEN SCIENCE

Table 1 from Reference 9 summarizes the position of microwave processing in the world of materials R/D *policy*. Given the new realities of the globalized economy and the change in the goals of modern materials technology [viz. cheaper, faster greener, Roy (3)]. The excitement in the field stems first from the fact that this research is 100% *relevant* to so many technologies. Second, this clear example of applications-driven work demands a high degree of interdisciplinarity, from electrical engineers, to plasma physicists and materials scientists. Table 2 lists the status of microwave science and technology today, all as preamble to the rest of the paper which describes the Penn State MRL research in the field over the last dozen years. This is described in reference (9) which also gives a detailed literature analysis, etc.

Table 1
Direct Materials Processing Using Microwaves

Global Technology Context
- The globalization of the world economy
- The role of technology and the new drivers in industry
- New goals for materials technologies: faster, cheaper, greener
- Microwave processing as perfect example of a relevant advance

Microwave processing is
- Applications driven
- Highly interdisciplinary
- Vigorous because so much is ***not*** understood

Table 2
Microwave Science and Technology

- Most lively and exciting part of ceramics research, because so much is NOT known
- "Uncertainty Principle": One cannot design apparatus easily because of field-workpiece interaction
- Interactions of microwave fields on surface (atomic) layers not understood
- Mass of unexplainable "microwave effects" on:
  * reaction products
  * reaction paths, mechanisms
  * reaction kinetics

The Present Innovation

The current paper presents a series of innovations which bear on this major new discovery and other key questions in the whole field. First, it extends the use of microwave sintering to new families which contain not only white ceramics and "semiconducting ceramics" but major composite families containing substantial amounts of metals. It obtains finer microstructures unattainable by conventional heating. Second, it specifically utilizes the device of reducing the variable valent oxides to trigger very high reaction rates in useful dielectric compositions which otherwise heat up very slowly, and unambiguously demonstrates an example of a "microwave effect" in new reaction paths in important ceramic materials synthesis.

Experimental

The main apparatuses used in our studies are reported in Ref 9. They normally operate in air, but they have been adapted to use an atmosphere of H_2 or N_2+H_2 as the ambient to protect samples from oxidation. Most significantly this one family of apparatus allows for *continuous* processing of small and large (up to 1m. long 10 cm. diameter) objects.

Microwave sintering of WC+Co composites

Hard metal composites due to their unique combination of hardness, toughness and strength, especially the tungsten carbide (WC) based composites, are universally used for cutting tools and drills: machining of wear resistant metals, grinding, and mining, geothermal, oil, and gas drilling. Such materials are required to possess highly abrasive and wear resistance properties. Conventional methods for sintering WC with Co as a binder phase involve high temperature and lengthy sintering cycles of the order of one day. These are very energy intensive. Consequently, the production cost of these materials is quite high. Moreover, in the conventional sintering method, the carbide specimen must be subjected to very high temperatures (up to 1500 °C) for long periods in order to achieve a high degree of densification/sintering. Such conditions unfortunately favor undesirable WC grain growth in the presence of Co-liquids. Consequently, the mechanical strength and hardness of the tool is diminished. It is a well known fact that finer microstructures provide superior mechanical properties and longer life of the product. Researchers have found that certain materials can be incorporated into the WC matrix to help prevent grain growth during the sintering process. Additives such as titanium carbide (TiC), vanadium carbide (VC) and tantalum carbide (TaC) have been used for this purpose. Unfortunately such additives deleteriously affect the mechanical properties of the tools, and add substantially to the overall cost of the tool. Use of nanometer size WC/Co powders has also been proposed to make these tools. But, the high cost of nanopowders, large scale production, and the difficulty of maintaining the nanostructure during conventional sintering are the main issues which have made this proposal unattractive to commercial manufacturers. In 1991, J. P. Cheng in a Ph.D. thesis [16] first showed that WC+Co composites could be sintered in a microwave apparatus. Gerdes and Willert-Porada[17] also reported the sintering of similar WC objects from normal size powders.

Processing and Characterization of WC/Co Test Samples

Since most of the drilling industry uses tungsten carbide with cobalt content varying between 6 to 12%, we selected initial compositions of WC with 6 and 12% Co only. Several WC test samples of 1/4-inch diameter were pressed uniaxially at a pressure of 300 MPa with 2 wt% PVA as a lubricant. The green densities were 7.78 and 7.28 gm/cc for two WC/Co compositions respectively. The samples were sintered in a single mode microwave furnace in ultra pure H_2 atmosphere at temperatures ranging from 1150° to 1350 °C and soaking time varying from 1 to 60 minutes. Near theoretical densities were achieved in 10-minute sintering time in a microwave field. It was found that for 6% Co the optimum temperature was 1350 °C to achieve highest density of about 14.8 gm/cc in 10 minutes, for 12% Co the corresponding conditions were 1300 °C and 10 minutes, and the density was about 14.6 gm/cc. These sintering conditions are much lower than normally used in a conventional heating of WC samples.

The samples were also characterized for their hardness by Vickers and Rockwell A methods. The highest hardness values achieved were about 93 Rockwell A for the 6% Co samples microwave sintered at 1350 °C for 10 minutes, and about 91 for the 12% Co sample processed in microwave at 1250 °C. These values are higher than normally achieved by conventional methods. Figure 1 shows the change in hardness vs. temperature and time respectively of the microwave processed samples. These samples were characterized for their microstructures by SEM, and the data showed very fine average grain size of <1μm.

WC commercial green bodies containing 12% and 6% Co were sintered in our microwave system. Such greenbodies were obtained from commercial suppliers and the properties of the sintered tools were clearly equal to or superior--especially in superfine microstructure--to conventional heating (see below). Most significantly, we showed (9) that with the use of microwave techniques for sintering WC-based tools using nanopowders, the grain growth can be reduced to a minimum, and this without adding any grain-growth inhibitors, and a very fine initial grain size can be retained and materials with higher strength can be synthesized (see Figure 2).

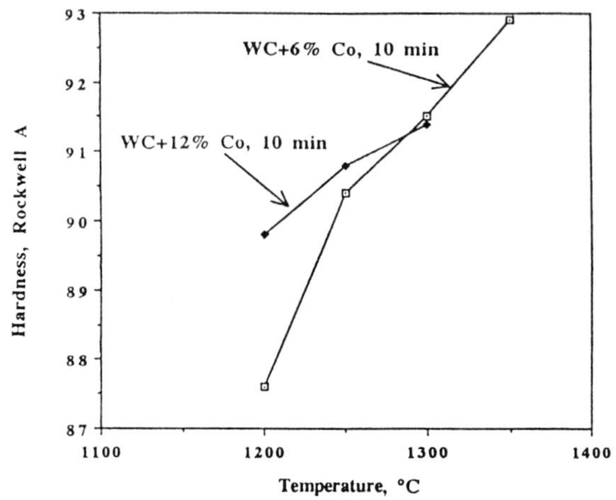

Figure 1. Hardness vs. sintering temperature of the microwave processes WV test samples.

Large numbers of WC (6-12%) samples were successfully sintered to "full" density in our microwave apparatus at 1250°-1320°C in 10-30 minutes. These conditions are radically lower than conventional sintering conditions. Fully dense WC+Co commercial samples can be obtained in 10 minutes and at a temperature of 1250°C (Co: 10%) and 1300°C (Co: 6%). WC+Co commercial cutting tools of a wide variety of special shapes and sizes have been sintered in the microwave field with faithful retention of size and shape. Sintering multi-samples batches in a total cycle time of 70 minutes including the dewaxing step has been routinely achieved. Figure 3 shows some of the commercial WC+Co tools processed in our microwave system. The larger bodies in each pair are in the green state and smaller ones are after microwave sintering. Table 3 provides a comparison between microwave and conventionally processed WC/Co cermets.

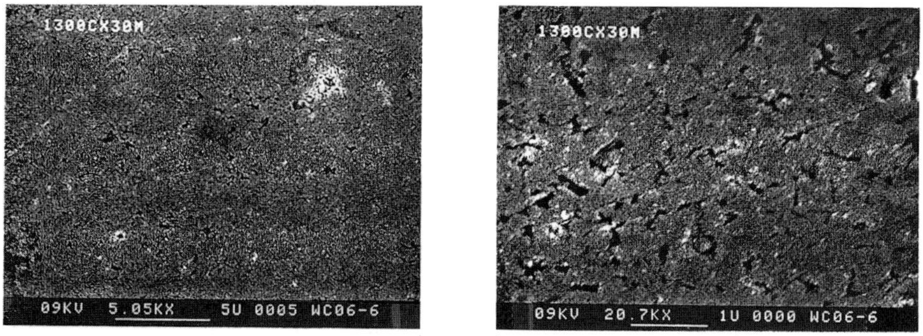

Figure 2. Microstructures of sintered WC-Co made by microwave sintering of nanopowder; note absence of any large Co-lakes.

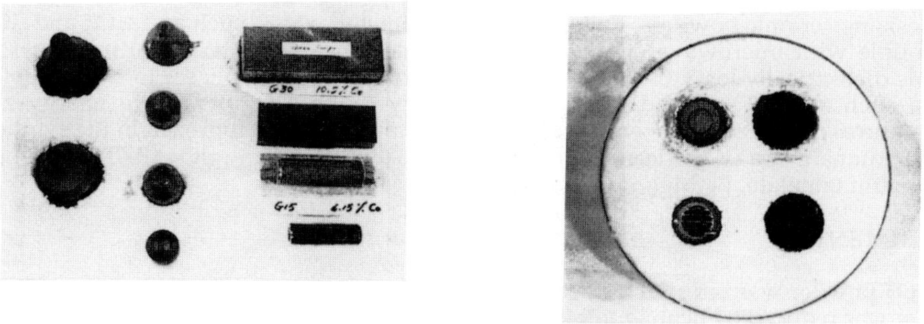

Figure 3. Typical commercial cutting tools of WC sintered in 15 minutes to full density and superior properties. In each case green bodies are the larger samples of the same shape. There is no distortion or slumping in any sample.

Table 3
Comparison of microwave and conventional processes for sintering WC-Co composite

Sintering Temperature (°C)	1300	1450
Total Cycle Time	90 min	12-24 hrs
Sintering Time (minutes)	10	60
Density (% T.D.)	99.8	99.7
Average Grain Size (μm)	0.6	2
Bending Strength (MPa)	1800	1700
Hardness (Rockwell A)	93	91

To summarize, it has been shown that microwave processing can sinter WC-Co tool bits and related parts with better properties, in about one-tenth the cycle time required by conventional means, and most significantly that it can help retain the benefits of a very fine microstructure when starting with very fine powders. Details have been provided in papers by Cheng et al. (10) and Gigl et al. (11).

Utilization of Reduced Oxide Powder in Materials Synthesis

The application of microwave processing to ferroelectric materials is not new, but in the past, little advantage had been found in speed. In the second innovation reported in this paper, we now have been able to use the concept of pre-reduction of phases such as TiO_2 and Ta_2O_5 to give a highly absorptive material in preparing batches for sintering various commercial electroceramic powders.

The main idea is to create a defect structure to make microwave coupling more efficient at room temperature. We have found that by using pre-reduced TiO_2 and Ta_2O_5, $BaTiO_3$, "PZT" and "BMT" can be synthesized at astonishingly low temperatures, between 300°C and 700°C in 5-12 minutes. Conventional methods for the synthesis of these phases require temperatures in the range of 900° to 1400°C and several hours of soaking time. This is for the first time, that *reduced oxide precursors and microwave processing* have been purposively used in synthesizing ceramic powders. Pure stoichiometric metal oxides, such as Ta_2O_5 andTiO_2, do not couple with microwave energy efficiently unless heated to temperatures where they become dielectrically lossy (>1000°C). By partially reducing these phases to oxygen defective states, such as Ta_2O_{5-x} and TiO_{2-x}, their ability to absorb microwave energy at low temperatures can be radically enhanced. We have employed this approach in the synthesis of barium titanate ($BaTiO_3$), lead zirconate titanate, $Pb(Zr_{0.52}Ti_{0.48})O_3$,(PZT), and barium magnesium tantalate, $Ba(Mg_{0.33}Ta_{0.67})O_3$, (BMT).

EXPERIMENTAL

TiO_2 powder was reduced by heating at 1150°C for 2 hours in a tube furnace and Ta_2O_5 powder was reduced by heating to 1450°C for 2 hours in a graphite furnace in forming gas (5% H_2-95%N_2). The precursors used in microwave synthesis by this route were $BaCO_3$ and TiO_{2-x} (for $BaTiO_3$), $PbCO_3$, ZrO_2 and TiO_{2-x} (for PZT) and $Ba(OH)_2.xH_2O$, MgO and Ta_2O_{5-x} (for BMT). The mixed powders were pressed into pellets and inserted in the 2.45 GHz ovens with appropriate insulation.

$BaTiO_3$

The X-ray diffraction data for the microwave products of $BaTiO_3$ powders using TiO_{2-x}, showed most surprisingly the formation of the hexagonal-$BaTiO_3$ phase. This appeared at a nominal pellet temperature of 300°C with no soak time. The formation of the tetragonal-$BaTiO_3$ phase increased with soak time and the reaction was completed at 700°C. The total time necessary for $BaTiO_3$ synthesis via this route was less than 12 minutes. Conventional synthesis of $BaTiO_3$ using the same reactant mixture occurs above 1300°C and always proceeds via formation of Ba_2TiO_4 first. The phase diagram for $BaTiO_3$ shows that the high temperature hexagonal phase of $BaTiO_3$ is only stable above 1400°C. The presence of this phase at 300°C and its disappearance by 700°C, with Ba_2TiO_4 *never* appearing indicates radically different reaction pathways from conventionally processed material. This reaction path difference is a convincing demonstration of some "microwave effect."

PZT

One of the major problems with conventional processing of PZT is the vaporization of PbO, which starts near 750°C. Apart from the obvious environmental hazard this poses, it also results in an incomplete reaction leaving large amounts of unreacted ZrO_2. Stoichiometric PZT mixtures, i.e. with no excess PbO, were reacted in a microwave field using TiO_{2-x} powder in the starting material. Analysis of the XRD patterns of the microwave processed PZT showed that the reaction was nearly complete at 600°C with only trace amounts of PbO and ZrO_2 detectable, indicating virtually complete reaction before PbO volatilization. By 900°C trace amounts of ZrO_2 remained, which may have been caused by imperfect mixing. Again, the *cubic* PZT phase was formed first at 600°C, followed by the nucleation and increase of the tetragonal PZT phase. By 900°C a 50-50 mixture of the cubic and tetragonal phases was evident. The total time required to achieve synthesis at 600°C in the microwave field was less than 8 minutes.

BMT

$Ba_3MgTa_2O_9$ the highest melting oxide known has become one of the most useful GHz frequency dielectrics. The reaction mixture using Ta_2O_{5-x} coupled and heated very rapidly until the temperatures reached 550°-600°C. At that temperature the sample ceased to heat, indicating weak microwave absorption. XRD analysis showed that while BMT was the major phase which had already formed, significant amounts of the reactants still remained. The failure to heat was probably due to the fact that pure BMT is a microwave transparent material exhibiting very low dielectric loss and as it formed the total absorption decreased. Moreover apparently the Ta_2O_{5-x} was oxidized during the partial reaction which may be caused by the tendency of the BaO to absorb excess oxygen and transform topotactically to

BaO_2 especially since we started with the hydroxide. As a result, microwave heating of the reaction mixture above 600°C required the use of a secondary coupler (SiC). By 1100°C the reaction was complete. By conventional heating, BMT synthesis requires at least 3 hours at 1400°C, and often much higher.

These three titanates and tantalates are among the most important ferroic materials in industry. It has been shown that using TiO_{2-x} or Ta_2O_{5-x} each material can be synthesized in far less time and at *much* lower temperatures than can be achieved by either microwave synthesis from the stoichiometric powders (TiO_2 or Ta_2O_{5-x}), or by conventional heating methods. Temperatures such as 300°C and 700°C for forming binary titanates are totally unprecedented and unexplainable by classical sintering models. Details have been provided by Mathis et al. (12).

3. MULTIPLE PULSED LASER PROCESSING: DIAMOND SYNTHESIS

At the Applied Diamond Conference held at NIST, Washington, DC in 1995, the QQC Corporation presented a poster session in which they announced a new method by which layers of diamonds were synthesized in the laboratory ambient, on a WC substrate shrouded by CO_2 and N_2, at a rate of about 1μm per *second* (i.e., 100-1000 faster than CVD). The entire diamond community reacted with firm disbelief although beautiful unimpeachable evidence in morphology, Raman spectra, etc., was presented. The present senior author visited P. Mistry at QQC head quarters and observed most remarkable achievements in metal and ceramic processing. Our laboratory (13,14) has now examined the diamond coated WC tools in some detail. Mistry et al. have also provided details on the process (15). Figure 4 shows the schematic of the set up and the key features are four **pulsed** lasers operating simultaneously including a high power (100W and up) 248nm excimer, YAG and CO_2 lasers. Figure 5 shows the typical diamond coating found on a WC substrate. Figure 6 shows our (PSU-MRL) section through the WC and diamond film and Figure 7 shows the remarkable feature of loss of Co from the upper 8μ of the carbide and hence a graded structure between it and the diamond. This accounts for the exceptionally strong bonds between the substrate and diamond, giving QQC tools superb performance data (15). Thus the applications of a radically new process are already established long before *any* science has been conceived for—indeed the diamond community in a remarkable demonstration of disbelief of the new for a year questioned the veracity of the inventors.

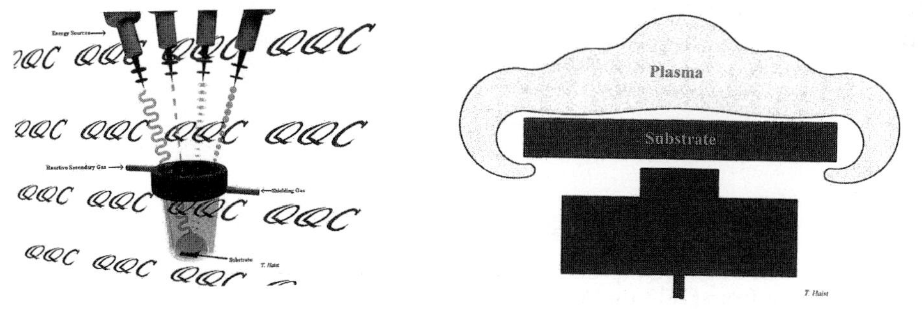

(a) (b)

Figure 4. (a) Schematic shown of pulsed laser array fired simultaneously, and (b) plasma surrounding workpiece.

But this application is not the only one for this remarkable process. The QQC company and the inventor, P. Mistry, have sold millions of dollars of metal parts which have been processed in a variety of ways. One example will have to suffice here. A section through a 3" diameter, 3" long steel punch for shaping aluminum soft drink and beer cans is shown in Figure 7. The steel punch has been transformed for the upper several mms to a martensitic phase, with same hard precipitate, appearing. A double (2-3μ) thin layer of martensite and one of titanium carbide the added cladding material. The data from our work have been summarized by Roy et al. (16). The performance of this clad, transformed tool was >10^2 times better than the similar tools which had not been so processed.

Figure 5. Typical QQC-process diamond coating (40μ) on a WC cutting tool.

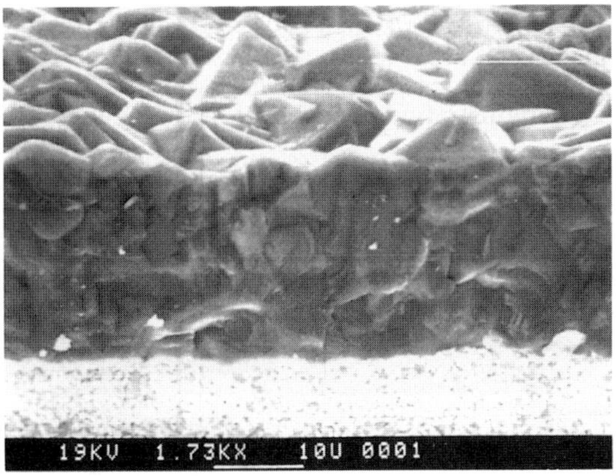

Figure 6. Section through QQC process diamond coating on WC. Microstructure of diamond is radically different from CVD, and it contains 1μm spheres of W.

The scientific understanding of the process developed empirically by Mistry et al. has only just been initiated, yet some two dozen radically new applications are already on the market. No better example of this thesis can be found. Genuine examples proven in the marketplace are first rejected as impossible by the very same community, which will, two-three years later, claim to have done the basic science which "led" to the self same technology.

4. ACKNOWLEDGMENT

We acknowledge the support of Dr. R. Pohanka of the Office of Naval Research for research on radically new methods of materials synthesis.

Figure 7. CoKd map (left) for section showing 8μm-Co depletion layer near the diamond interface. with WQ.

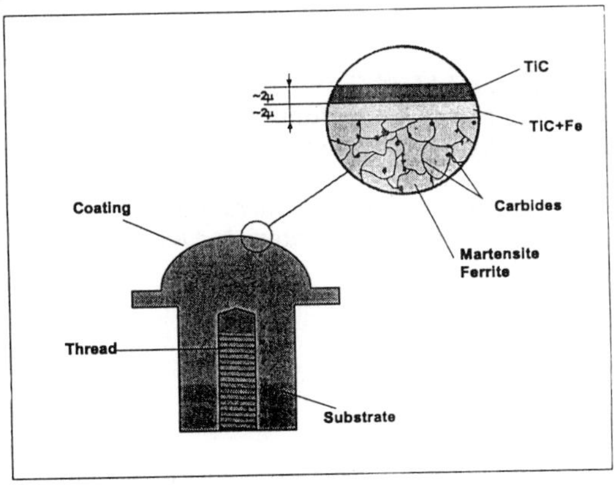

Figure 8. Section through QQC laser processed steel punch, which has been transformed to martensites and clad with a 2μ layer of TiC -- all in the solid state retaining net shape.

5. REFERENCES

1. "The Nature and Nurture of Technological Health: Ceramics and Civilization, High-Technology Ceramics—Past, Present, and Future," W.D. Kingery (ed.), American Ceramic Society 11, 351-370 (1987).

2. "Telestic (Purposeful) and Atelestic (Curiosity-Driven) Research in the Post Supercollider World," *Hybrid Microelectronic Materials*, pp. 1-14 (1996).
3. "The New Materials/Market Development Cycle: A Perspective on the Historical Changes," Proceedings of the Professor Anthony Kelly Symposium, University of Surrey, Surrey, England 1-29 (September 22, 1994).
4. *Lost at the Frontier—U.S. Science and Technology Policy Adrift* (with Deborah Shapley), Philadelphia: ISI Press, 223 pp. (1984).
5. R. Roy and E.F. Osborn, "The System Lithium Metasilicate-Spodumene-Silica," J. Am. Chem. Soc. 71, 2086-2096 (1949).
6. O. Mazurin and E.A. Porui Koshits, "Phase Separation in Glass," Elsevier, New York (1984), see. p. 11.
7. (a) R. Roy, "Metastable Liquid Immiscibility and Subsolidus Nucleation," J. Am. Ceram. Soc. 43, 670-671 (1960); see also (b) R. Roy, "Phase Equilibria and the Crystallization of Glass. Symp. on Nucleation and Crystallization in Glasses and Melts," Ed. Reser et al., Am. Ceram. Soc. 39-45 (1962).
8. R. Roy, "Aids in Hydrothermal Experimentation: II. Methods of Making Mixtures for Both 'Dry' and 'Wet' Phase Equilibrium Studies," J. Am. Ceram. Soc. 49:145-146 (1956).
9. R. Roy, D. Agrawal, J.P. Cheng and M. Mathis, in *Microwaves: Theory and Application in Materials Processing IV*, "Microwave Processing: Triumph of Applications-Driven Science in WC-Composites and Ferroic Titanates," American Ceramic Socieity (1997).
10. J. P. Cheng, D. K. Agrawal, S. Komarneni, M. Mathis and R. Roy, "Microwave Processing of WC-Co Composites and Ferroic Titanates," *Mat. Res. Innovat.* 1, 44-52 (1997).
11. P. Gigl, D., E. Breval, J. Cheng, D. K. Agrawal, R. Roy, "Structure Properties of Microwave sintered Cemented Tungsten Carbide Materials," Presented at the First World Congress on Microwave Processing, January 5-9, Orlando, Florida (1997).
12. M. D. Mathis, R. Roy, D. K. Agrawal, R. H. Plovnick, and R. M. Hutcheon, "Microwave Synthesis of Ceramic Powders Using Reduced or NonStoichoimetric Precursors," Presented at the First World Congress on Microwave Processing, January 5-9, Orlando, Florida (1997).
13. A. Badzian, R. Roy, P. Mistry and M. C. Turchan in A. Paoletti and A. Tucciarone (eds.), The Physics of Diamond, IOS Press, Amsterdam (1997, in press).
14. A. Badzian, B.L. Weiss, R. Roy, T. Badzian, W. Drawl, P. Mistry and M.C. Turchan, *Diamond Relat. Mater.* (1997, in press).
15. P. Mistry, M. Turchan, S. Liu, G. Granse, T. Baurmann and M. Shara, "A Revolutionary Diamond Synthesis Technique: The QQC Materials Deposition Process," *Innovations in Materials Research* 1(2):193-207 (1996).
16. P. Mistry, M. Turchan, R. Roy, S. Gedevanishvili and E. Breval (in press, 1997).

Materials Science and Engineering Serving Society
S. Sōmiya, R.P.H. Chang, M. Doyama and R. Roy (Editors)

Green Materials: Industrial Ecology

R. A. Laudise

Bell Laboratories, Lucent Technologies, 700 Mountain Avenue, Murray Hill, New Jersey 07974, USA

In the 19th Century (and to this day in some developing countries) belching smokestacks were viewed as a sign of progress. Industrialization, indeed, should be viewed as a sign of progress. It leads almost inevitably to higher living standards, greater longevity and better quality of life. In the 19th Century when industrial production was relatively small the earth could be viewed as a virtually infinite sink capable of absorbing waste products endlessly without harm. By about the middle of the 20th Century events such as the near extinction of the California Condor (described by Rachel Carson in the book "Silent Spring") mercury poisoning in Japan, Rhine River pollution, etc., taught us that industrial activities could adversely effect not just local ecosystems but the entire planet (e.g. the ozone hole, global warming). The response was largely regulation of emissions: clean up the smokestack, the effluent pipe, the landfill. The number of "command and control" laws grew exponentially. The cost to economic efficiency began to exceed the benefits.

Lately we have begun to understand the implications, for pollution and its prevention, of population growth and increased industrialization in the 3rd World and in the vibrant Pacific Rim economies. By the middle of the 21st Century population will increase at least 50% and probably then stabilize as most countries have relatively high standards of living. Industrial production will probably increase at least 5 fold as the newly industrialized nations come on stream. If present end-of-pipe pollution control is about as effective as now (or increases somewhat) ~10X more pollution will occur. Large parts of the world will suffer ecodisasters like that seen already in some parts of the former Communist Block.

What to do? How will we achieve a transition to sustainability? Many people see the solution in redesigning production processes so as to be initially green, choosing materials so as to avoid problems and designing for reuse and recycling. The view that industrial products and processes should be compatible with the natural ecosystem is often called Industrial Ecology. A case can be made that this approach is not an option – it is a *necessity* for the 21st Century.

Industrial Ecology (Design For Environment/Green Design):

- Uses benign processes to make benign products
- Gets beyond the smokestack, waste pipe and landfill
- Minimizes wastes and turns them into by-products
- Turns junk into recycled scrap
- In the Quality sense
  - – Gets out of the response mode
  - – Does it right the first time
  - – Realizes that good industrial ecology can pay for itself
- Will be a source of competitive advantage

Total quality management in production and manufacture emphasizes that if we can find a metric and continuously improve it and if we look for the root causes of a problem we are well on the way to process and product improvement. These principles apply equally well to green design. The root cause of our environmental problems is bad materials and bad process choices – we will need to make changes at square one. Environmental metrics are not hard to come by: toxic content in effluent gas and waste water, ratio of solid waste/useful product, etc. The goal should be zero discharge. Quality and environmental responsibility pays. For ab initio improvement at the corporate level:

- Industrial hygiene costs are reduced
- End-of-pipe costs are reduced
- Landfill costs are reduced
- Torts, superfund, compliance and regulation costs are reduced
- Process, materials and energy efficiency are improved
- Product differentiation results in improved customer sales
- International competitiveness is improved
- Brand value is enhanced
- Employee satisfaction is improved
- Community acceptance of factories is improved.

Already "eco labels" have been shown to enhance sales, while international corporations are learning that their products must comply with the most environmentally strict market in which they want to compete. Laws are being changed so as to prevent pollution costs being externalized from the producer to the larger society. For instance "take back" laws ensure design for recycling rather than design for dumping with landfill paid by the taxpayer.

Total lifecycle must be considered in product design. For instance, improved fuel efficiency even at the cost of using 'environmentally troublesome materials" may lead to an even cleaner product for total life cycle for automobiles and airplanes.

Another important green trend is dematerialization. A few chips of Si do what a whole mainframe computer did a decade ago. Less materials are used, less energy is consumed and therefore the product is surely greener.

Another important green trend is functional substitution. Because of the materials revolution – Si chips, optical fibers, etc. – the cost of communication is so low that communication can replace commuting. 4-5 million people in the U.S. now commute to work. Enablers of telecommuting include:

- Conference Calls
- Speakerphones
- Voice Mail
- Fax
- E-mail
- Low Bandwidth TV
- Internet
- Home Pages
- Cellular / Laptops
- PC/Smart Terminals
- Search Engines / Intelligent Agents

It is interesting to point out that in the real world even in "long term" research laboratories more effort is still spent in end-of-pipe problems by far than on "ab initio clean" problems. Perhaps the largest single group of industrial ecology researchers is that funded by the Lucent Technologies Foundation (originally AT&T Foundation) Industrial Ecology Fellowship program. Since 1993 these fellowships have been awarded to senior researchers in universities for research on generically clean processing, green materials substitution, and policy changes to incent green behavior including law, accounting, economics and management. Curricula development including new course design has also been included in these fellowships. Beginning in 1997 the U.S. National Science Foundation has become co-sponsor of the Fellowships with a doubling in number and a new title "NSF/Lucent Technologies Industrial Ecology Research Fellowships". Further information or an application can be obtained at the NSF web site. The Fellowships (each for two years at $50,000/year) awarded in 1997 include:

1. Green Modularity: Ecology and Product Retirement, Prof. John Kaplan, University of Alabama – The research seeks to identify how to incorporate into the process of mechanical design the concept of modular units that can be disassembled and recycled or reused at the end of a product's life.
2. Nonflowing Chemical Processing for Thin Film Manufacturing, Prof. Carol McConica, Colorado State University – A study of a technique to eliminate the unnecessary flow of chemicals during processing of thin films manufactured for integrated circuits.
3. Modeling the Industrial Ecosystem, Prof. Thomas E. Graedel, Yale University – A mathematical investigation of the flows of materials, energy and capital in an industrial econosystem, analogous to models of biological ecosystems.
4. Life Cycle Assessment/Design Methodology for Reinforced Commingled Recycled Plastic Lumber (CRPL), Prof. John Engblom, Florida Institute of Technology – Analysis of structural form and potential life cycle of synthetic lumber, and development of a method to design this product with computer technology.
5. Removal of Organic Films and Contaminants from Surfaces Using Elevated Pressure, Elevated Temperature Water, Prof. Dennis W. Hess, Georgia

Institute of Technology – Development of a new, ecologically superior approach to cleaning and conditioning surfaces of contaminants during fabrication of semiconductors.

6. Life Cycle Design of Building Integrated Photovoltaic Systems, Prof. Gregory A. Keoleian, University of Michigan – A comprehensive assessment of the full energy, environmental and economic benefits of equipping a building with solar energy technology compared to conventional systems for generating electricity.
7. Formalization of Disassembly Process to Support Serviceability and End-of-Product Life Options, Prof. Allada Venkat, University of Missouri-Rolla – An in-depth analysis of a formal process to design products that can be disassembled into reusable, recyclable or benignly disposable components at the end of a product's useful life.
8. In-Situ Generation of Hazardous Reactants for Chemical Synthesis. Prof. George W. Roberts, North Carolina State University – A study of a new, environmentally benign approach to the problem of using reactive and hazardous chemicals in manufacturing: generating and consuming hazardous chemicals in a single reactor.
9. Models and Instruction for Life Cycle Material Content Decisions, Prof. Julie Ann Stuart, Ohio State University – A study of a polymer degradation approach to designing products with reusable materials. The study emphasizes integration of environmentally conscious engineering lessons into the student engineering curriculum.
10. Environmental and New Technology Adoption in the U.S. Steel Industry, Prof. Timothy Considine, Pennsylvania State University – Integrated assessment of new recycling technology available to the steel industry: life cycle analysis of steel making and pollutant by-products; scrap recycling; investment in new steel-making technologies; the impact and tradeoffs of pollution control policies and environmental goals.
11. Models and Tools for End-of-Life Product Management, Prof. Winston Knight, University of Rhode Island – Analysis and development of analytical computer tools to be used during early product design to evaluate the ease of disassembly and recycling of products, with the goal of more efficient methods of bulk recycling.
12. Pollution Control in Fuel Cell Applications Using Ceramic Candle Filter for Cleaner Power Generation, Prof. Ziaul Huque, Prairie View A&M University – The researcher seeks to develop a method of removing particulates and toxic pollutants from coal gas, for use in fuel cells. Research will include developing and testing a working prototype of a ceramic filter.
13. Environmental Process Diagnosis and Improvement Tool, Prof. René van Berkel, IVAM Environmental Research, Univ. of Amsterdam (Netherlands) – Research at the University of Amsterdam involves developing computer software to help identify, evaluate and implement environmental improvements to manufacturing production processes and products.

14. An Environmental Justice Template of the Industrial Ecology Paradigm, Prof. Ely A. Dorsey, Howard University – Researchers will examine fairness issues involved in how sites are chosen for new manufacturing facilities and pollution cleanup.
15. High Solids, Water Based Tape Casting, Profs. Daniel J. Shanefield and W. Roger Cannon, Rutgers University – Aims to eliminate volatile solvents used to prepare ceramic tape - a large source of environmental impact in the ceramic industry.
16. "Point Source" Metals Recovery Via Spouted Bed Electrolytic Reactors, Prof. Joseph M. Calo, Brown University – A new design for electrolytic recovery of metals used in electroplating, with strong prospects for reducing costs.
17. Focusing on Ecology Within Concurrent Engineering Framework, Prof. Susan E. Carlson Skalak, University of Virginia – Researchers propose developing software to help incorporate environmental factors into engineering planning practices.
18. Design for Disassembly to Support Virtual De-manufacturing, Prof. Rajit Gadh. University of Wisconsin-Madison – Developing computer-aided product design tools which address issues of ease of manufacturing and ease of disassembly and recycling.

Ab initio environmental impact will almost certainly become a central activity in materials science and engineering in the 21st Century. Companies and countries which ignore this truth will very likely pay both severe economic and quality of life consequences.

Many texts and papers review industrial ecology concepts. An excellent general instruction is Graedel and Allenby [1].

REFERENCES

1. T. E. Graedel and B. R. Allenby, Industrial Ecology, Prentice Hall, Englewood Cliffs, New Jersey, USA, 1995.

Materials Science and Engineering Serving Society
S. Sōmiya, R.P.H. Chang, M. Doyama and R. Roy (Editors)

Ceramics for the Next Millennium

Koichi Kugimiya
R&D Division, Matsushita Electric Ind. Co., Ltd.
3-1-1 Yagumonakamachi, Moriguchi, Osaka 570, Japan
Tel: 81-6-906-9029 Fax: 81-6-903-0996

Abstract

Ceramics are human friendly in its nature as the history shows for the past billions of years. For the last hundred years, the electro-ceramics have contributed not only to improve quality of our life and to enrich our culture and daily livings, but also to realize ever developing intellectual demands. But now we face difficulties and pollution is a big issue.

For the future, ceramics can contribute far more to the world ecology and give us intellectual satisfaction. Adding to the time consuming big projects lasting a few decades, we should share fair amount of research efforts to E-Tech's; more familiar technologies around our living. Each E-Tech is small but adds up to a huge energy saving so as to reduce power plants into a half. E-Tech's should have more attention from the academic world and supports from governments. The most essential is an efficient information system connecting researchers worldwide.

1. THE PAST

Ceramics and electro-magnetics have been embedded in earth from its birth. Thunderstorms were madly rough on rocks crusted over the earth dynamo. The earth, a great natural ceramics, was borne 4.6 Million years ago. Since then, ceramics have been nature friendly. After a while, human interaction with ceramics became intimate. The interactions, together with electromagnetism are listed in Table 1. As well known, magnetite and apatite are embedded into many living organism. Magnetite as a bio-compass is found in some microbes, brains of pigeons, bees and salmon, and apatite as bones and teeth:basic frame of our body.

The oldest stone tool, the first man processed natural rock ceramics, was found in Ethiopia at the age of Homo Rudolfensis. And further down the years, earth wares or fired clay pots which were a few steps closer to recent ceramics were found at Fukui cave in Japan. Even mirrors were made of stones in the old days. Finely polished flat obsidian(volcanic glass) were found inside of tomb in Turkey, 6000 years BC. It is believed that the stone mirror was used by necromancers for incantation in the old days(Fig.1). Improved earth wares, porcelain or china, became abundant world wide and found everywhere in the world. They have enriched the culture and were very important to every day life. They were so valuable and popular that a manufacturing process of porcelain was painted on a wall of a pyramid, 2000BC(Fig. 2).

Static electricity and magnetism were known in the east and west, old days. Their mysterious attraction was expressed as they attract each other like a mother loves her son. It was even astonishing to learn that in Iraq they already used batteries, the famous Baghdad cell, to electroplate gold(Fig. 3). It was more than 2000 years before Gasner in Germany and Okui in Japan had introduced dry cells for the convenient application. Okui applied dry cells to electric clocks, his major business.

Technology and science had slowly picked up and contributed to civilization but for many hundred of years, alchemists were general images of scientists. Recent modernization have improved the image of poor scientists a little better, but they(including us) still have some nature of Jekyll and Hyde.
Along with modernization, ceramics were used widely as mainly insulators in the beginning. Many companies were founded about 1900 as we see many companies inaugurate 100 anniversary these days. After radios and TV's were invented, widespread application and requirements were piled up, and so-called electro-ceramics grew up today to a very big industry; 2 billion dollars in Japan alone. Adapting many requests, ceramics had improved their own properties and now have almost any property, electromagnetic, optic and many others. Even they can be superconductive. Ceramics have been used various way in our social life and contributed to enrich our daily living, materials/intellectual satisfaction, world wide as other science and technology have done.

2. THE PRESENT

Since application of ceramics are very wide and spread over almost every sectors of electric and electronic devices, many kind of atomic elements are used to meet requirements. Then, some contains elements harmful to environment such as lead. They are sometimes major components and sometimes additives of very small amount. Although ceramics parts are very tiny, the total amount become very large due to the wide application. They are found in every house and office, every car, and anywhere in the world. This wide spread distribution poses laborious difficulty to collect and properly dispose of devices containing ceramics.

Another big environmental issue is a big energy saving without inconvenience of general public in a good comfortable life. Actually in Japan(Fig. 4), the shares of transportation and office/housing sectors, each occupy 1/4, are kept increasing due to frequent cargo and passenger car traffics, and also due to a rich living style, while energy consumption of industry is declining because of the great efforts by the industries and the government. In the growing private sector, big energy eaters in Japan(Fig. 5) are refrigerators, lighting/TV, and cooling. They all add up to almost 2/3 of the grand total although many of us are not aware of the fact. Among them, recent popularity of air conditioning poses a big headache to power plants, because they all turn on at almost the same time of hot afternoons of a few hot days. To meet for this very short seasonal requirement, extra power plants were built and almost a half are idle for the most of years. If a peak power consumption could be somehow averaged out, it will make the power plant operation more efficiently and cut down about 1/3 of power plants in Japan alone.

Also note that 10% efficiency improvement of refrigerators and lighting or displays will result in the saving of several nucleus power plants. It was also reported that vender machines of canned drinks in Japan, 2 million units in total, amount to 1 nucleus power plant of 1 million KW. Another less-known fact is that waiting or idling powers like clock, phone, TV and VTR account for close to 10% of the total house power consumption. Although each consumes very small power, all of them are kept working 24 hours a day continuously and add up to a awful big share. These are facts just in Japan but apply everywhere in the world in the same manner.

To save big energy with an advantage of ceramics, there are actually many big government projects worldwide competing better efficiency on transportation and power plants; ceramics for gas turbine engine, fuel cell, nucleus plants and superconductors. They all are really huge projects involving many high-tech's and need big facilities. More over they were on the projects for a few decades and will need awful more decades to complete and to really prove the effectiveness. We do need some effective projects: much closer to our living and much sooner to realize.

3. THE FUTURE

Ceramists should serve to our society more than before, effectively dedicating our technologies, knowledge, systems and networks we have built up for many years. For many years to come, the most important issues for general public or for their living are still mental/intellectual satisfaction, health and ecology. For ecology, there are many challenging subjects to us; saving energy, creating clean energy, reducing unfavorable elements. Reduction of unfavorable elements from ceramic materials have been carried out in variety of materials for some years now. Surely we have to continue these efforts. Since there are too many ceramic materials, important is the effective distribution and exchange of data and information throughout the world to excite and accelerate developments. Some new means for the swift information exchange are necessary, fi, an internet data system.

The issues related energy saving and creation in a public domains may be the most important subjects left for the world cooperation. As stated above, the share of the office/housing sector is steadily increasing and the efforts to reduce energy is now left mostly up to the public efforts. A scattered and small efforts are now under way and thus only a limited success is expected. I believe that now is the right time for Ceramists to act and to take a lead for improvement.

The peak cut or shift technologies are not well noticed but truly are one of the most important technologies as described above. The technologies do not save or create energy but total energy efficiency is surprisingly improved. Required are technologies somehow energy be sifted from day to night, winter to summer. Efficient thermo-accumulators(heat cans) and better heat insulators are of course essential. To cut down the peak electric demand at summer time, small solar cell power generation is also very effective. Each houses can be self supported or partially supported by Sun, specially hot summer days.

Recent developments of phosphor or fluorescent materials are promising for even better materials. For a long time, the improvement of fluorescent materials have been almost forgotten. We surely need intensive and focused efforts to improve the efficiency of the materials.

New aspects are reported on so-called Honda-Fujishima(H-F) effect which generates Hydrogen by UV irradiation. Hydrogen will be effective to store energy for a long time, from summer to winter, for instance. It is recently proved that some chemical reactions are accelerated and chemical modulation of Anatase made wider light spectra effective for better efficiency. Now in Japan some products using the H-F effect are introduced.

Thermoelectric materials like Peltier devices are making little progress for many years too. We badly need new materials to recover wasted heat effectively.

At a glance, contribution of those materials above appear to be very small and tedious but really is as effective as big projects like gas turbine projects of billion dollars worth. Furthermore those materials are familiar and can be studies without heavy installment. Call it E(Easy, Economical, Exciting and Excellent)-tech. Anybody with appropriate knowledge can join the race and make contributions. As being proved at the fever period of the high temperature superconductive oxides research 10 years ago, extraordinary progress will be made in a short period, once number of excited and eager researchers(professional or amateur) would join. Quick worldwide distribution of information, technologies and new findings, big or small, is essential. With internet, we should now perform far better than before.

4. SUMMERY

The demands for energy are enormously growing and carbon oxide accumulation is accelerating even faster. We may exceed the unknown limit of control and start the nature to wildly evolve carbon oxides stored as many forms in nature. Many may not aware of, but our precious time is running out.

Ceramists have been granted for the eminent contribution to our society for long time and now they are expected to do more in future in many ways.

We should extend our research efforts not only to a few big projects associated with big money but also to, in a sensible share, E-Tech's improving familiar materials like Phosphor/Fluorescent /Optic Materials, Heat shifters/Cooling/Heating/Heat Insulator, and Clean Energy /Low thermal energy/Solar Energy. The E-Tech's may not appear brilliant scientific issues to scientists but should yield the most effective energy saving. As greater a number of devoted researchers becomes and better communication is, sooner the astonishing and exciting results should come out. Of course, to built the efficient system, appropriate government supports and funding are very helpful.

Table 1 Brief History of Ceramics and Electromagnetic

4.6 billion	Birth of Earth --- Formation of Natural Ceramics
3.8 billion	Birth of living cells
	Inclusion of Magnetite and Apatite
4 million	Apelike Australopithecus(Lucy and her kin)
2.5 million	Homo
	Stone tools/Ethiopia
100 thousand	Homo Sapience Sapience
15 thousand	Earth wares/Fukui cave, Japan
	Clay furnaces/Vestonice, Moravia
6 thousand	Stone mirrors/Catal Fuick, Turkey
7 century	Magnetism: Magnetite and iron attraction
	Static electricity: Amber attraction/Thales, Greek amber=electrum(Persian)/electrica(Greek)
4 century	Compass/China
3 century	Baghdad battery/Baghdad, Iraq
B.C.	
A.D.	Ages of alchemists
1752	Thunder and kite/Franklin
1776	Electro-medicare:Elekittel/Gennai Hiraga, Japan
1882	Edison bulb company
1885	Dry Cell/Gasner(Germany), Okui(Japan) for clock
1887	Tokyo bulb and power company
1900+-	Ceramic companies for electric application
1918	Foundation of Matsushita
1920	Radio broadcasting/USA
1930	Kato-Takei Ferrite
1935	Titania:beauty powder
1936	TV broadcasting/UK
	Developments of Electronics and Electro-ceramics
1986	High-Temperature-Superconductive Ceramics Fever
1997	The 3rd Okinaga Symposium on Materials and Engineering Serving Society

Theban period : BC 3000～BC 1700

Fig. 2 Egyptian Process

Fig. 1 Stone Mirror

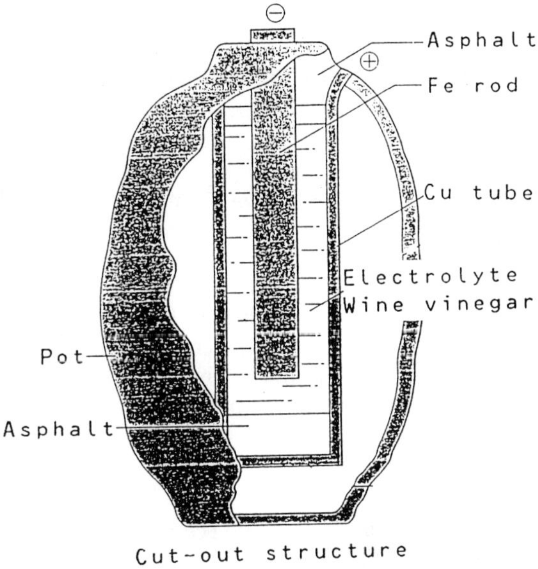

Fig. 3 Baghdad Cell

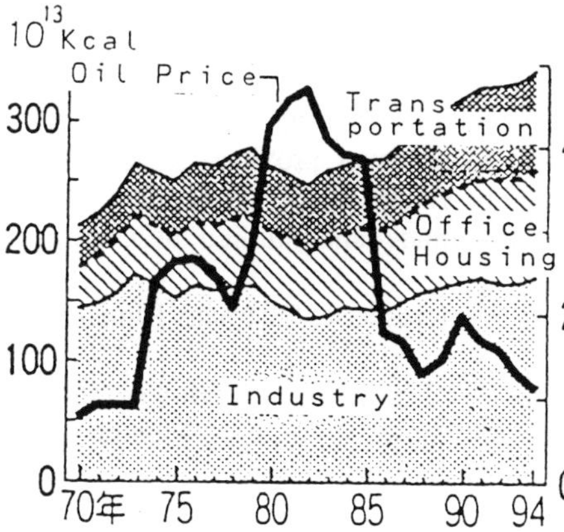

Fig. 4 Energy Consumption in Japan

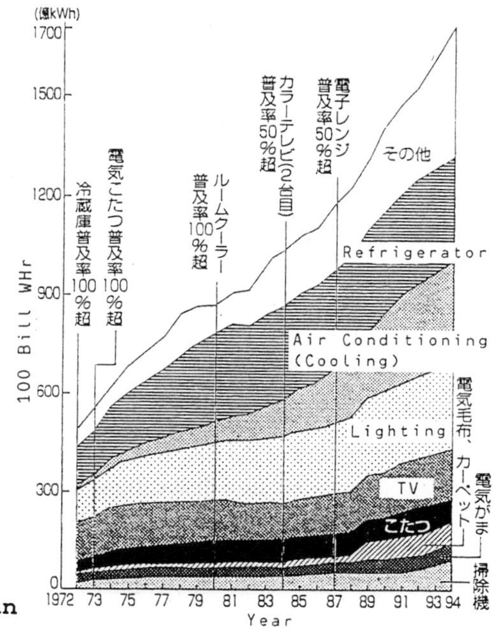

Fig. 5 Electric Power Consumption Home Use in Japan

Materials Science and Engineering Serving Society
S. Sōmiya, R.P.H. Chang, M. Doyama and R. Roy (Editors)

The way to the resolution of 0.1 nm

Kazuo Ito

JEOL Ltd., 1-2 Musasino 3-chome Akishima Tokyo 196-5885 Japan

Abstract

The resolution of electron microscope has been improved from 10 nm at the beginning stage to the practical limit of 0.1 nm during the past 60 years. The progress up to 1 nm was made almost entirely in Germany from the year of 1937 to 1954. However, 1 nm to 0.1 nm was not so easily made, and took more than 40 years until JEOL produced and installed a 0.1 nm resolution microscope in 1994 at the Max-Planck Institute in Stuttgart , Germany. The history of the progress of the resolution and the technical background are described briefly.

1. PROGRESS OF RESOLUTION

I would like to show the progress of resolution taking some examples in the field of crystallography. In table I and table II, chronological events related to the development of electron microscope are shown, world wide and in JEOL, respectively. The first epoch making achievement appeared in J. W. Menter's paper reporting that the crystal lattice planes were resolved with Siemens Elmiscope I. At that time, the resolution was 1 nm.

Table I Historical events showing the development of electron microscope

1937-1939:	The first electron microscope, exceeding the resolution of optical microscope, was made in Siemens Company, Germany, in 1937 by B. von Borries and E. Ruska. [1]
1939:	The first scanning microscope was made by M. von Ardenne. [2]
1948:	The theoretical resolution limit and the image contrast of electron microscope were studied by O. Scherzer. He concluded that by use of an electron microscope with resolution of 0.2nm, which was provided to be possible theoretically, even carbon atoms would be visible. [3]
1948-1953:	The scanning electron microscope was made and studied at Cambridge University. The resolution reached 10 nm.
1954:	The first high resolution (1 nm) electron microscope, "Siemens Elmiskop 1," was introduced to the market.
1956:	For the first time, the crystal lattice planes with spacing 1.2 nm were resolved and also dislocations in crystal were directly observed by J. W. Menter. [4]
1968:	A.V. Crew announced the scanning transmission microscope of 0.5nm resolution and obtained the single atom image with it. [5]
1982:	The scanning tunneling microscope was invented by G. Binnig, et. al. [6]

Table II JEOL LTD. Historical events showing the development of the resolution

1947: The first electron microscope DA-1 was produced and commercialized, the resolution of which was not better than 10 nm. (Fig.4)

1971: JEM-100B, which made the world standard of high resolution far better than 1 nm, was introduced to the market after continuous efforts of improvement and remodeling for more than 20 years. One of the produced electron microscope was installed at J.M. Cowley's laboratory in Arizona State University, Phoenix, U.S.A., where S. Iijima displayed beautiful crystal structure images. [7]

1973: JEM- 100C was used by H. Hashimoto for imaging single Th atoms for the first time.[8]

1980: JEM-200CX, with 200kV accelerating voltage and a resolution better than 0.24nm, was installed at Sir Hirsch's laboratory in Oxford, U.K., where they imaged dislocations in Si, making an epoch. [9]

1984: JEM-4000EX, with 400kV accelerating voltage and 0.19nm resolution, was made and installed at H. Hashimoto's laboratory in Oosaka University, Japan.

1994: The highest resolution electron microscope, 0.1 nm resolution and 1,250kV accelerating voltage, was installed at Max-Planck-Institut, Stuttgart, Germany[10] (Fig .5)

Since JEM-100B, JEOL has been thought to be the manufacturer of high resolution electron microscopes. Many efforts have been made in JEOL Ltd. to improve the resolution, and more details of crystal structure have been revealed. In obtaining all of the following examples, our experiences were fully utilized. In 1971. S. Iijima reported that his electron microscope image coincided exactly with the structure determining by X-ray diffraction. (Fig.1) In 1973, H. Hashimoto showed the image of isolated Th atoms taken with the transmission electron microscope. (Fig.2)

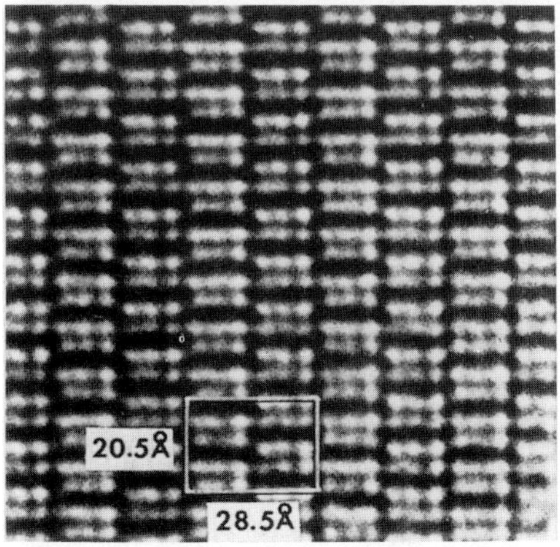

Fig.1 Structure image of $Ti_2Nb_{10}O_{29}$.

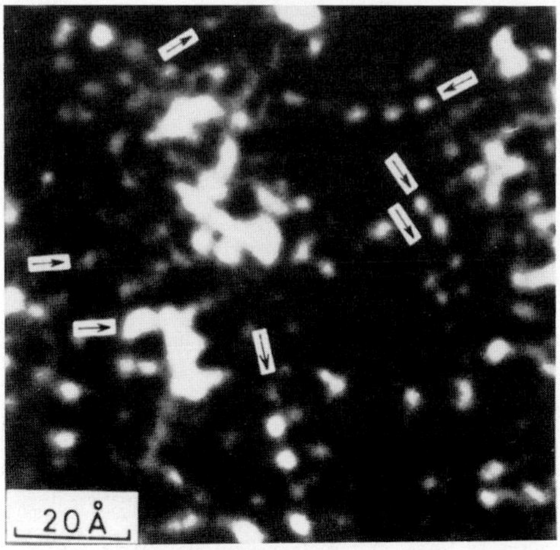

Fig.2 Dark field image of Th atoms.

Then, in 1980 Sir, Hirsch et al. of Oxford University, U.K. obtained the image of beautiful dislocations of Si with JEOL's 0.24nm resolution-200kV electron microscope.(Fig.3)

We continued efforts to improve the resolution, and in 1984 we made the 0.19nm resolution- 400kV electron microscope, which was installed at H. Hashimoto's laboratory in Osaka University, Japan. Finally, in 1994, JEOL produced and installed the 0.1 nm resolution-1250kV electron microscope at Max-Planck Institute, Stuttgart, Germany. This 0.1 nm is considered to be the practical limit of resolution at this moment.

For comparison we show the first electron microscope which we produced in 1947 and the 1250kV electron microscope in Fig.4 and Fig.5, respectively.

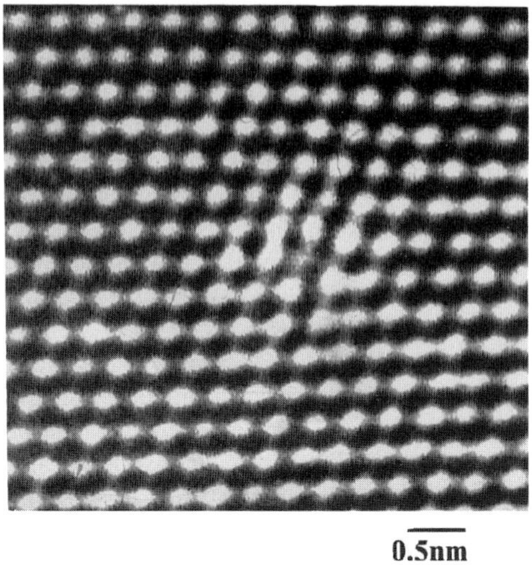

Fig.3 60° dislocation in Si

Fig.4 The first TEM DA-1

Fig.5 Ultra-high voltage electron microscope JEM-ARM-1250

2. RESOLUTION LIMIT OF ELECTRON MICROSCOPE

Since we do not have any effective mean to correct the spherical aberration of objective lens, we have to rely on a high accelerating voltage to improve the resolution. Calculation shows the practical limit of resolution to be as follows:

100kV : 0.3 nm, 200kV : 0, 19 nm, 400kV : 0. 15 nm, 1250kV : 0. 10 nm.

Here, I would like to show the influence of resolution on the image. Fig.6 is the computer simulation images of β-Si_3N_4 for three electron microscopes with 0.19, 0.15 and 0.1 nm resolutions. The image improvement by resolution progress is obvious.

FIg.7 shows comparison of the image of Cu grain boundary taken with a 0.17 nm resolution-400kV electron microscope and that taken with 0.1 nm resolution-1250kV electron microscope at Max-Planck Institute. The left photograph exhibits twins clearly, but the irregular contrast of dots, which must correspond to the atoms location, is observed. However, the right photograph shows the clear arrangement of dots without irregular contrast as a result of 0.1 nm resolution.

JEM-2010 d = 0.19 nm	**JEM-4010 d = 0.15 nm**	**JEM-ARM1250 d = 0.1 nm**
Cs = 0.5 mm Cc = 1.1 mm	Cs = 0.7 mm Cc = 1.6 mm	Cs = 1.6 mm Cc = 4.1 mm
δf = 42 nm	δf = 40 nm	δf = 41 nm

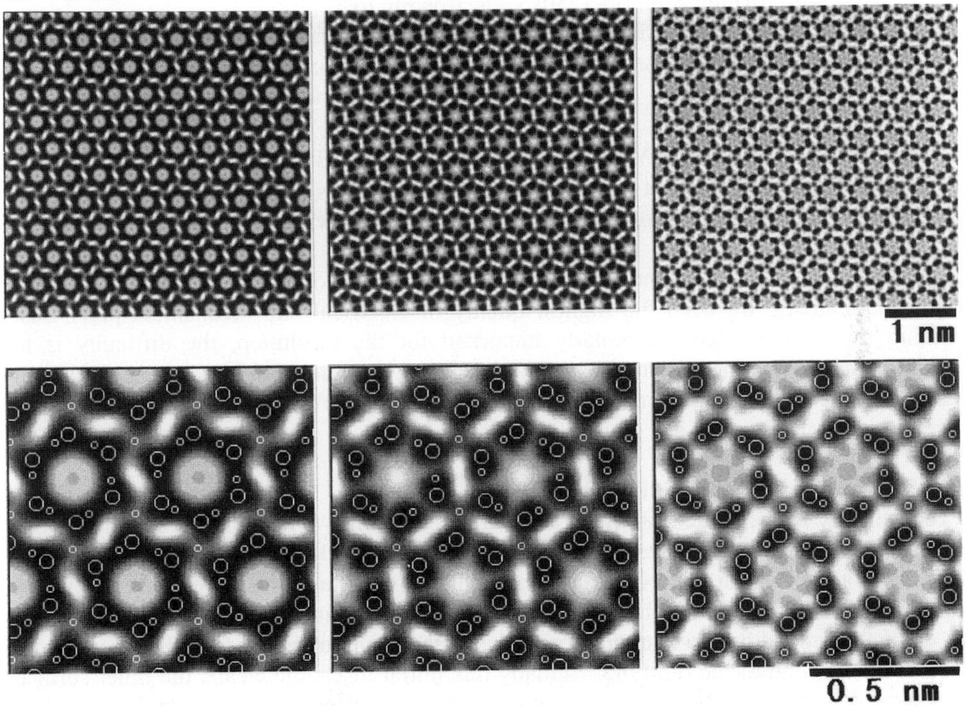

Fig.6 Computer simulated images of β-Si3N4 for three microscopes with different resolution
Lower three are the enlargements(4 times) of the above.
Large circle : Si atomic column position ; small circle : N atomic column position.

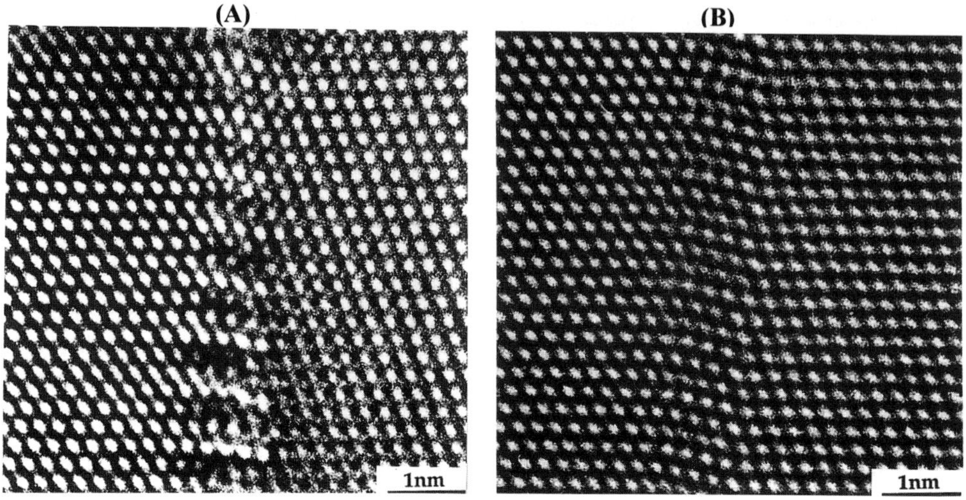

Fig.7 Comparison of images taken with 0.17nm resolution-400kV electron microscope and with 0.1nmresolution-1250kV electron microscope.
JEM-ARM 1250 can be able to obtain the atomic arrangement (B) without any strange image contrast, which is seen in the left photo (A) in the best image taken with JEM-4000EX. Photograph : Cu grain boundary

3. TECHNICAL PROBLEMS BEHIND THE RESOLUTION IMPROVEMENTS

I would like to show what kinds of technical problems exit behind the resolution improvement, and to mention some reasons why we needed such a long year to reach the 0.1 nm resolution. Just imagine that we are observing the atom arrangement under a magnification of 10 million times or more. One microscope consists of about 30 thousands parts. Undoubtedly, we need the highest quality of electrical and mechanical parts. Just counting the main matters particularly important for the resolution, the difficulty is in getting a sufficiently enough stability of high voltage, keeping an atom scale mechanical stability of specimen stage, protecting the specimen from the outside vibration in the atom scale, etc.

3.1 Stability of the accelerating voltage

Theoretically we need the stability of one part per million. We have the Cockcroft-Walton circuit to create the high voltage and use the negative feedback system for stabilization. We need a special precaution to prevent discharge in the high voltage area. Even with maximum precaution the most common phenomena in the high voltage microscope is that discharges, big or small, depending the situation, occur when the high voltage is first switched on the electron microscope in the morning. Actually this initial discharge cleans the inner surface of electron gun, though it might causes some damage to insulators. We devised a so-called alternating current conditioning, meaning of which comes from the operation that the alternating high voltage is applied to the electron gun at first, instead of direct current. This is a very effective way to clean the electron gun.

3.2 Mechanical stability of specimen stage

Here is another mission impossible: we must exchange the specimen easily and quickly, and then we must control smoothly the x-y movement and the tilting angle along the x-y axes. The latter is necessary for the observation of a crystal under the exact incident beam direction. Then, once the observation field is chosen, this field stays still in the range of 0.1 nm during the exposure time (about 10 seconds). Remember, we observe the image under the magnification of several hundred thousand times, and further enlarge 20 times by a TV system, resulting in a magnification of 10 million times or more. In the early electron microscope, the image tends to move jumpingly and drifts for some time after the operator stops the movement of specimen stage. This is the world of electron microscope, i. e. the world of careful design and very skillful workmanship.

3.3 Precaution for anti-vibration

We must carefully check the installation site, especially vibration condition of the site. We should have the best anti-vibration mount on which the electron microscope is placed. And, we should also have specimen stage sufficiently strong against vibration. The remaining vibration of several Hz with amplitude of one micron on the anti-vibration mount can be reduced to relative displacement of 0.1 nm on the specimen stage with the resonance frequency of several hundred Hz of the specimen stage,. Here again, we see that the 0.1 nm resolution is really the limit.

4. CONCLUSION

I have described the progress of the resolution up to 0.1 nm and explained why this is the practical limit which we can realize with actual present-day technology. However, this does not mean that no progress may be expected in the future. Perhaps it is senseless to go toward a higher voltage to achieve a higher resolution. A recent paper [11] shows that the spherical aberration can be corrected by using the non-symmetrical lens. Therefore, the resolution of 0.12nm can probably be achieved with 200kV electron microscope. It is nice to hear that there is always a new frontier ahead.

5. ACKNOWLEDGEMENTS

The author is indebted to Dr. S. Iijima, Prof. H. Hashimoto, Prof. Sir Hircsh, Dr. J. L. Hutchison and Prof. M. Ruhle for helpful suggestions and allowing to use their photos.

REFERENCES

1. B.v. Borries and E. Ruska: Wiss. Veroff. Siemens-Werke, 17-1,99 (1938)
2. M. von Ardenne: Z. Tech. Phys. 19, 407-416 (1938)
3. O. Scherzer: J. Applied Phys. 20 20-29 (1949)
4. J.W. Menter: Proc. Roy. Soc. A 236, 119-135 (1956)
5. A.V. Crew: J. Electronmicros. (Japan): 28 Supplement S9-S15 (1979)
6. G. Binning et. al.: Phys. Rev. Letters 49, 57-60 (1982)
7. S.Iijima: J. Appl. Phys. 42, 5891-5892 (1971)
8. H. Hashimoto et. al.: Electronmicros. 22, 123-134 (1973)
9. P.B. Hirsch et al.: Electron Microscopy 1 146-149 (1980)
10. M. Ruhle et al.: Ultramicroscopy 56 1 -10 (1994)
11. H. Rose: Optik 85, 19-24(1990)

Materials Science and Engineering Serving Society
S. Sōmiya, R.P.H. Chang, M. Doyama and R. Roy (Editors)
1998 Elsevier Science B.V.

Recent Advances in Alumina

A. J. Perrotta
Alcoa Technical Center, Alcoa Center, PA 15069-0001

ABSTRACT

Alumina continues to play an important role in the materials industry with the Bayer process still providing alumina after 100 years of continuous processing. Higher quality alumina, derived through improvements in refining of bauxite to alumina and also post-processing of Bayer alumina, is being produced for aluminum production, ceramic applications, catalytic processing, and feedstocks for alumina and alumina-bearing products. In addition to these traditional applications of alumina, which are continuously being improved, alumina is being studied as part of a new frontier on nanophase materials and nanophase technology. A recent advance in this area is the formation of nanosized corundum, which is in contrast to the transitional aluminas that have been traditionally used in the above applications that required nanosized alumina.

INTRODUCTION

This paper reviews the recent advances in alumina from one perspective and, certainly, does not cover all the advances especially those at high temperatures. These advances have been developed through the whole spectrum of alumina technology starting with the principal alumina production process, the Bayer process. The principal product from the Bayer process is the production of smelting grade alumina for aluminum production. The starting material for the alumina-based industry, and also the precursor for the smelting grade alumina, is gibbsite. The gibbsite production derived by the caustic digestion of bauxite, and subsequent precipitation from supersaturated sodium aluminate solution is still the principal source of alumina in the materials industry. Post processing of Bayer alumina gives alumina for ceramic application, catalytic processing, and feedstocks for alumina-bearing products. In addition, alumina is being researched as part of the new frontier on nanophase material, which are leading to new products with improved properties. This paper will review, from one perspective, some of the work leading to improvements in alumina quality and the utilization of alumina which includes some of the basic research on nanoaluminas and associated phases.

DISCUSSION

Bayer Alumina - Serious research and implementation of this research has been done on the sodium oxalate problem in the processing of bauxite for the production of alumina. Sodium oxalate, coprecipitated with gibbsite in the Bayer process, gives lower alumina quality by the incorporation of sodium impurity in the gibbsite product and, also the generation of fines in the product. Work on the removal of sodium oxalate from Bayer streams has been done recently by the author and co-workers (1-3). Much of this work has been done using the concept of treating the streams with materials such as lime or magnesia that react with alumina-bearing streams to form compounds known as layered double hydroxides. The layered double hydroxides are structurally composed of aluminum-substituted calcium or magnesium hydroxides that have positively charged layers which intercalate inorganic or organic anions between the layers. This process is done insitu by the reaction of lime or magnesia with soluble aluminum in the Bayer stream to form the positively charged layers and the concurrent intercalation of the anions to form the layered double hydroxide. These are ideal structures for removal of oxalate, and other impurities from Bayer streams since these impurities normally exist in the anionic state and are charge compensated by the sodium ions in the strong caustic solution. The structure is shown in Figure 1; it can accommodate large anionic species between its layers (4).

Figure 1: Layered double hydroxides.

$Ca_2\,Al(OH)_7 \cdot 3H_2O$ - Hydrocalumite

Other cations: - Mg^{2+} Al^{3+}*; Fe^{2+}-Fe^{3+}; $Ni^{2+}Al^{3+}$
- Li $Al_2(OH)_6CO_6$**; $[Al_2(OH)_6] \cdot Li_2C_2O_4$***

Pillaring Anions*: $Mo_7O_{24}^{6-}$, $Cr_2O_7^{2-}$, SO_4=, PVA† in Hydrocalumite, C_2O_4=, C_2H_5COO-

---

$[Ca_2Al(OH)_6+]$ Hydroxide layer

OH-, H_2O Anions, neutral guest molecules

$[Ca_2Al(OH)_6+]$ Hydroxide layer

The effect of calcium hydroxide on oxalate reduction is given in Table 1 (1). Treatments at room temperature in long-term runs up to 93 hr show a significant reduction in sodium oxalate in the alumina stream. This reduction in oxalate content is correlated with the formation of the layered double hydroxide, hydrocalumite, as determined by X-ray diffraction analysis of the filtered solid product. Similarly, the use of magnesia also decreases the oxalate concentration as shown in Figure 2. Relatively small doses of magnesia result in a fairly sizable reduction in the oxalate content of the alumina stream. In addition, as shown in Figure 3, the liquor color is very much enhanced giving a much whiter alumina hydrate for low temperature applications (e.g., polymeric filler, fire retardant).

Table 1
Effect of $Ca(OH)_2$ on Oxalate Reduction (Long Duration Runs with Pt. Comfort Spent Liquor)

Run No.	$Ca(OH)_2$ g	Pt. Comfort Liquor (mL)	Temp. (°C)	Time (hr)	Sodium Oxalate (g/L)	TOC (g/L)	XRD Results
26-STDA		X			3.3	7.7	
26-STDB		X			3.3	7.7	
26-1A	20	250	R.T.	93	0.8	7.8	maj. split peak HC, (C)
26-1B	20	250	R.T.	93	0.8	7.6	maj. Split peak HC, (C)
22-STD		X			3.4	7.9	
22-1	10	250	R.T.	93	0.7	8.2	maj. HC trace CH2, C
22-1B	10	250	R.T.	93	1.0	7.6	
31-STD		X	R.T.		3.4	7.7	
31-1	10	250	R.T.	93	1.0	7.6	

*HC=Hydrocalumite, CH_2=$Ca(OH)_2$, C=Calcite ()=Very Small amount, Split Peak=More than one form

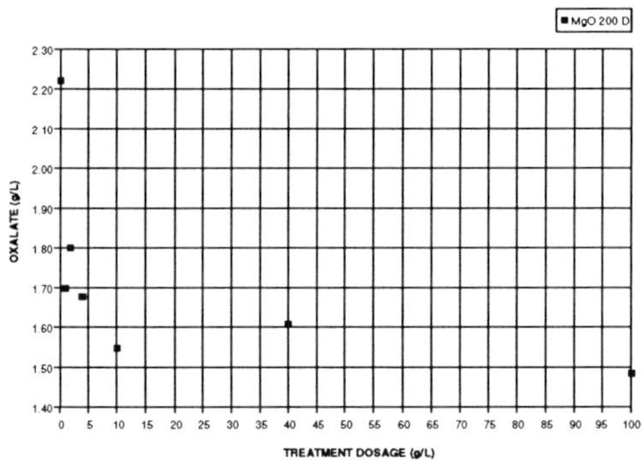

Figure 2: Oxalate in solution after oxalate precipitation.

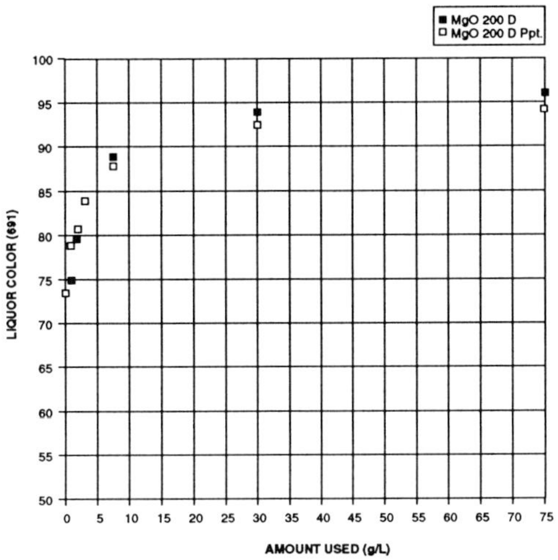

Figure 3: MgO treatment of spent liquor.

The treatment of Bayer process lake water (3) for removal of carbonate, oxalate, sulfate and even some monovalent chloride ions is shown in Figures 4 and 5. The caustization of the lake water is nearly complete after one treatment with nearly complete carbonate removal of the carbonate ion. The divalent oxalate and sulfate ion concentrations are also significantly reduced; the reduction of the sulfate ion also helps to give a higher quality alumina product. Interestingly, there is also a 50% reduction in the amount of total organic carbon.

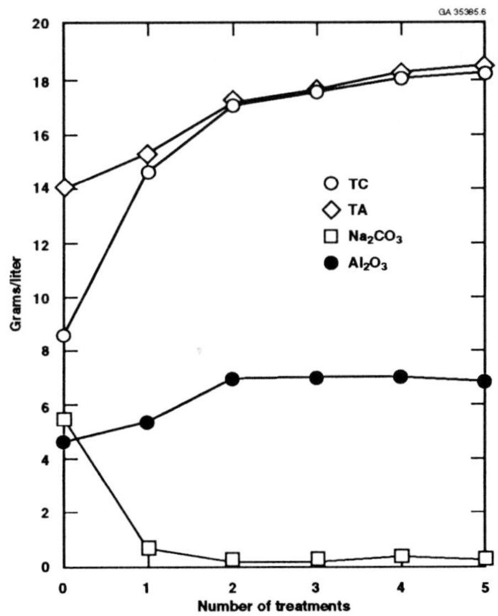

Figure 4: Bayer refinery lake water treatment using 32 g/L hydrocalumite for 1 hr at R.T.

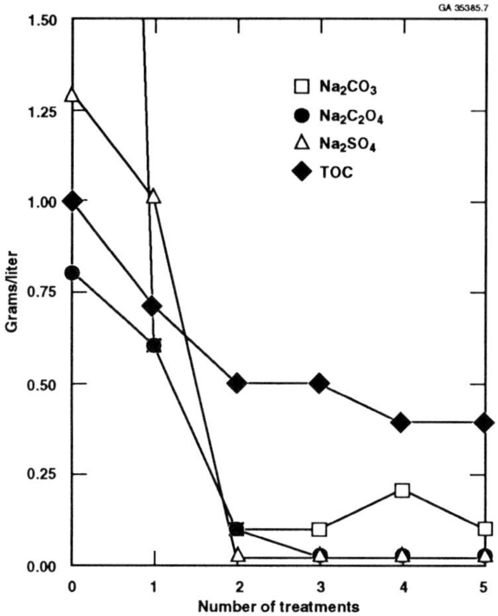

Figure 5: Bayer refinery lake water treatment results using 32 g/L hydrocalumite for 1 hr at R.T.

High Surface Area Corundum - Historically, the formation of anhydrous aluminas with high surface areas lie in the transitional alumina series (e.g., gamma, eta alumina) formed by heat treatment of boehmite, bayerite and alumina gels. Although diaspore is the thermodynamically stable phase relative to boehmite, metastable boehmite is usually encountered and the boehmite-diaspore transition is exceedingly sluggish. Seeding of boehmite with diaspore seed crystals allows the hydrothermal conversion of boehmite to diaspore (5). Diaspore formed by seeding bayerite or boehmite with natural Chinese diaspore ore (>30%) in 1N NaOH gave pure diaspore after 35 days (6). Dehydration to corundum at 530°C resulted in a surface area of 66.2 m^2/g (7).

Recently, much higher surface areas for corundum have been achieved using a diaspore precursor and are shown in Figure 6. Surface areas from the dehydration of fine diaspore to corundum are in excess of 160 m^2/g. The diaspore synthesis giving these high surface area corundum is a totally "green" synthesis without the addition of any alkali or acid to speed up the conversion of boehmite to diaspore (8). Much work remains to be done to make this conversion reaction into a continuous process for the production of high surface area corundum.

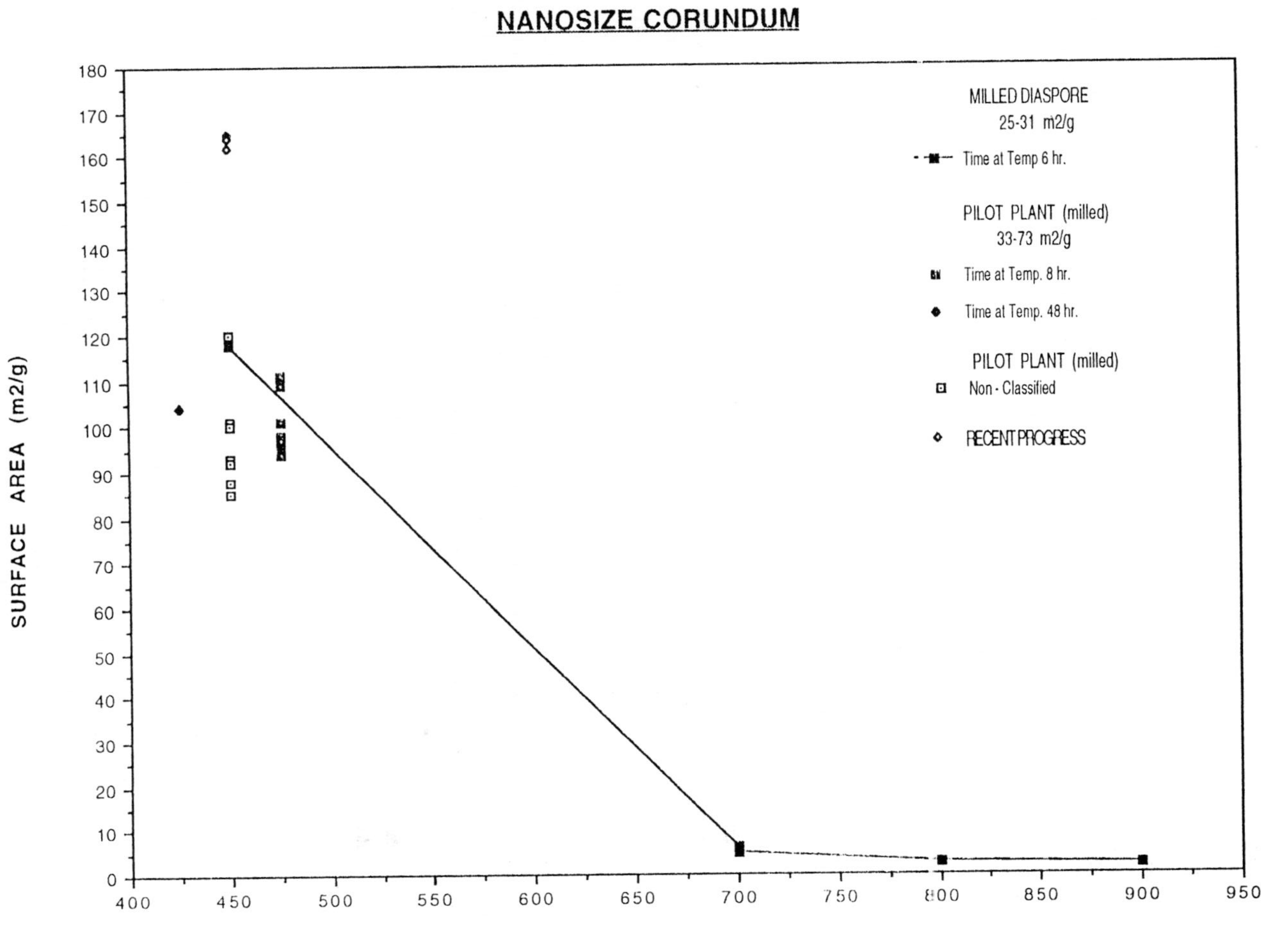
NANOSIZE CORUNDUM
MILLED DIASPORE
25-31 m2/g
Time at Temp 6 hr.
PILOT PLANT (milled)
33-73 m2/g
Time at Temp. 8 hr.
Time at Temp. 48 hr.
PILOT PLANT (milled)
Non - Classified
RECENT PROGRESS
SURFACE AREA (m2/g)
HTT (C)
0
10
20
30
40
50
60
70
80
90
100
110
120
130
140
150
160
170
180
400
450
500
550
600
650
700
750
800
850
900
950

Figure 6

In contrast to using the topotactic transition of diaspore to corundum at 500°C, a new sol-gel route has been discovered giving the formation of alumina gels that form corundum around 500°C (9). This route utilizes the low temperature hydrolysis (~0°C) of an alkoxide such as aluminum-sec-butoxide to form a sol which is subsequently aged at low temperature. The aged sol is then stirred at elevated temperature (~80°C) to form a gel which is then dried and heated to elevated temperatures to form the corundum. There appears to be a correlation between aging time and the temperature such that longer aging times (~30 days) result in lower temperatures for corundum formation. This crystallization behavior is the first sol-gel-derived one that forms corundum at a temperature as low as 500°C and is similar to that of diaspore which, in contrast, has been synthesized under hydrothermal conditions. More work is needed to increase the weight fraction of corundum formed at 500°C using the sol-gel route which would then approach the more nearly complete topotactic conversion of diaspore to corundum.

Mesoporous Alumina (Alumina Molecular Sieves) - Activated aluminas, used for decades in hydrodesulfurization, the Claus process, dehydration of alcohols to give olefins, have recently been extended in their syntheses by the use of organic surfactants to give pure alumina mesoporous materials (10). The development of these new mesoporous aluminas was stimulated by the new high silica mesoporous materials (MCM-48, MCM-41) which gave, after burnout of the surfactants, narrow pore size distributions in the mesoporous range of 15-100 Å and specific surface areas as high as 1000 m^2/g (11). One new mesoporous alumina has a narrow pore size distribution centered at 20 Å with specific surface areas as high as 710 m^2/g. Three new mesoporous alumina molecular sieves (MSU 1-3) were formed by the hydrolysis of tri-sec-butoxyaluminum at ambient temperature in the presence of nonionic polyethylene oxide surfactants; these were calcined at 500°C. Pore sizes ranged from 40 to 80 Å with surface areas ranging from 425 to 535 m^2/g (12,13). These latter molecular sieves have a greater thermal stability which is thought to be due to less structural change of the alumina during calcination because of the use of neutral surfactants (14). This work will be continued in a joint project between Michigan State University and Alcoa.

Phase Transitions and Thermal Stability - Recently, the work on diaspore discussed earlier has led to the discovery of a new transitional alumina reaction sequence. Using the nomenclation of Misra and Wefers (15), this gives a new alpha transitional series. Diaspore, formed hydrothermally, and dehydrated under vacuum gives an intermediate structure (16,17) whose X-ray diffraction data is given in Table 2. Interestingly, although diaspore and corundum have only octahedral coordination of aluminum, the intermediate phase has a significant amount of tetrahedrally coordinated aluminum in the structure as determined by Al-27 nuclear magnetic resonance experiments. This intermediate phase, denoted α'-Al_2O_3, has relatively high surface areas as shown in Table 3. The presence of water at elevated temperature rapidly converts the intermediate phase to corundum as shown by the insitu high temperature X-ray diffraction data (Figure 7). The presence of the tetrahedral aluminum species derived from the hexagonal close packed diaspore structure may provide a new catalytic support.

Table 2
Cell Dimensions and Indexing of X-Ray Powder Data for α'Al_2O_3

a = 9.566(5), b = 5.124(2), c = 9.156(4), β = 91.76(3)
Error of Fit = 2.04

h	k	l	2theta $d_{calc.}$	$D_{obs.}$	I/Io
1	1	0	19.64	19.64	54
1	0	$\bar{2}$	21.25	21.26	74
2	1	0	25.46	25.46	42
2	1	1	27.50	27.49	45
2	1	$\bar{2}$	31.82	31.66	4D*
0	2	0	35.00	34.99	61
1	2	1	37.70	37.72	74
2	1	$\bar{3}$	38.71	38.66	51
4	0	2	43.20	43.21	100
4	2	0	52.30	52.29	31
3	0	$\bar{5}$	57.21	57.22	58
4	2	$\bar{4}$	66.54	66.55	36
4	2	4	68.19	68.18	45

Table 3. Crystal Sizes (Å)

Nano-Alumina	Surface Area (BET)	X-ray Line Broadening
Diaspore 23733-21-1A (TP)	237 Å (73.4 m^2/g)	
Diaspore 23733-21-1A (BP)	458 Å (38 m^2/g)	
α-Al_2O_3 23733-21-1B from TP 21-1A 425°C/8 hr	145 Å (104 m^2/g)	175 Å (86 m^2/g)
α-Al_2O_3 23733-21-1C from TP 21-1A 475°C/8 hr	138 Å (109 m^2/g)	175 Å (86 m^2/g)
α'-Al_2O_3 23733-18-1L 400°C, 138 hr Vacuo (from 18-A)	154 Å (97.7 m^2/g)	97 Å (155 m^2/g)

Investigations of the stability of aluminas in the nanocrystalline state has given a thermodynamic basis for some rethinking of the equilibrium relationships among the alumina phases in the nanocrystalline region. Recently, new data from adsorption microcalorimetry, thermogravimetric analyses, BET adsorption experiments, and high temperature solution calorimetry data show that α-Al_2O_3 becomes energetically stable with respect to γ-Al_2O_3 at specific surface areas >125 m^2/g (18,19). From molecular dynamics simulations, the calculated enthalpy of alumina polymorphs as a function of specific surface area is shown in Figure 8. A crossover to γ-Al_2O_3 becoming energetically the stable phase occurs at about 125 m^2/g. The experimentally determined enthalpies shown in Figure 9 agree quite well with the calculated enthalpies giving a crossover to the energetically stable γ-Al_2O_3 at about the same specific surface area. This recent thermodynamic data is the first conclusive experimental evidence that surface energy differences can favor the formation of a particular polymorph.

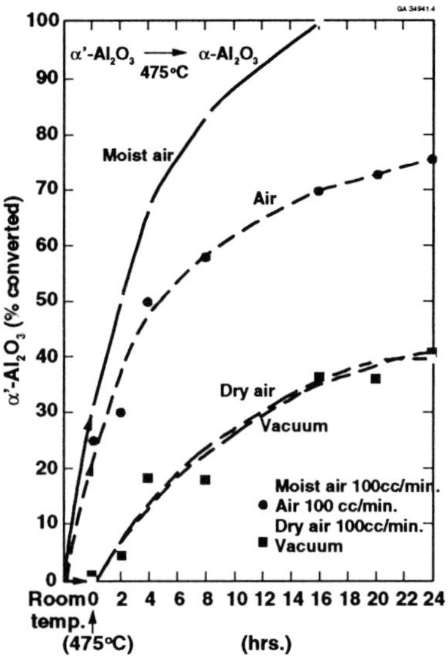

Figure 7

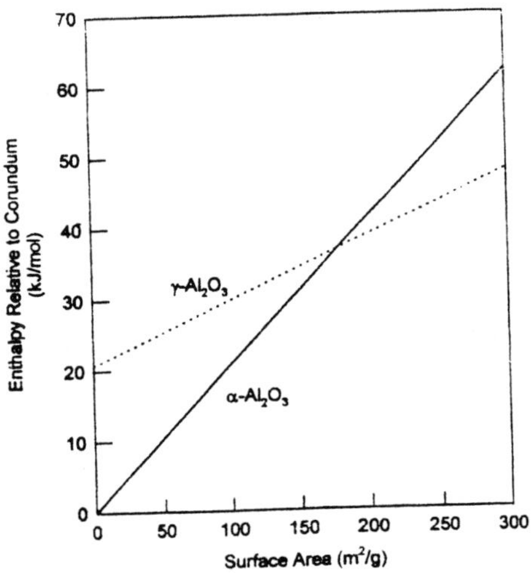

Figure 8

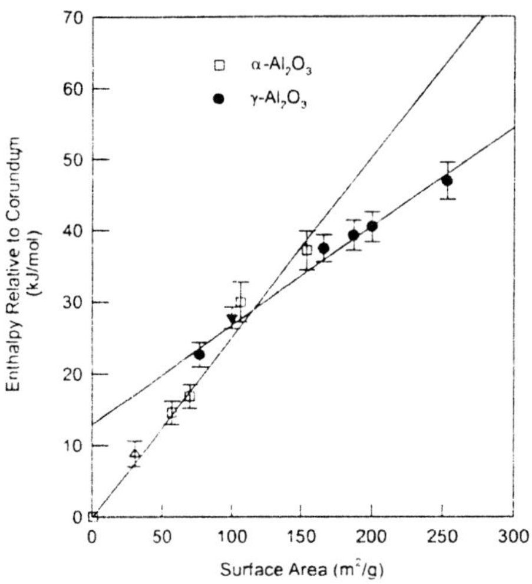

Figure 9

References

1. A. J. Perrotta and F. Williams, "Hydrocalumite Formation in Bayer Liquor and Its Promotional Effect on Oxalate Precipitation," Light Metals 1995, edited by J. Evans, The Minerals, Metals & Materials Society, 1995, 77.
2. A. J. Perrotta and F. Williams, "Layered Double Hydroxide Formation in Bayer Liquor and Its Promotional Effect on Oxalate Precipitation," Light Metals 1996, edited by Wayne Hale, The Minerals, Metals & Materials Society, 1996, 17.
3. A. J. Perrotta, F. Williams, L. Stonehouse, "Layered Double Hydroxides for Treatment of Bayer Process Lake Water," Light Metals 1997, edited by Reidar Huglen, The Minerals, Metals and Materials Society, 1997, 37.
4. C. Misra and A. J. Perrotta, "Composition and Properties of Synthetic Hydrotalcites," Clay and Clay Minerals, 1992; also U.S. Patent 5,075,089 (1991).
5. G. C. Kennedy, "Phase Relations in the System Al_2O_3-H_2O at High Temperatures and Pressures," Am. Jour. Sc., 1959, vol. 257, 563.
6. T. Tsuchida and K. Kodaira, "Hydrothermal Synthesis and Characterization of Diaspore," J. Mat. Sci., 25 (1990) 4423.
7. T. Tsuchida, "Preparation and Reactivity of Acicular α-Al_2O_3 from Synthetic Diaspore," Solid State Ionics (1993) 464.
8. A. J. Perrotta, Private communication, 1997.
9. K. Kamiya, J. Yotani, R. Semba, J. Matsuoka and H. Nasu, "Sol-Gel Preparation of Alumina Gels Forming α-Al_2O_3 around 500°C," Journ. Cer. Soc. Japan, 104, [7] (1996), 685.
10. F. Vaudry, S. Khodabandeh and M. E. Davis, "Synthesis of Pure Alumina Mesoporous Materials," Chem. Mater. (1996), 8, 1451.

11. C. T. Kresge, M. E. Leonowicz, W. J. Roth, J. C. Vartuli and J. S. Beck, "Ordered Mesoporous Molecular Sieves Synthesized by a Liquid - Crystal Template Mechanism," Nature (1992) [359] 710.
12. S. A. Bagshaw and T. J. Pinnavaia, "Mesoporous Alumina Molecular Sieves," Angew. Chem. Int. Ed. Eng. (1996), 35, No. 10, 1102.
13. S. A. Bagshaw, E. Prouzet and T. J. Pinnavaia, "Templating of Mesoporous Molecular Sieves by Nanionic Polyethylene Oxide Surfactants," Science, September 1995 [269] 1242.
14. P. Tanev and T. J. Pinnavaia, Private communication, June 1997.
15. C. Misra and K. Wefers, "Oxides and Hydroxides of Alumina," Alcoa Technical Paper 19, 1987.
16. A. J. Perrotta and R. B. Minnick, "Transition Alumina and Method for Its Production," U.S. Patent 5,334,366, August 2, 1994.
17. A. H. Carim, G. S. Rohrer, N. R. Dando, S. Y. Tzeng, C. L. Rohrer and A. J. Perrotta, "Conversion of Diaspore to Corundum: A New Alpha-Alumina Transformation Sequence," submitted for publication to Journ. Amer. Cer. Soc.
18. J. M. Michele, A. Navrotsky and A. J. Perrotta, "Effects of Increased Surface Area and Chemisorbed H_2O on the Relative Stability of Nanocrystalline γ-Al_2O_3 and α-Al_2O_3, J. Phys. Chem. B, Vol 101, No. 4 (1997) 603.
19. M. Michele, A. Navrotsky, A. J. Perrotta and A. Auroux, "Surface Energies and Phase Stability in Nanocrystalline Aluminas" Science (accepted for publication).

Materials Science and Engineering Serving Society
S. Sōmiya, R.P.H. Chang, M. Doyama and R. Roy (Editors)

Crystallization process of amorphous aluminas to α-alumina

Zenbe-e Nakagawa[a], Taisuke Aosaki[b], and Naoya Enomoto[c]

[a]Mining College, Akita University, 1-1, Tegata Gakuen-cho, Akita 010, Japan

[b]Present Address: Kao Cooperation

[c]Materials and Structures Laboratory, Tokyo Institute of Technology, 4259, Nagatsuta, Midori, Yokohama 226, Japan

Amorphous aluminas were prepared through the spray-pyrolysis method and the calcination method of aluminum nitrate. The crystallization processes of these materials to α-alumina were compared with those of crystalline aluminum hydroxide, bayerite, and the various heat-treated materials. Amorphous aluminas contain five coordination number of Al and the phase transition temperatue from γ- to α-alumina is lower than that of crystalline materials.

1.INTRODUCTION

Crystallization processes of crystalline aluminum hydroxides have been extensively studied for a long period[1]. On the contrary, amorphous aluminas and/or gels are collected simply as "amorphous mateirals" and their crystallization process to α-alumina is less investigated. Amorphous materials would be more reactive than the crystalline ones and should be more important in various fields as a source of alumina.

In this paper, amorphous aluminas were prepared through the spray-pyrolysis method and the calcination method of aluminum nitrate. As their crystallization process and phase transition process are considered to be strongly affected by the local structure around Al atom in the materials, the coordination state of Al was focused.

2.EXPERIMENTAL PROCEDURES

Spray pyrolysis of aluminum nitrate was conducted by spraying the ethanol-water 1:1 solution of 0.2mol/l Al^{3+} concentration into a furnace kept at 600°C (sample SPA600). Calcination procedure of aluminum nitrate was to soak at 400°C for 12h in a furnace (sample CAL400). Two materials were amorphous by X-ray diffraction test. A reference material is a crystalline aluminum trihydroxide, bayerite (sample RH40). Samples were annealed at 500, 700, and 900°C for 12h in air, in vacuum of 10^{-3}Pa, and in moisture at vapor pressure of about $5x10^4$Pa. Crystalline phases in the heat-treated specimens were examined by X-ray diffractmetry(XRD). The crystallization and phase transition temperatures were determined as

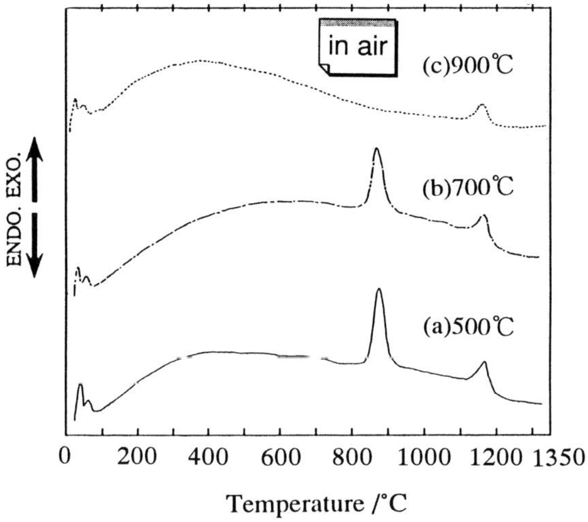

Figure 1 DTA curves of the sample SPA600 heat-treated at vaious temperatures in air

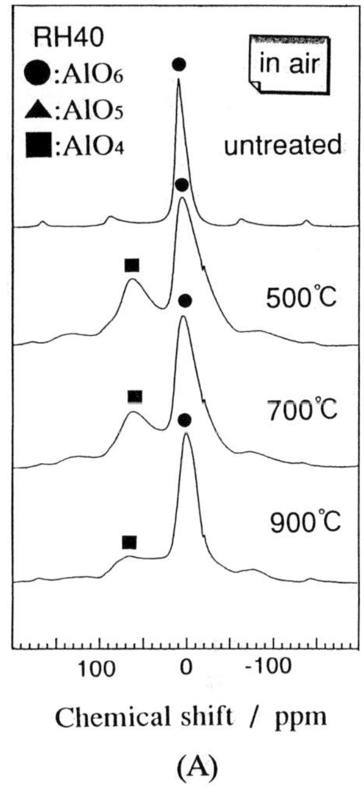

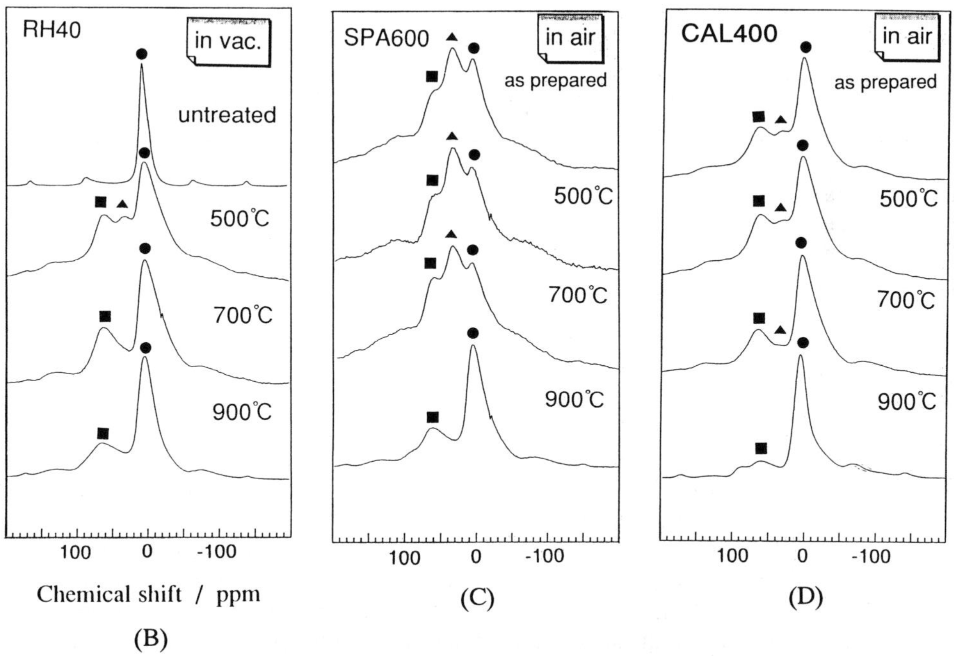

Figure 2 Changes in NMR spectra of the samples heat-treated at various temperatures in air or in vacuum

Table 1 Summary of DTA and XRD data for the sample RH40

Sample		RH40		
Treatment Condition		XRD Phase	Tγ	Tα
Untreated /As Prepared		Bayerite	350~620°C	1270°C
In Air	500°C	γ	——	1269°C
	700°C	γ	——	1267°C
	900°C	γ + θ	——	1270°C
In Vacuun	500°C	γ + Amor.	749°C	1107°C
	700°C	γ	——	1208°C
	900°C	γ	——	1197°C
In H_2O Vapor	500°C	γ	——	1253°C
	700°C	γ + θ	——	1255°C
	900°C	θ	——	1257°C

Tγ: γ Phase Transition Temperature
Tα: α Phase Transition Temperature
γ : γ -Type Al_2O_3 θ : θ -Type Al_2O_3 α : α -Type Al_2O_3

Table 2 Summary of DTA and XRD data for the samples SPA600 and CAL400

Sample		SPA600			CAL400		
Treatment Condition		XRD Phase	Tγ	Tα	XRD Phase	Tγ	Tα
Untreated /As Prepared		Amor.	878°C	1166°C	Amor.	817°C	1094°C
In Air	500°C	Amor.	877°C	1167°C	Amor.	822°C	1095°C
	700°C	Amor.	875°C	1167°C	Amor.+ γ	817°C	1094°C
	900°C	γ	——	1169°C	γ + α	——	1079°C
In Vacuun	500°C	Amor.	878°C	1173°C	Amor.	824°C	1087°C
	700°C	Amor.	880°C	1170°C	Amor.+ γ	833°C	1087°C
	900°C	γ	——	1173°C	γ + α	——	1079°C

Notation of the symbols is the same in Table 1.

those of the top of the exothermic peak on DTA curves which were measured at the heating rate of 10°C/min.

Coordination state of Al atoms to oxygen atoms was determined by MAS-(Magic Angle Spinning) NMR measurement. Aluminum sulfate hydrate was used as a standard sample of Al-six-coordination. The values of the chemical shift of AlO_4, AlO_5, and AlO_6 are 50～80, 30～40, and -10～20ppm, respectively.

3.RESULTS AND DISCUSSION

One example of DTA curves is shown in Figure 1 and summary of DTA and XRD data is listed in Tables 1 and 2. An amorphous material has two exothermic peaks on the DTA curve based on the crystallization to γ-phase and the phase transition to α-phase, though a crystallized material has only one peak based on the phase transition to α-phase.

In the reference material, RH40, the heat-treatments in vacuum and in moisture decrease the transition temperature and the difference is large for the treatment in vacuum. The heat-treatment at 500°C in vacuum produced an amorphous material, which causes to decrease the phase transition temperature. In this case, γ-phase transforms directly to α phase via no θ-phase.

The phase transition temperature of two amorphous materials prepared from aluminum nitrate are lower than that of RH40 in air, and there are no temperature diferences under their heat-treatment conditions between in air and in vacuum. Among them, the temperature of CAL400 is lower than that of SPA600. Similarly, the crystallization temperature of CAL400 is also lower than that of SPA600. These results suggest that high crystallization temperature produces high crystallinity of γ-phase and that the high crystallinity increases the phase transition temperature.

Figure 2 shows NMR spectra of the samples heat-treated at various temperatures in air and/or in vacuum. In as-prepared materials, amorphous materials, SPA600 and CAL400, contain AlO_4 and AlO_5, while bayerite, RH40, contains only AlO_6. The specimen from RH40 heat-treated at 500°C in vacuum also contains AlO_5. Therefore, amorphous aluminas are infered to be characterized to contain AlO_5. These materials transform direcly from γ- to α-phase and have low phase transition temperatures. If the five coordination number of Al forces to firm an amorphous structure, the degree of amorphous state is to be high in SPA600 specimen which has a high crystallization temperature. High crystallization temperature enhances crystallinity of γ-phase and increases the phase transition temperature to α-phase. The low crystallization temperature and the low transition temperature of the specimen from RH40 heat-treated at 500 °C in vacuum support these ideas.

CONCLUSIONS

Amorphous aluminas are characterized to contain five coordination number of Al and their crystallization process is strongly influenced by the existence of AlO_5.

REFERENCE

1. W.H.Gitzen, ed., "Alumina as a Ceramic Materials", Am.Ceram.Soc., (1970)

Materials Science and Engineering Serving Society
S. Sōmiya, R.P.H. Chang, M. Doyama and R. Roy (Editors)

Optical nonlinearity of ZnO composite nanoparticles with different interfacial chemical environments

Rong-yao Wang, Xiao-chun Wu, Bing-suo Zou, Li Wang, and Ji-ren Xu

Optical Physics Lab., Institute of Physics, Chinese Academy of Sciences, P.O.Box 603, Beijing 100080, P.R.China
FAX:+86-10-62562605, E-mail:zoubs@aphy02.iphy.ac.cn

ZnO nanoparticles with anionic surfactant DBS (dodecyl benzene sulfonate) or cationic surfactant CTAB (cetyltrimethyl ammonium bromide) modifying surfaces were synthesized by microemulsion method. The sign and magnitude of both the real and imaginary parts of the third-order optical nonlinearity $\chi^{(3)}$ of these ZnO composite nanoparticles with different chemical environments were measured by single-beam Z-scan technique using 790 nm, 85 fs duration laser pulse. The ultrafast nonlinear response time (~ 250 fs) measured by time-resolved pump-probe technique at excitation wavelength of 647 nm suggest that the optical nonlinearity below band-gap originate mainly from a rapid electronic polarization process or virtual process such as the optical Stark effect.

1. INTRODUCTION

Semiconductor nanocrystals which exhibit large and fast third-order nonlinearities have been intensively studied in the last decades. The search for new optical materials with strong optical nonlinear behavior has motivated experiments with surface chemical modification, because the large ratio of surface atoms to bulk atoms in semiconductor nanoparticles results in strong influence of the surface states on their optical properties. The synthesis of nanocomposite materials by means of surface chemistry is a new direction in the study of physical properties of nanometer-sized materials[1]. Various organic function molecules are used to modify the surface of nanoparticles in order to improve the propeties of nanoparticles[2-4]. Recently, we have shown that surface modification of CdO nanoparticles with different surfactant molecules play an important role in their optical properties[5]. In this work, we report that under proper thermal treatment surface modification of ZnO nanoparticles with surfactant molecules result in the formation of a new ZnO-surfactant composite nanoparticles. The large and fast optical nonlinearity below band-gap occurring in these new organic-inorganic composite nanoparticles were examined by Z-scan and pump-probe technique.

2. EXPERIMENTAL

Nanometer-sized ZnO were prepared by the water/oil-type microemulsion method[6]. The oil phase is constituent by xylene. The surfaces of ZnO nanoparticles were modified by anionic surfactant DBS or cationic surfactant CTAB. The colorless ZnO/DBS or /CTAB organosols absorb light of wavelengths shorter than 370nm (Fig.1). It mainly reflects the novel nature of

ZnO nanoparticles. During evaporating the solvent the color of samples transformed from colorless to red. Then the red samples were redissolved in xylene for optical studies. Their average radii of 4 nm and the standard deviation of ±0.8 nm were measured with the high-resolution transmission electron microscopy. The concentration of the samples are about 10^{-3} M.

Single beam Z-scan technique [7] was used to measure the third-order optical nonlinear susceptibility of the red ZnO/DBS or /CTAB composite nanoparticles. This technique is based on the principle of spatial beam distortion and relies on the fact that the sample itself acts as a thin lens with varying focal length as it moves along the axis of a focused laser beam. Laser pulses, generated from a Ti: Sapphire laser (Spectra-physics TSA-10) with wavelength of 790nm, pulse width of 85 fs and repetition rate of 82 MHz, were focused to a waist radius about 36 μm, which was measured with a scanning knife edge. The thickness of the sample was 1 mm, which was smaller than the Rayleigh length Z_0 of the focused laser beam (Z_0 =5 mm). The linear transmittance S was 0.07 for the closed-aperture and 1 for the open-aperture, respectively. The transmittances with closed- and open- aperture were measured in the far field as the sample was moved through the focal plane to determine the sign and magnitude of both the real and imaginary parts of the third-order optical nonlinearity $\chi^{(3)}$ of ZnO composite nanoparticles.

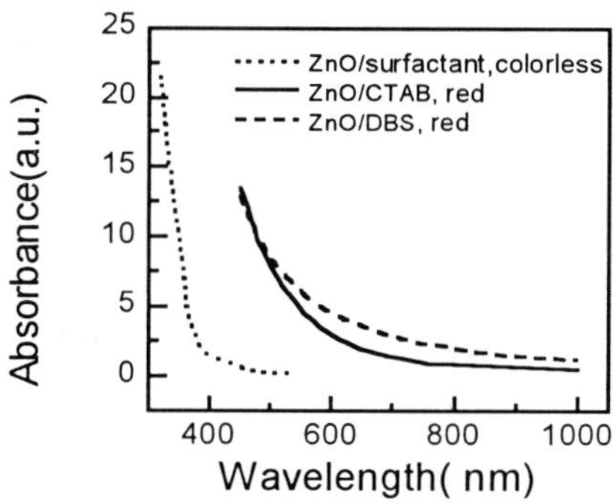

Figure 1. Absorption spetra of ZnO/DBS and ZnO/CTAB nanoparticles in xylene at room temperature. The absorption spectrum of colorless ZnO nanoparticles coated with surfactants is also plotted.

3. RESULTS AND DISCUSSIONS

The absorption spectra of red nanocomposite ZnO/DBS(sample 1) and ZnO/CTAB(sample 2) in xylene are shown in Fig. 1, from which we derive that the optical gap for either sample 1 or sample 2 is ~2.3eV(540 nm), which has a large absorption edge red-shift compared to the case of colorless samples (the optical gap energy~3.2eV). Being considered the catalysis of ZnO nanoparticles and the changes of the packing parameter of surfactant molecules at high temperature[8], it may be reasonable to assume that small particle ZnO colloids form charge transfer complexes with DBS or CTAB at intercfacial layer that result in the appearance of the new optical gap in the visible region[4]. But the detailed structures of ZnO composite nanoparticles are not clear yet.

Figure 2 shows the normalization transmission with open-aperture as a function of the sample position along the lens axis. With increasing the excitation intensity, the two-photon absorption process in two samples were enhanced but showed different intensity dependence.

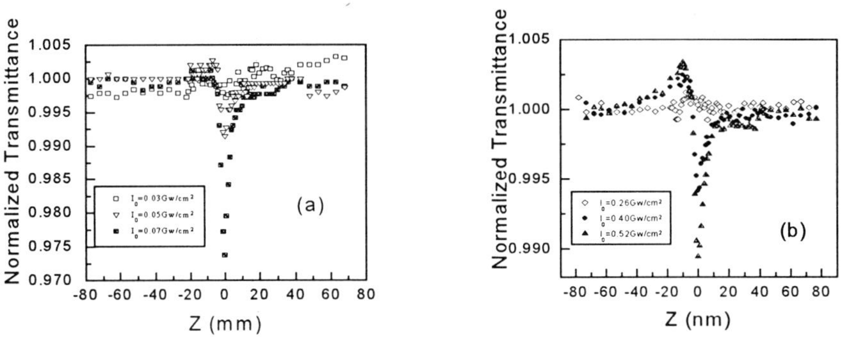

Figure 2. Z-scan open-aperture transmission curves at a various of the excitation intensities I_0 (a) ZnO/DBS (b) ZnO/CTAB.

The normalized transmittance for open-aperture condition was given in Ref.[7] by

$$T(z) = \frac{1}{\pi^{1/2} q(0,0,z)} \int_{-\infty}^{\infty} \ln\left[1 + q(0,0,z)\right] \exp(-\tau^2) d\tau \qquad (1)$$

The only unknown parameter in Eq(1) is β in q(0,0,z)

$$q(0,0,z) = \frac{\beta I_0 (1 - e^{\alpha L})}{(1 + z^2/z_0^2)\alpha} \qquad (2)$$

Where, α is the linear absorption coefficient, β is the two-photon absorption coefficient, L is the sample thickness. I_0 is the intensity of the laser beam at the focus (z=0), τ is the laser pulse width. A plot of T(0) as a function of $q(0,0,0)=\beta I_0 L_{eff}$, where $L_{eff} = [1-exp(-aL)]/a$, can be used to determine β directly. We derived experimentally that $\beta_1 \approx 11.6$ cm/Gw for ZnO/DBS and $\beta_2 \approx 0.45$ cm/Gw for ZnO/CTAB.

In Fig.3 the normalization transmission curves for the closed-aperture as a function of the sample position are plotted. To quantify the nonlinear refractive index γ, the ananlytical procedure reported in Ref.[7] was adopted. We chose the experimental data at low excitation intensities and divided the closed-aperture Z-scan by the open-aperture ones performed at the same excitation intensity. Then the differences between the peak and valley, ΔT_{p-v}, in the resultant Z-scan are measured. The nonlinear refractive index γ can be evaluated according to following equation:

$$\gamma = \frac{1.414\ \lambda}{2\pi L_{eff}} \cdot \frac{\Delta T_{p-v}}{0.406\ (1-s)^{0.25}\ I_0} \tag{3}$$

The experimental values are $\gamma_1 \approx -5.2\times10^{-16}\ m^2/W$ (-3.8×10^{-10}esu) for ZnO/DBS and $\gamma_2 \approx -2.2\times10^{-17}\ m^2/W$ (-1.5×10^{-11}esu) for ZnO/CTAB, respectively. The estimated error in these measurements is ±30%, which originates mainly from the uncertainty of the waist measurement and the fluctuation of laser energy.

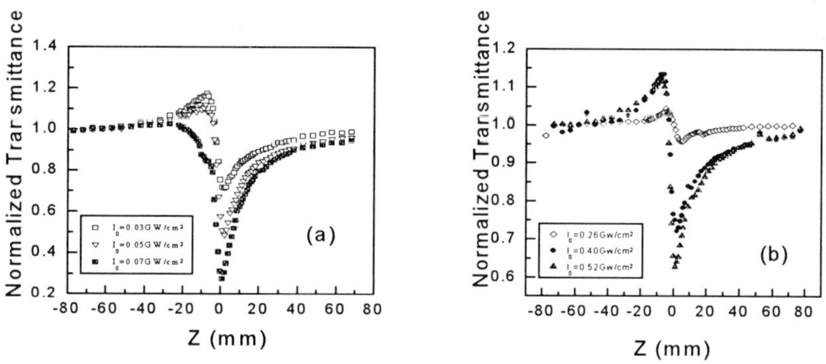

Figure 3. Z-scan closed-aperture transmission curves at a various of the excitation intensities I_0 (a) ZnO/DBS (b) ZnO/CTAB.

It can be concluded that the nonlinear refraction change in sample 1 is larger relative to sample 2. Meanwhile, the third-order optical nonlinearity (including two-photon absorption coefficient, β, and the bound electronic nonlinear refractive index, γ) of ZnO composite nanoparticles were examined to be interfacial chemical environment dependence. These two kinds of ZnO composite nanoparticles have the same core structure. The interfacial layer is composed of the surface dangling bond of ZnO nanoparticle and the polar head of surfactant. Therefore, we suggest that the differences in optical nonlinearity should be ascribe to the different charge transfer complexes formed in ZnO/DBS and ZnO/CTAB composite nanoparticles.

Because the normalized incident photon energy $\hbar\omega/E_g$ for λ = 790 nm is about 0.7, which satisfies the two-photon absorption condition $1/2E_g < \hbar\omega < E_g$, two-photon absorption and optical Stark effect play an important role on the large optical nonlinearity, both of which are laser pulse duration dependent. But the contribution to nonlinear refraction from two-photon absorption is less important because of small two-photon absorption. The separation between the transmittance peak and valley in resultant Z-scan is $Z_{p-v} \approx 1.7Z_0$, which indicates the overwhelming effects of the third-order nonlinear process. Furthermore, the negative sign of the nonlinear refraction is consistent with the Optical Stark effect [9]. In addition, because of high repetition rate of laser pulse, some slow nonlinearities might be involved. For example, because ZnO composite nanoparticles have apparent long tail state (Fig. 1), these surface trapped states or band-tail states may be indirectly excited [10]. With rapid relaxation of the photocarriers to the deep states in about several ns, the long lifetime of the deep trapped states would cause a nonnegligible contribution to the total nonlinearity [11]. As for the thermally-induced nonlinearity, it is positive for excitation wavelength below the band-gap[12]. We think that it can be neglected because the absorption coefficients of samples at 790nm are very small ($\alpha_1 \approx 0.1 cm^{-1}$,$\alpha_2 \approx 0.08\ cm^{-1}$) and any changes of scan curves did not occur when repeated two-direction scans were performed.

In our another time-resolved pump/probe experiment, 647 nm, 120 fs laser pulse was used to determine the temporal characteristics of the optical Kerr nonlinearity in the same samples with different concentration. This part of work will be reported elsewhere. With reference of CS_2, Both ZnO/DBS and ZnO/CTAB nanocomposites exhibit an ultrafast nonlinear response (~250 fs), and no significant slow effects are observed. The measured optical nonlinear refraction is ~ 10^{-11} esu for ZnO/DBS and ~10^{-12}esu for ZnO/CTAB, respectively. The difference in magnitude of the nonlinear refraction is consistent with the result at 790 nm excitation. These experimental data at two excitation wavelengths in the nonresonant region indicate that the dominant mechanism of the ultrafast nonlinear optical response for ZnO composite nanoparticles with surface chemical modification is a rapid electronic polarization process or virtual process such as the optical Stark effect.

4.CONCLUSIONS

In summary, The third-order optical nonlinearity of ZnO composite nanoparticles originate mainly from the bound electronic nonlinearity when they are excited with 790nm and 647nm femtosecond laser at room temperature. The large and fast nonlinear optical response in ZnO/DBS composite nanoparticles indicates a potential application of special chemical modification to gain a new optical nonlinear materials.

REFERENCE

1. G.A.Ozin, Adv. Mater., 4 (1992) 612.
2. Torimoto, K.Maeda, T.Sakata, H.Mori, and H.Yoneyama, Chem. Lett., 977 (1994)
3. H.Matsumoto,T.Matsunage, T.Sakata, H.Mori, and H.Yoneyama, Langmuir, 11 (1995) 4283.
4. T.Rajh, D.M.Tiede, M. C. Thurnauer, J. Non-crystalline Solids, 205-207 (1996) 815.
5. X.C.Wu, R.Y.Wang, B.S.Zou, P.F.Wu, Li.Wang, J.R.Xu, Appl.Phys.Lett.,71 (1997) (to be published)
6. X.C.Wu, B.S.Zou, J.R.Xu, B.L.Yu, G.Q.Tang, G.L.Zhang, and W.J.Chen, Nanostruct.Mater., 8 (1997) 179.
7. A.A.Said, M.Sheik.Bahae, D.J.Hagan, T.H.Wei, J.Wang, J.Young, and E.W.Van Stryland, J.Opt.Soc.Am., B9 (1992) 405.
8. A.N. Maitra, K.L.Mittal, P.Bothorel (eds.), Surfactants in solution, Plenum Press, New York, 1986.
9. D.Cotter, M.G.Burt, and R.J.Manning, Phys. Rev.Lett., 68 (1992) 1200.
10. H.Hang, S.W.Koch, Quantum Theory of the optical and electronic properties of semiconductors, World scientific, Singapore, 1990.
11. H.Ma, A.S.L.Gomes, Cid.B.de Araujo, J. Opt. Soc.Am., B12 (1992) 2330.
12. R.K.Jain , M.B.Klein, and R.A.Frisher (eds.), Optical Phase Conjugation, Academic, Newyork, 1983.

Materials Science and Engineering Serving Society
S. Sōmiya, R.P.H. Chang, M. Doyama and R. Roy (Editors)

Ultraviolet light emission properties of ZnO single crystals♣

N.OHASHI, T.SEKIGUCHI*, H.HANEDA**, Y.TERADA, T.OHGAKI, J.TANAKA**

T.TSURUMI and O.FUKUNAGA♉

Department of Inorganic Materials, Tokyo Institute of Technology,
2-12-1 O-okayama, Meguro, Tokyo 152 JAPAN.

* Institute for Materials Research, Tohoku University, Sendai 980-77, Japan

** National Institute for Research in Inorganic Materials, Namiki, Tsukuba 305, Japan.

Luminescent properties of zinc oxide(ZnO) single crystals were studied by means of cathodoluminescence. The effects of impurity on luminescence properties were discussed by measuring secondary ion mass spectra. Effect of oxygen vacancy on CL spectra were examined by characterization of single crystal annealed under various conditions.

1. Introduction

Zinc oxide (ZnO) having wide band gap of 3.3eV is one of the candidates for ultraviolet (UV) optical devices. Recently, Reynolds et al.[1] fabricated a UV laser device made of ZnO thin films and Tang et al.[2] fabricated a UV laser with high emission efficiency. An advantage of ZnO as UV optical materials is that the exciton binding energy in ZnO (about 60 meV[3]) is much larger than that in the other semiconductor materials; for example, that in GaN is 20 meV. The drawback on the use of ZnO as UV emitting material is that the occurrence of visible light emission due to defects and/or impurities decreases the UV emission efficiency.

In the present study, single crystals of ZnO were grown by a flux method and their luminescence properties were measured by means of cathodoluminescence (CL). The

---

♣ This study is partly supported by Special Coordination Fund 'Frontier Ceramics Project' from Science and Technology Agency of Japan.
♉ Present Address: Advanced Ceramics Technology Co. Ltd., Higashi Oume, Oume, Tokyo 192 Japan.

position dependence of CL spectra and depth profiling of impurity concentration were compared to elucidate the effects of impurities and defects upon emission properties.

2. Experimental procedure

Single crystals of ZnO were prepared by a flux method using PbF_2. Purity of the source ZnO was 99.999% or 99.9% and that of PbF_2 99.9 or 99.5%. A mixture of PbF_2 and ZnO, Pb:Zn=8:2 in molar ratio, filled in a Pt crucible was melted at 1040-1100°C in oxygen gas flowing condition and then slowly cooled to 950°C in 30-40 h. The grown crystals were then annealed in oxygen gas at 800-1200°C for 6-12 h.

Chemical composition of the grown crystals were analyzed by induction coupled plasma atomic emission spectra (ICP) method. Depth profiles of impurities were measured by secondary ion mass spectra (SIMS) from the $(000\bar{1})$. Cathodoluminescence (CL) spectra were measured at 30-300 K using a scanning electron microscope equipped with optical spectrometer. Position dependence of CL spectra was measured on the fractured surface of the crystal. The detail of the CL apparatus is written elsewhere[4].

3. Result and discussion

Impurities detected by SIMS and/or ICP method was Pb, F, Li, Al, F and/or Cu. Concentration of Cu and Al was estimated to be less than 0.1 mol% of Zn by ICP analysis and signals corresponds to Pb, F, Li, Al and F were not detected by ICP analysis. As for the $(000\bar{1})$ face region, SIMS analysis suggested that impurity concentration steeply decreased upto 3 μm and that impurity concentration was almost constant in the depth range over 3 μm. In this paper, the depth range of 0-3 μm and that of over 3 μm are, hereafter, abbreviated by "surface" and "bulk" region, respectively.

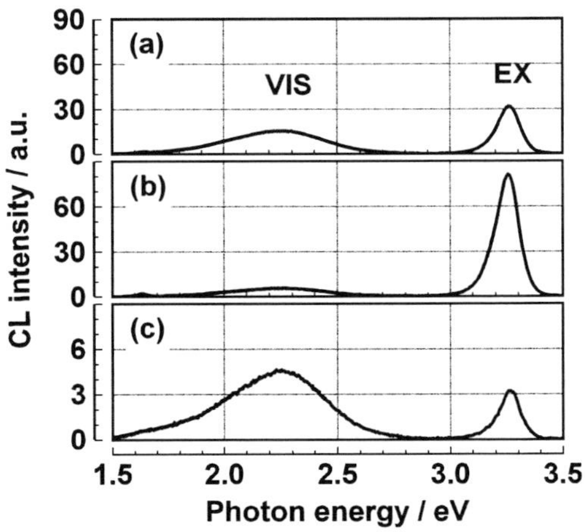

Fig. 1 CL spectra for as-grown ZnO crystal grown from ZnO powder with 99.9% purity. Spectra were collected from fractured surface. (a) $(000\bar{1})$ face with oxygen topmost layer, (b) bulk and (c) (0001) face with zinc topmost layer.

Figures 1 show the typical CL spectra of ZnO single crystal. Two CL peaks were observed, i.e. excitonic UV luminescence at 3.3 eV and broad emission band in the visible light region of 1.5-2.8 eV. The excitonic luminescence and visible light emission are abbreviated by 'EX' and 'VIS' region, hereafter. In the as-grown crystals, profiles of CL spectra depend on the position in the crystal. Both EX and VIS intensity increased with the increase of the depth from the surface at surface region, while VIS and EX intensity decreased and increased, respectively, with the depth from the surface in the bulk region.

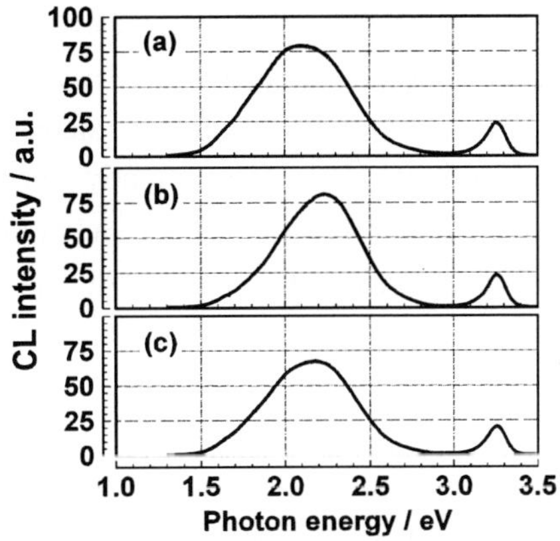

Fig.2 CL spectra for ZnO single crystal annealed in O_2 gas at 1000°C. The crystal applied was grown from ZnO with 99.9% purity. (a) $(000\bar{1})$ face, (b) bulk and (c) (0001) face.

Figures 2 show CL spectra of the ZnO annealed in the oxygen gas atmosphere; these spectra were obtained from annealed surface without cleaving. As indicated in these figures, the intensity of VIS emission increased and that of EX emission decreased by annealing in oxygen gas. Furthermore, spectral profile in VIS region was changed by annealing; in fact, lower energy component of VIS emission (1.5-2.0 eV) appeared in the annealed crystal. The intensity of the lower energy component was high in the surface region as indicated in Fig. 2(a). It seems that strong position dependence of CL intensity observed in the as-grown sample was suppressed by annealing but the profiles of CL spectra still depended on the position even in the annealed crystals. The manner of position dependence of CL spectra is common among the as-grown and annealed crystals, indicating that the position dependence of CL spectra is brought by position dependence of impurities. In fact, intensity of VIS emission from ZnO crystal grown from ZnO source of 99.9% purity was higher that that grown from ZnO source of 99.999% purity.

In the previously published papers, e.g. ref. 5, the VIS emission was ascribed to the ionized oxygen vacancy ($V_O^{\bullet}$) giving g=1.96 ESR signal. It was reported that intensity of both VIS emission and g=1.96 ESR signal increase simultaneously by forming gas annealing[5]. In the present study, intensity of VIS emission increased by annealing in oxygen gas. In general, oxygen vacancy concentration decreases with the increase in oxygen partial pressure on the annealing. On the crystal growth procedure in this study, evaporation of PbF_2 and ZnO was observed and it is consequently considered that partial pressure of oxygen gas during the crystal growth was less than 1 bar. The effect of annealing was also confirmed by Hall measurements as reported elsewhere[6]: carrier concentration determined by Hall effect slightly decreased by annealing in O_2 gas, suggesting that concentration of oxygen vacancy and/or zinc interstitial decreased by annealing. Therefore, it is considered that the increase of VIS emission intensity by annealing is not due to the increase of oxygen vacancy concentration. Assuming that VIS emission is originated by $V_O^{\bullet}$, the increase of VIS emission intensity by annealing means that concentration of $V_O^{\bullet}$ increases by annealing in spite of the decrease in total concentration of oxygen deficiency. One of the other possible ways to explain the increase of VIS emission intensity by annealing is to consider the change in chemical state of impurity by annealing. The higher energy component of VIS emission is assignable to intra-atomic transition of Cu ion[3] and the lower energy component to donor-to-acceptor transition[7], since position dependence of CL spectra seems to correspond with depth profile of Cu and Li concentration. It is expected that chemical state of Cu and Li is modified by structural reconstruction and/or oxidation via annealing, since Fermi level sifts by decrease of oxygen vacancy concentration by annealing and the impurity level in band gap sifts by structural reconstruction around impurity ions. It is plausible that impurity concentration is one of the important factors to discuss the CL spectra of ZnO. In conclusion, it is worth to note that the controlling of the chemical state of impurities and/or defects is important to increase the UV emission efficiency of ZnO crystal as well as purification of the crystals.

References

1 D.C. Reynolds, D.C. Look and B. Jogai, *J. Solid State Commun.* **99** (1996) 873.

2 Z.K. Tang, P.Yu, G.K.L.Wong, A.Ohtomo, H.Koinuma, Y.Segawa, M.Kawasaki, *J. Solid State Commun.* **103**(1997) 459.

3 D.M.Kolb and H.-J.Schulz, *Current Topics in Materials Science*, vol. 7. pp226, (ed. E.Kaldis), North-Holland Publishers Co., Amsterdam, 1981.

4 T.Sekiguchi, N.Ohashi, Y.Terada; *Jpn. J. Appl. Phys.* **36** (1997) L289

5 K.Vanheusden, W.L.Warren, C.H.Seager, D.R.Tailant, J.A.Voigt and B.E.Grande, *J. Appl. Phys.* **79** (1996) 7983

6 T.Ohgaki, Graduation thesis, Tokyo Institute of Technology (1997), *in Japanese*.

7 H.-J.Egelhaaf and D.Oelkrug, *J. Crys. Growth.*,**161**(1996)190

Materials Science and Engineering Serving Society
S. Sōmiya, R.P.H. Chang, M. Doyama and R. Roy (Editors)
1998 Elsevier Science B.V.

Mechanical Properties at High Temperature of High Purity Mullite Ceramics

Hiroshi OHNISHI, Katsumi MAEDA, Saburo KOSE and Toshio KAWANAMI
NIKKATO CORP. 3-2-24, Orionocho, Sakai 590 Japan

1. Abstract

The mechanical properties at high temperature, which were the flexural strength and creep resistance of high-purity mullite ceramics were investigated.

The Al_2O_3/SiO_2 ratio of mullite samples were 72wt% Al_2O_3/28wt% SiO_2 and 75wt% Al_2O_3/25wt% SiO_2. Some samples contained a little ZrO_2 due to prepared using ZrO_2 media. All samples were fabricated using the Sol-Gel method and sintered near theoretically. The strength at R. T. depended not on the Al_2O_3/SiO_2 ration, and that increased as the grain size decreased. On the other hand, the ratio of the strength at high temperature for at R. T. was influenced not only on the Al_2O_3/SiO_2 ratio, but also on the grain size. And also, the degradation ratio decreased with increasing the amount of alkalis and ZrO_2. The strain-rate of the composition with Al_2O_3-rich, little alkalis and ZrO_2, and larger grain size was smaller. The fracture surface of the composition with lower strain-rate indicated the small slow crack growth region. The Al_2O_3/SiO_2 ratio, the amount of alkalis and ZrO_2, and the grain size dependence of the strain-rate showed similar tendency to those of the ratio of strength at high temperature for at R. T..

Key words: mullite, Al_2O_3/SiO_2 ratio, alkalis, ZrO_2, strain-rate

2. Introduction

Mullite is not a stoichiometric composition, as shown in the phase equilibrium diagrams for Al_2O_3-SiO_2 system reported by Aksay et al.[1] and Prochazka et al.[2]. Therefore, the mechanical properties at high temperature should change as a function of Al_2O_3/SiO_2 ratio. Ohnishi et al. revealed that mullite ceramics with 74-76wt% Al_2O_3 exhibited smaller degradation of strength and high creep resistance[3], and reported impurity dependence of mechanical properties for near stoichiometric composition was investigated[4]. They, however, have been investigated the influence of the relationship between Al_2O_3/SiO_2 ratio and impurity on the mechanical properties.

The purpose of this work, the influence of impurity of mullite ceramics prepared using the Sol-Gel method on the mechanical properties at high temperature were investigated.

3. Experimental Procedure

Mullite powder was prepared using the Sol-Gel method. An Al_2O_3 sol and a SiO_2 sol were mixed in two ratios which were 72wt% Al_2O_3/28wt% SiO_2 to 75wt% Al_2O_3/25wt% SiO_2, and subsequently calcined. The powder was ground by a ball mill using high-purity Al_2O_3 and zirconia media (NIKKATO CORP. OSAKA, Japan). The reason to use ZrO_2 media was to improve sinterbility. The ground powder was isostatically pressed at 98 MPa and sintered at 1600 to 1750°C in air. Phase identification was performed by X-ray diffraction. The flexural strength was measured at R. T. and 1500°C in air in 3-point bending on a 30mm span using 3x4x40mm test specimens at cross-head speed of 0.5mm/min. Creep tests were conducted at 1500°C in air using 4-point bending on a 30mm lower span and a 10mm upper span using 3x10x45mm test specimens. The strain-rate was calculated using the equations derived by Hollenberg et al.[5].

The slow crack growth was evaluated by observation of the fracture surface of specimen, which was fractured at 1500°C in air with Vickers indentation at 98 N by 3-point bending on a 30mm span at cross-head speed of 0.05mm/min. The microstructure were observed by scanning electron microscopy (SEM) after etching, and the average grain size was determined by the intercept method[6].

4. Results and Discussion

X-ray diffraction analyses revealed that all samples were composed mullite crystal.

Figure 1 show some representative microstructures of samples. Sample B was annealed sample A to obtain coarser grained sample. Samples A, B and C with near stoichimetric composition contained enlonged and granular grains mixture. On the other hand, samples D and E with Al_2O_3-rich composition were composed granular grains. For samples A, B and D, a little glassy phase was observed at grain boundary, and samples C and E had little glassy phase as reported by Kanzaki et al.[7].

Table 1 shows the chemical compositions, density, flexural strength and average grain size of each sample. The ratio of strength means the degradation ratio of strength at 1500°C for at R. T..

Table 1 Chemical compositions, density, flexural strength and average grain size of each sample

Sample		A	B	C	D	E
Al_2O_3/SiO_2 (wt%)		71.8/27.0	←	71.1/28.6	74.3/25.1	74.8/25.0
Alkali (wt%)		0.05	←	0.03	0.05	0.03
ZrO_2 (wt%)		0.40	←	tr	0.40	tr
Density (g/cm^3)		3.12	3.12	3.12	3.10	3.12
Flexural strength (MPa)	R.T.	303	215	292	270	192
	1500°C	144	127	216	189	143
Ratio of strength (σ 1500°C/σ R.T.)		0.48	0.59	0.74	0.70	0.74
Average grain size (μ m)		4.4	10.7	2.9	7.2	11.0

Samples A, B and D contained a little ZrO_2, because these samples were fabricated by grinding using ZrO_2 media. In addition, these samples contained larger amount of alkalis, which had a tendency to form glassy phase at grain boundary. All samples had 98% of theoretical density and greater. The strength at R. T. increased as the grain size decreased. Though Al_2O_3/SiO_2 ratio of sample A was the same as that of sample B, the ratio of strength of sample A was smaller than that of sample B due to smaller grain size of sample B. Sample C with almost the same Al_2O_3/SiO_2 ratio as sample A and B showed lower degradation of strength, in spite of smaller grain size than samples A and B. The ratio of sample D, which had the same amount of alkalis and ZrO_2 as sample A, was smaller than samples A and B, because of Al_2O_3-rich composition as reported Ohnishi et al.[3]. However, the ratio of sample D with Al_2O_3-rich composition exhibited higher than sample C with. around the stoichiometric composition. Sample E showed the highest ratio of strength, because it had Al_2O_3-rich composition, little alkalis and ZrO_2, and larger grain size.

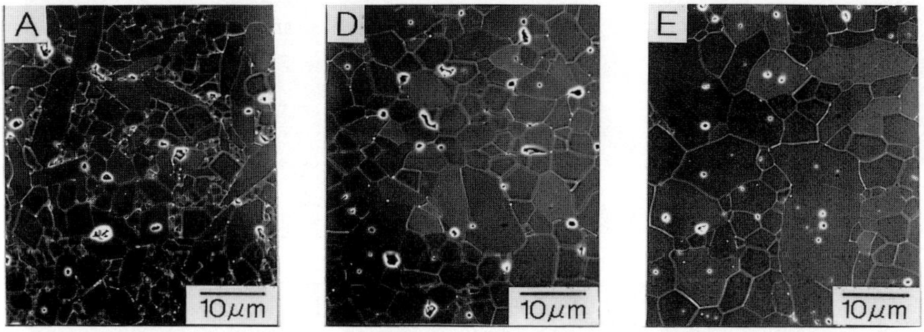

Figure 1. Typical scanning electron micrographs of samples after eching

Figure 2 shows the log-log plots of steady-state creep rates vs. stress from 5 to 120 MPa at 1500°C for each sample. The strain-rate of sample E, which had only mullite crystal and little amount of glassy phase and ZrO_2, was lower than other samples. On the other hand, sample A, which had near stoichiometric composition, large amount of alkalis and ZrO_2 and small grain size, indicated higher strain-rate.

Figure 3 shows the SEM micrographs of the fracture surface of samples A and E with Vickers indentation, which was fractured at 1500°C. Remarkable intergranular fracture was observed in the slow crack growth region, and the transgranular fracture is predominant in the out side of the region.

The slow crack growth region of sample A was larger than that of sample E. These results showed the strain-rate was larger as the slow crack growth was larger. In the same composition, sample C showed lower strain-rate than samples A and B. In addition, the strain-rate of sample C was lower than that of sample D with Al_2O_3-rich composition. But, the stress-exponent, n, of sample C was higher that of samples A, B and D even at lower stress. Moreover, the stress-exponent of samples A, B and C increased further at high stress. These increasing exponent indicated that the creep mechanism changed to dislocation creep or interface-reaction controlled deformation[8]. The Al_2O_3/SiO_2 ratio, the amount of alkalis and ZrO_2, and the grain size dependence of the strain-rate showed similar tendency to those of the ratio of strength at high temperature for at R. T..

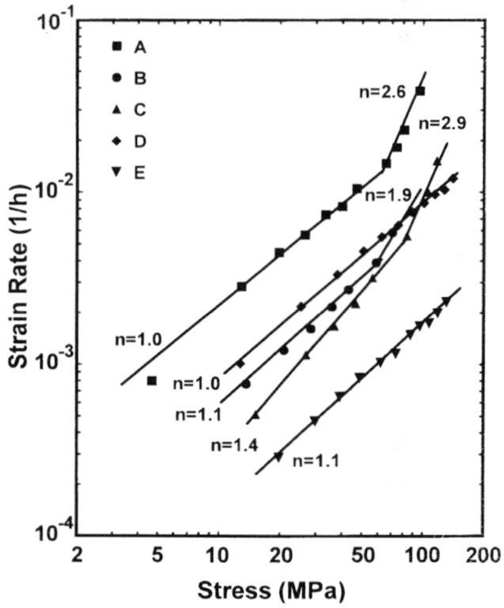

Figure 2. Steady-state strain vs. stress at 1500°C for each sample

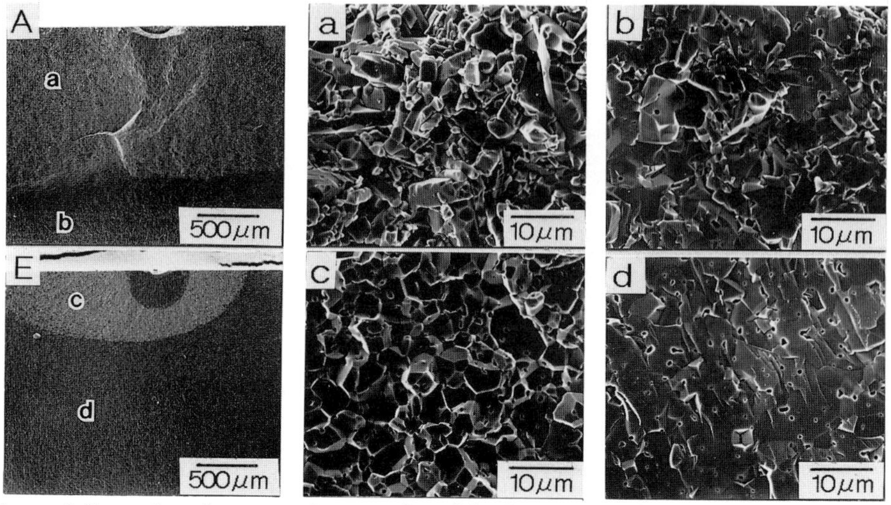

Figure 3. Scanning electron micrographs of the fracture surface of sample A and E with Vickers indentation

5. Conclusion

The mechanical properties at high temperature of high-purity mullite with 72wt% Al_2O_3/28wt% SiO_2 and 75wt% Al_2O_3/25wt% SiO_2 ratio using powder fabricated by the Sol-Gel method were investigated.

1. The strength at R. T. depended not on Al_2O_3/SiO_2 ratio and the amount of alkalis and ZrO_2, but on the grain size.
2. In the same amount of alkalis and ZrO_2, the ratio of strength of 75wt% Al_2O_3/25wt% SiO_2 composition was higher than near stoichiometric compositions. However, the near stoichimetric compositions with little amount alkalis and ZrO_2 showed higher ratio of strength than 75wt% Al_2O_3/25wt% SiO_2 compositions with larger.
3. The difference of the strain-rate of each sample showed the same tendency of the ratio of strength. The strain-rate of 75wt% Al_2O_3/25wt% SiO_2 composition with little amount of alkalis and less ZrO_2 indicated the lowest of all.
4. The slow crack growth region of sample, which indicated lower strain-rate, was smaller, while that showed higher strain-rate was larger.

Reference

[1]I. A. Aksay and J. A. Pask, *J. Am. Ceram. Soc.*, **58** [11-12] 507-12 (1975)
[2]S. Prochazka and F. J. Klug, *J. Am. Ceram. Soc.*, **66** [12] 874-80 (1983)
[3]H. Ohnishi, K. Maeda, T. Nakamura and T. Kawanami, "Ceramic Transactions Volume 6" Edited by Somiya et al. The American Ceramic Society, Inc. pp. 605-612 (1990)
[4]H. Ohnishi, T. Kawanami, A. Nakahira and K. Niihara, *Nippon Seramikkusu Kyokai Gakujutsu Ronbunshi,* **98** [6] 541-47 (1990).
[5]G. W. Hollenberg, G. R. Terwilliger and R. S. Gordon, *J. Am. Ceram. Soc.*, **54** [4] 196-99 (1971).
[6]R. L. Fullman, J. Metal., [3] 447-52 (1953).
[7]S. Kanzaki, O. Abe, M. Ohashi and H. Tabata, Processing of 2nd International Symposium of Ceramic Components for Engines, (1986) pp. 625-631.
[8]Y. Okamoto, H. Fukutome, K. Hayashi and T. Nishikawa, *J. Eur. Ceram. Soc.*, **6** 161-68 (1990).

Materials Science and Engineering Serving Society
S. Sōmiya, R.P.H. Chang, M. Doyama and R. Roy (Editors)

New theory on fracture toughness of polycrystalline alumina in consideration of fracture toughness of grain and grain boundary

Junichi Tatami[a], Kouichi Yasuda[b], Yohtaro Matsuo[b], Shiushichi Kimura[c]

[a]Reserch Fellow of the Japan Society for the Promotion of Science
2-12-2 Ookayama, Meguro-ku, Tokyo 152, Japan

[b]Department of Inorganic Materials, Faculty of Engineering, Tokyo Institute of Technology
2-12-2 Ookayama, Meguro-ku, Tokyo 152, Japan

[c]Institute of Inorganic Synthesis, Faculty of Engineering, Yamanashi University
7, Miyamae-cho, Kofu-shi 400, Japan

Fracture toughness of polycrystalline alumina was formulated as a function of grain size and fracture toughness of grain and grain boundary focusing on crack path. The fracture toughness of grain and grain boundary were derived from the fracture toughness of single crystal alumina and alumina bicrystal. Comparison between the theory and the experimental results showed that the theory explained the grain-size dependence of fracture toughness of polycrystalline alumina.

1. INTRODUCTION

In order to improve fracture toughness of ceramics, many theories have been proposed. However, they can not be applied to the material whose a crack extends in a different way which is assumed a priori in the theories. In our previous study, we directly observed crack propagation behavior of ceramics in scanning electron microscope[1]. As a result, it was shown that the elementary processes of crack propagation of polycrystalline ceramics were composed of transgranular fracture and intergranular fracture. Although the fracture toughness of polycrystalline ceramics should be related to the fracture toughness of grain and grain boundary and crack path, such a theory has not been carried out. Purpose of this study is that a new theory of the fracture toughness of polycrystalline ceramics is framed in consideration of the fracture toughness of grain and grain boundary. In addition, the fracture toughness of grain and grain boundary is obtained by the measurement of fracture toughness of single crystal and alumina bicrystal. Finally, the validity of this theory is investigated by comparison between the theory and the experimental result of the fracture toughness of polycrystalline alumina.

2. THEORY ON FRACTURE TOUGHNESS OF POLYCRYSTALLINE ALUMINA

In this theory, as shown in Figure 1, it is assumed that a two-dimensional polycrystal consists of grain and grain boundary and that crack arrives at any location on the grain boundary in the y-direction with equal probability. The released energies are calculated when a crack propagates transgranularly and intergranularly. On the assumption that the path which has the smaller released energy occurs as the actual crack path, the expected value of the fraction of transgranular fracture and the critical energy release rate of polycrystalline ceramics are derived as follows,

$$\hat{f} = \frac{d - 2y_C}{d} \tag{1}$$

$$\hat{G}_{IC} \equiv G_{IC}^{\ grain}\hat{f} + G_{IC}^{\ boundary}\left\{1 + \frac{p}{6d}y_C^{\ 2} + \frac{q}{2d}y_C\right\}\left(1 - \hat{f}\right) \tag{2}$$

where $G_{IC}^{\ grain}$ and $G_{IC}^{\ boundary}$ are the critical energy release rate of grain and grain boundary respectively. y_c can be expressed as follows:

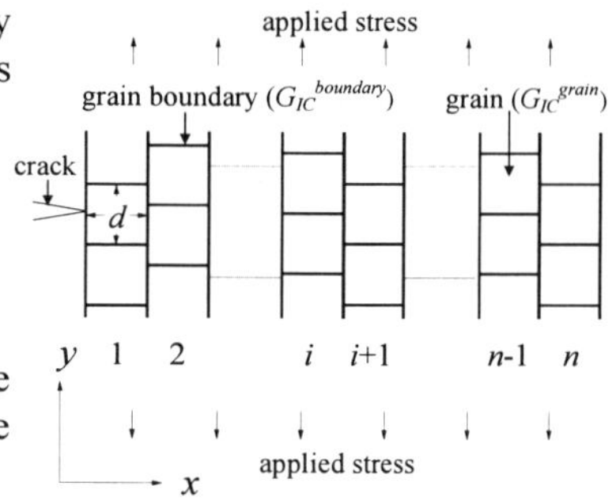

Figure 1. Model of microstructure

$$y_C = \begin{cases} \dfrac{-q + \sqrt{q^2 + 2\left(\dfrac{G_{IC}^{\ grain}}{G_{IC}^{\ boundary}} - 1\right)pd}}{p} & (d \geq d^*) \\ \dfrac{d}{2} & (0 < d < d^*) \end{cases}$$

. Here, $$d^* = \frac{8\left(\dfrac{G_{IC}^{\ grain}}{G_{IC}^{\ boundary}} - 1\right) - 4q}{p} \tag{3}$$

p and q is estimated by linear approximation of the apparent critical energy release rate during crack deflection ($\widetilde{G}_{IC}^{\ deflection}(t)$) as follows:

$$\frac{\widetilde{G}_{IC}^{\ deflection}(t)}{G_{IC}^{\ boundary}} \doteqdot p'\left(\frac{t}{2a}\right) + q = pt + q \tag{4}$$

where t is deflection length and a is crack length. Consequently, the fracture toughness of polycrystalline ceramics was formulated as the fraction of transgranular fracture (namely crack path), grain size and the fracture toughness of grain and grain boundary.

3. FRACTURE TOUGHNESS OF GRAIN AND GRAIN BOUNDARY OF POLYCRYSTALLINE ALUMINA

3.1. Fracture toughness of grain of polycrystalline alumina

The fracture toughness of single crystal alumina was measured on four crack propagating planes and two crack propagating directions for each plane by controlled surface flaw method and single edge V-notched beam method. Figure 2 shows the experimental results on critical release energy rate of single crystal alumina. The fracture toughness of single crystal alumina varied from 16 Jm^{-2} to 65 Jm^{-2} depending on the crystal orientation. The fracture surfaces of $\{2\bar{1}\bar{1}0\}$, $\{01\bar{1}0\}$, $\{01\bar{1}2\}$ planes were cleavage, but that of (0001) plane was uneven and consisted of $\{01\bar{1}2\}$ cleavage planes.

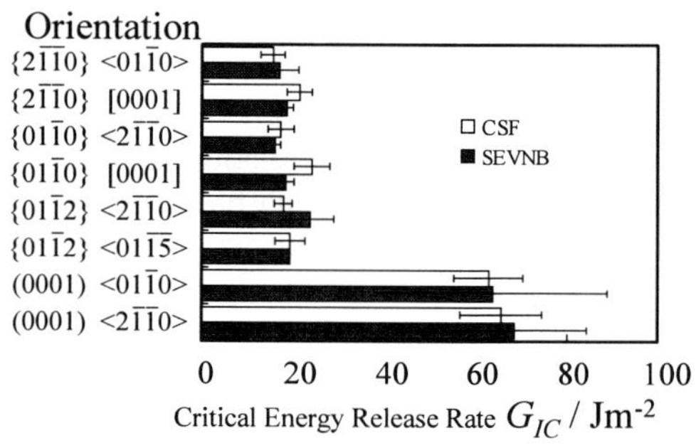

Figure 2. Crystal orientation dependence of the crystal energy release rate of single crystal alumina

The fracture toughness of non-cleavage plane, like (0001) plane, is analyzed on assumption that the composition of cleavage planes which has the smallest released energy occurs as the actual fracture surface. As a result, the fracture toughness of grain was formulated as follows,

$$G_{IC}^{grain} = \min\left[\sum\left\{\left(\frac{K_I^{\infty}}{K_I^{tip,i}}\right)^2 \times G_{IC}^{cleav,i} \times \alpha_i\right\}\right] \quad (5)$$

where the first term in brace is the effect of decrease in apparent stress intensity factor by crack deflection, which is evaluated by elastic anisotropy of single crystal alumina and the geometrical relationship between cleavage plane and main crack. The second term is the fracture toughness of cleavage plane and the third term is the effect of increase in fracture surface area. By substituting experimental results for $G_{IC}^{cleave,i}$, as shown in Figure 3, the stereographic projection of the fracture toughness of grain of polycrystalline alumina was calculated. As a calculation, it was shown that the fracture toughness of grain of polycrystalline alumina varied from 18Jm^{-2} to 108Jm^{-2} depending on the crystal orientation, and its average was 42Jm^{-2}.

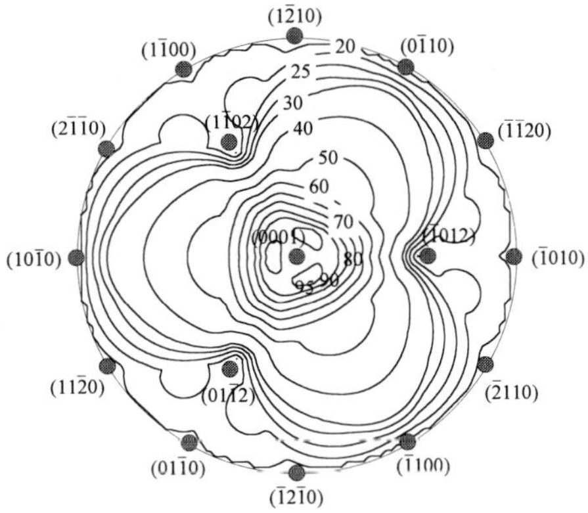

Figure 3. Stereographic projection of the critical energy release rate of single crystal alumina.

3.2. Fracture toughness of grain boundary of polycrystalline alumina

Alumina bicrystals were fabricated by joining two alumina single crystals at 1900°C for 24 hours in Ar flow. The fracture toughness was measured by single edge V-notched plate tension test. The critical energy release rate of (0001) Σ=3 twist boundary of alumina was 16 Jm^{-2} and larger than energy for creating new surfaces, $2\gamma_s^{(0001)\Sigma=3}$(=10Jm^{-2}) which was calculated by surface energy and boundary energy. It seems that not only energy for creating new surfaces but also plastic dissipative energy contributes to the critical energy release rate of (0001) Σ=3 twist boundary of alumina. Based on this concept, the critical energy release rate can be expressed as follows

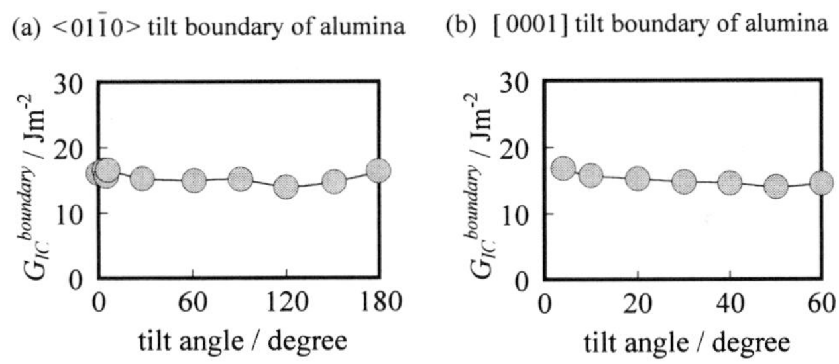

Figure 4. The dependence of the tilt angle on the critical energy release rate of grain boundary.

$$G_{IC}^{boundary} = \left(2\gamma_{surface} - \gamma_{boundary}\right) + 2\gamma_{plasticity} = (1+\chi)\gamma_{surface}\left(2 - \frac{\gamma_{boundary}}{\gamma_{surface}}\right) \quad (6)$$

Here, it is assumed that the plastic dissipative energy is proportional to the energy for creating new surfaces. χ is the proportional constant. From substitution 0.38 which was experimental result [2] for $\gamma_{boundary} / \gamma_{surface}$ in Equation (6) on (0001) Σ=3 twist boundary of alumina, we obtained 10 Jm^{-2} as the value of $(1+\chi)\gamma_{surface}$. By substituting the value of $(1+\chi)\gamma_{surface}$ of (0001) Σ=3 twist boundary of alumina and the boundary energy on the previous experimental results [2-3] into Equation (6), as shown in Figure 4, the relationship between $G_{IC}^{boundary}$ and tilt angle of grain boundary was obtained. As a result, it was shown that $G_{IC}^{boundary}$ is not so dependent on the tilt angle. Furthermore, their average is 15Jm^{-2} in both cases. Consequently, in this study, the average of the fracture toughness of grain boundary of alumina was regarded as 15Jm^{-2}.

4. COMPARISON BETWEEN THEORY AND EXPERIMENTAL RESULT ON FRACTURE TOUGHNESS OF POLYCRYSTALLINE ALUMINA

Polycrystalline Al_2O_3 with average grain size from 1.8μm to 50μm was fabricated by hot pressing, and fracture toughness was measured by controlled surface flaw method and chevron notched beam method. As shown in Fig. 5, the experimental results showed that the fracture toughness of polycrystalline alumina increased with an increase in grain size. Solid curves are theoretical curves which were estimated by Equation (2) using the fracture toughness of grain and grain boundary. Comparison between the theory and the experimental results showed that the theory explained the grain-size dependence of fracture toughness of polycrystalline alumina. Consequently, it was shown that the theory on the fracture toughness of polycrystalline ceramics in consideration of the fracture toughness of grain and grain boundary was valid.

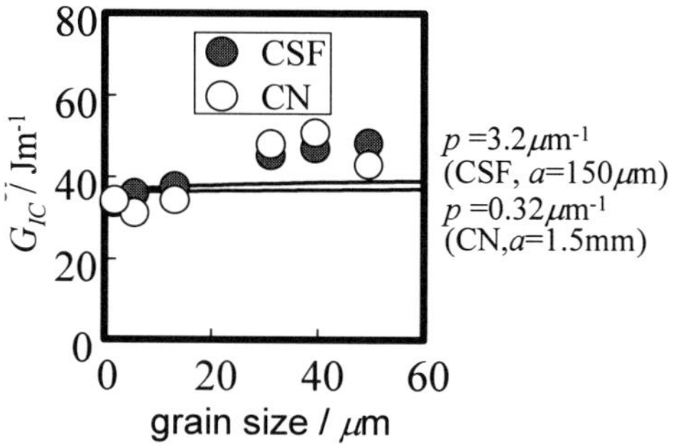

Figure 5. The grain size dependence of the fraction of transgranular fracture. The solid curves are theoretical ones.

5. CONCLUSIONS

Fracture toughness of polycrystalline alumina was formulated as a function of grain size and fracture toughness of grain and grain boundary focusing on crack path. The fracture toughness of grain and grain boundary were derived from the fracture toughness of single crystal alumina and alumina bicrystal. Comparison between the theory and the experimental results showed that the theory explained the grain-size dependence of fracture toughness of polycrystalline alumina.

REFERENCE

1. K. Yasuda, J. Tatami, Y. Matsuo and S. Kimura, J. Mater. Sci. Soc. Jpn, 33 (1997) 238.
2. J. F. Shackelford and W. D. Scott, J. Am. Ceram. Soc., 51 (1968) 688.
3. I. Sawaki, The graduation thesis of Tokyo Institute of Technology (1980).

Materials Science and Engineering Serving Society
S. Sōmiya, R.P.H. Chang, M. Doyama and R. Roy (Editors)

Preparation of high purity magnesium oxide and magnetic ferrite from chromium containing residue

I. Mathematical model and experiments under atmospheric carbonate leaching

Tao Qi* **Yi Zhang** **Yuehua Guo** **Zuohu Li**

Institute of Chemical Metallurgy, Chinese Academy of Sciences, Beijing, P.R. China 100080
Fax: (86)-10-62561822, E-mail: yizh@home.icm.ac.cn

Abstract A practical and economical new technology is developed to treat the chromium containing residue which consists of carbonate leaching of the residue — extraction of valuable magnesium — chromium detoxification — preparation of magnetic ferrite. In this paper effects of reaction temperature, CO_2 partial pressure, liquid to the residue ratio, addition of ammonium carbonate and applying ultrasonic field on the yield of magnesium in carbonate leaching experiments are investigated under the atmospheric condition. The results calculated by the unreacted shrinking core model (USCM) show good agreement with the experimental data. The new technology will bring about high environmental and economic benefits.

1. INTRODUCTION

Chromate production is usually a source of severe pollution among the basic chemicals industries. The highly poisonous chromium containing residue discharged from the traditional roasting plant ranks the top of the pollution source of solid waste and presents potential threat to underground water, rivers and marine area. Although there are a number of reports and patents for treating chromium containing residue, few economic and rational projects are presented to treat the residue at the very source. The general method is to fill it in land or stack it after the hexavalent chromium is detoxified. It is also used to manufacture ceramic, cement and brick et al.

The purpose of this papers is to present a new method to extract the valuable magnesium compound in the residue and prepare magnetic ferrite with the treated residue by reduction roasting, so that zero discharge of chromium containing residue is achieved. The new technology is believed to be a significant move to comprehensive utilization of chromium containing residue.

2. MODELING OF CARBONATE LEACHING

The unreacted shrinking core model (USCM) is widely used for fluid reactions because of the advantages of simplicity[1]. In this paper USCM is utilized to describe the carbonate leaching of chromium containing residue. Although the residue particles are porous, magnesium is distributed in the particles evenly. The carbonate leaching reaction can be represented as

$$1/2Mg(OH)_2(s)+H_2CO_3(l) \longrightarrow 1/2\ Mg(HCO_3)_2\,(l)+ H_2O(l) \quad (1)$$

* Author to whom correspondence should be addressed.

The basic equation in the shrinking core model for a spherical particle is derived from a differential mass balance for the fluid reactant diffusing through the ash layer formed around the particle due to reaction[2]

$$D_e\left(\frac{\partial^2 C_A}{\partial r^2}+\frac{2}{r}\frac{\partial C_A}{\partial r}\right)=\varepsilon\frac{\partial C_A}{\partial t} \tag{2}$$

The initial conditions and boundary conditions are[2]

$$t=0, C_A=0 \tag{3}$$

$$r=R, D_e\frac{\partial C_A}{\partial r}=k_f(C_{AO}-C_A) \tag{4}$$

$$r=r_c, D_e\frac{\partial C_A}{\partial r}=k_S C_A \tag{5}$$

At the surface of the shrinking core and ash layer there is the following relation

$$-\frac{\partial r}{\partial t}=\frac{1}{2}k_S\frac{C_A}{C_{BO}}=\frac{bD_e}{C_{BO}}\cdot\frac{\partial C_A}{\partial r} \tag{6}$$

The assumption that there is no reactant accumulation and convection in ash layer leads to a great simplification of equation (2)

$$D_e\left(\frac{\partial^2 C_A}{\partial r^2}+\frac{2}{r}\frac{\partial C_A}{\partial r}\right)=0 \tag{7}$$

The analytical solution of equation (7) basing on the initial and boundary conditions is[2]

$$t=\frac{2R^2C_{BO}}{C_{AO}}\left[\frac{1}{2k_S R}(1-\xi_c)+\frac{1}{6D_e}(1-\xi_c)(1+\xi_c-2\xi_c^2)+\frac{1}{3k_f R}(1-\xi_c^3)\right] \tag{8}$$

When the overall rate is controlled solely by the diffusion of the liquid reactant within the ash layer, k_s ,k_f>>D_e, equation (8) is reduced to

$$t=\frac{R^2C_{BO}}{3D_e C_{AO}}\left[1-3\xi_c^2+2\xi_c^3\right]=t^*\left[1-3\xi_c^2+2\xi_c^3\right] \tag{9}$$

where C_{A0} is the concentration of liquid reactant; C_{B0} is molar density of solid reactant; D_e is the effective diffusion coefficient within ash layer; R is the diameter of solid particles; t is the reaction time; t^* is the complete conversion time of the solid reactant, $t^*=R^2\cdot C_{B0}/(3D_e\cdot C_{A0})$; $\xi_c=r_c/R$ is the dimensionless radius. The fractional conversion of solid reactant is $X_b=1-(r_c/R)^3=1-\xi_c^3$. The concentration of liquid reactant(HCO_3^{-1}) is equal to its equilibrium concentration

$$C_{AO}=\sqrt{K_a\cdot P\cdot H} \tag{10}$$

where K_a is the first order dissociation constant of carbonic acid acid; P is the partial pressure of CO_2; H is the Herry constant of CO_2 dissolved in water.

3. EXPERIMENTAL

Atmospheric carbonate leaching was conducted in a 1L bubble column with a sintered glass distributor and a thermostat to control the reaction temperature within ± 0.5 ℃. The partial pressure of carbon dioxide was controlled by regulating the flow rate of CO_2 and N_2 with rotameters. Gaseous CO_2 of 99.5% purity and N_2 of 99.9% purity were from Praxair Company. After slurry of the residue and water was stirred with ultrasonic for 5 minutes, it was filled into the column and heated to the desired temperature. A small amount of $FeSO_4$ was added to slurry to reduce Cr^{6+} in the residue. Slurry samples were collected at selected intervals through a sampling valve and filtered. The magnesium concentration in the solution was analyzed by

Atomic Absorption Spectrometry (AA) after the solution is diluted to a certain concentration range. Iron and magnesium in the residue are measured by AA, too. Slurry is filtered after carbonate leaching. Magnesium carbonate light (MCL) is precipitated while the solution is heated at 102 ℃ for 10 minutes.

4. RESULTS AND DISCUSSION

4.1. The composition of chromium containing residue

The residue used in experiment was obtained from molten salt oxidizing of chromite in NaOH medium without burden. The typical composition of chromium containing residue is 14.2% Mg, 40% Fe, 0.59% Cr ,0.35% Si, 0.60% Al and 1.41% ΣNa. The X-ray diffraction of chromium containing residue indicates that $Mg(OH)_2$ and $1.08Na_2O{\cdot}Al_2O_3{\cdot}\,1.68SiO_2\,{\cdot}1.81H_2O$ are crystalline but Fe_2O_3 is amorphous. The Electron Scanning Micrography (SEM) shows that the chromium containing residue pellets are porous and consist of grains about 1-10μm.

4.2. Effect of reaction temperature on the yield of magnesium

Fig.1 shows that the yield of magnesium decreases with the increasing of reaction temperature because of decomposition of $Mg(HCO_3)_2$ with temperature above 26 ℃ at the ordinary atmosphere. The decomposition temperature of $Mg(HCO_3)_2$ is raised when the CO_2

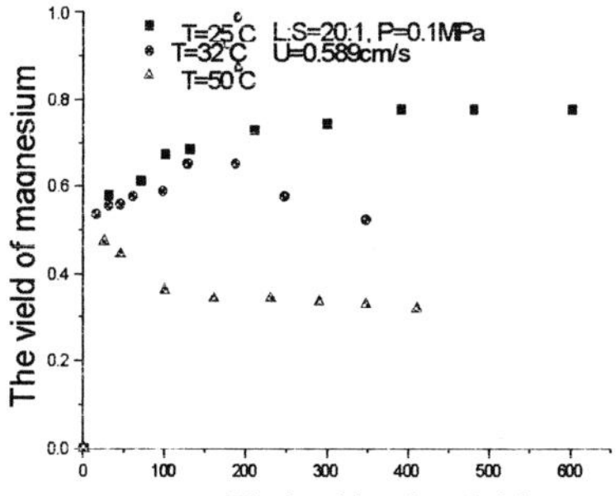

Fig. 1 Effect of reaction temperature on the yield of magnesium

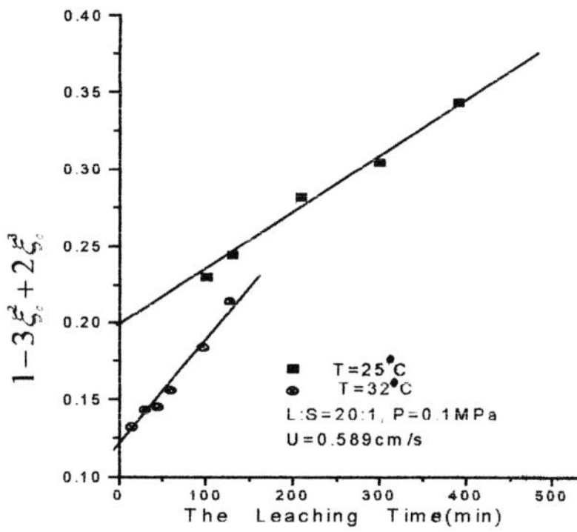

Fig. 2 The calculated results of USCM

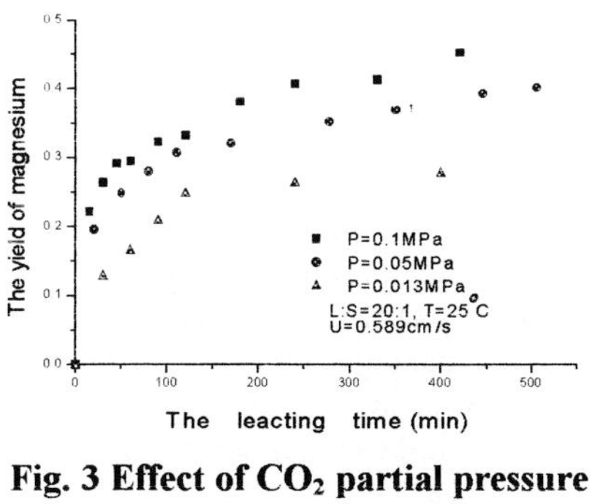

Fig. 3 Effect of CO_2 partial pressure on the yield of magnesium

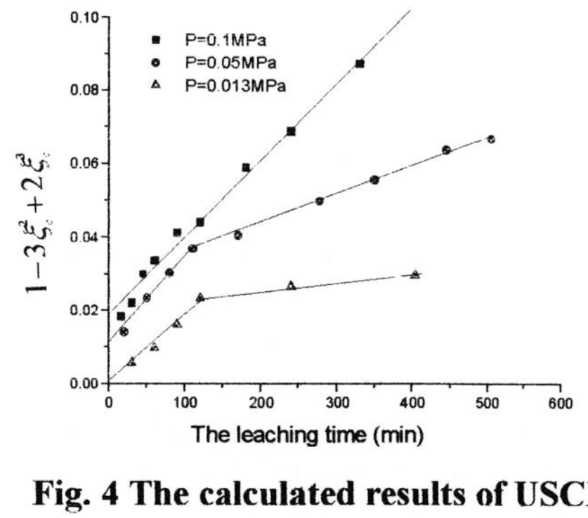

Fig. 4 The calculated results of USCM

partial pressure in the reactor is increased. Fig. 2 shows that the calculated results of equation(9) fits the experimental data well. In model calculation, the fractional conversion of solid reactant $X_b=C_{Bi}\cdot V/(C_{B0}\cdot W)$, where C_{Bi} is the molar concentration of magnesium at i time; V is the total volume of leaching solution; W is the mass of the residue.

4.3. Effect of CO_2 partial pressure on the yield of magnesium

The CO_2 partial pressure predominates the reactant concentration of HCO_3^- and the equilibrium concentration of $Mg(HCO_3)_2$. Fig.3 indicates that the higher the CO_2 partial pressure, the greater the yield of magnesium. As shown in Fig.4, the calculated results of equation (9) show good agreement with the experimental data for the different CO_2 partial pressure, but there are apparent breakpoints at P_{CO2}=0.05MPa and 0.013MPa.

4.4. Effect of liquid to chromium containing residue ratio on the yield of magnesium

The liquid to chromium containing residue ratio is an important factor for the yield of magnesium and energy consumption. Fig.5 indicates that the higher liquid to solid ratio results in the higher yield of magnesium.

4.5. Effect of ammonium carbonate and ultrasonic field on the yield of magnesium

In order to improve the reaction rate, addition of ammonium carbonate and application of ultrasonic field are used to enhance the leaching process as shown in Fig. 6. Although the reactant concentration HCO_3^- is increased in ammonium carbonate solution, the yield of magnesium decreases because the pH and ionic strength are higher than that of no ammonium carbonate solution.

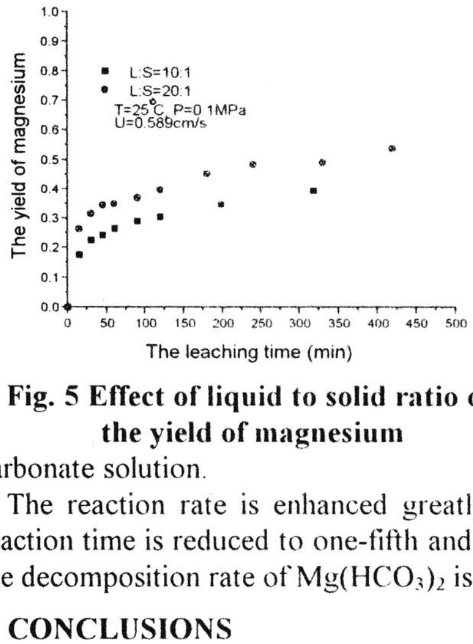

Fig. 5 Effect of liquid to solid ratio on the yield of magnesium

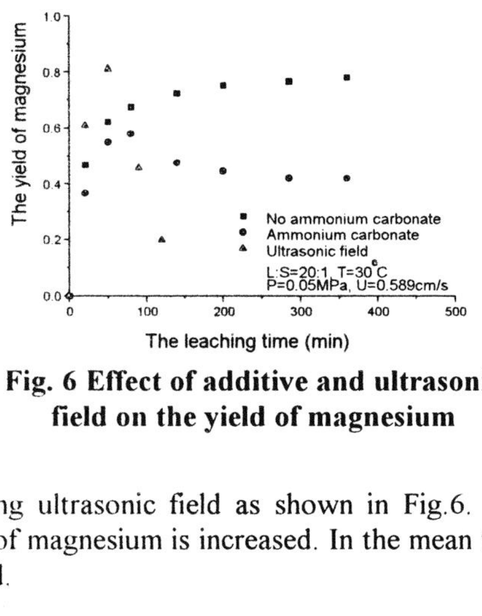

Fig. 6 Effect of additive and ultrasonic field on the yield of magnesium

The reaction rate is enhanced greatly by adding ultrasonic field as shown in Fig.6. The reaction time is reduced to one-fifth and the yield of magnesium is increased. In the mean time the decomposition rate of $Mg(HCO_3)_2$ is developed.

5. CONCLUSIONS

The carbonate leaching is an effective method to extract magnesium from chromium containing residue under ambient conditions and the overall rate is controlled by the diffusion of the reactant within particles. The yield of magnesium is increased with decreasing of the reaction temperature and increase of liquid to chromium containing residue ratio. The high CO_2 partial pressure and application of ultrasonic field result in a high yield of magnesium. The calculated results based on USCM show good agreement with the experimental data.

[1] Lindman N. and Simonsson D., Chem. Eng. Sci., 1979, **34**, 31
[2] Yang Shouzhi, Engineering Chemistry & Metallurgy(in Chinese), 1984, **5(4)**, 1

Materials Science and Engineering Serving Society
S. Sōmiya, R.P.H. Chang, M. Doyama and R. Roy (Editors)

Preparation of high purity magnesium oxide and magnetic ferrite from chromium containing residue

II. Pressurized carbonate leaching and preparation of magnetic ferrite

Yuehua Guo **Tao Qi*** **Yi Zhang** **Zuohu Li** **Jiashu Liu**

Institute of Chemical Metallurgy, Chinese Academy of Sciences, Beijing, P.R. China 100080
Fax: (86)-10-62561822, E-mail: yizh@home.icm.ac.cn

Abstract On the basis of atmospheric carbonate leaching, a new pressurized process is proposed to treat chromium containing residue. Effects of reaction time, reaction temperature, and CO_2 pressure on the yield of magnesium have been studied. Preparation methods of magnesium oxide and magnetic ferrite were investigated.

1. INTRODUCTION

Chromium containing residue, one of the important hazardous solid waste, has caused severe pollution. In recent years, many processes have been proposed to treat it. However, in most of the methods, chromium containing residue is just stacked up or used for building materials burden. No valuable element is recovered from the residue.

In this paper a new process has been developed to extract magnesium compound from the residue by pressurized carbonate leaching and prepare magnesium oxide and magnetic ferrite by thermolysising and reduction roasting, respectively. The residue can be not only detoxified, but also changed into new products.

2. EXPERIMENTAL

Chromium containing residue was obtained by molten salt oxidizing of chromite in NaOH medium, which was quite different from the traditional method of gas-solid roasting. Slurry of the residue and water was fed into a 1L autoclave made of stainless steel with a magnetic stirrer. Gaseous CO_2 of 99% purity was pumped into the autoclave to a certain pressure. The reaction temperature and stirring velocity were controlled automatically. Slurry samples were collected at certain intervals through a sampling valve and filtered. The leaching solution was heated to 102℃ for 10 minutes. The magnesium carbonate light (MCL) was precipitated from the solution. Magnesium oxide was produced by roasting MCL at 900℃ for 5 hours. The mud cake was washed with distilled water and then roasted in the hydrogen and steam atmosphere. Magnetic ferrite was acquired through magnetic screening. The flow diagram of the new process is shown in Fig.1.

---

* Author to whom correspondence should be addressed .

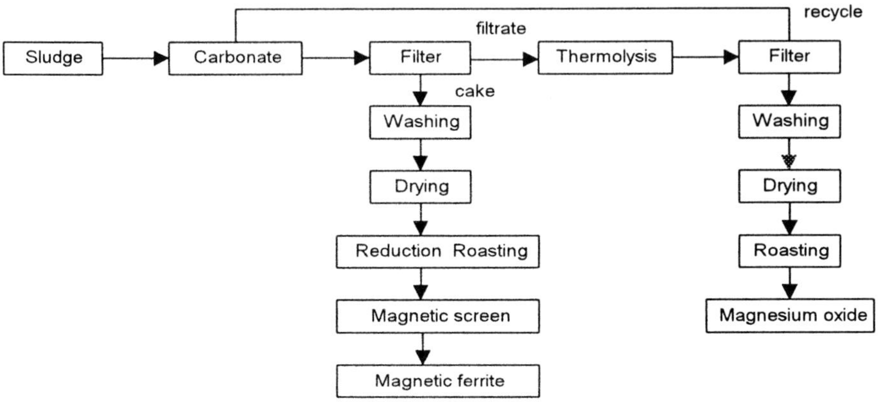

Fig.1 The flow diagram of the new process

3. RESULTS AND DISCUSSION

3.1 Carbonate leaching of chromium containing residue

Reaction mechanism

X-ray diffraction patterns and chemical analysis show that main components in the residue were ferric oxyhydroxide as an amorphous phase and crystalline magnesium hydroxide. Because of the different pH range for dissolving $Fe(OH)_3$ or $Mg(OH)_2$, the pH of the leaching solution is the key factor to separate iron and magnesium. In carbonate leaching process the pH value decreased gradually from 11 to 6.5. Magnesium was dissolved into the solution, but ferrum remained in the residue. Therefore, the pressurized carbonate process is one of the effective way to separate ferrum and magnesium. The reaction mechanism of magnesium hydroxide and carbon dioxide are involved the following steps

$$CO_2(g) = CO_2(aq) \quad (1)$$

$$CO_2(aq) + H_2O = HCO_3^-(aq) + H^+(aq) \quad (2)$$

$$Mg(OH)_2(aq) = Mg^{2+}(aq) + 2OH^- \quad (3)$$

$$Mg^{2+}(aq) + 2HCO_3^-(aq) = Mg(HCO_3)_2(aq) \quad (4)$$

$$Mg^{2+}(aq) + 2HCO_3^-(aq) = MgCO_3(aq) \cdot H_2O + CO_2(g) \quad (5)$$

Leaching kinetics

As shown in Fig.2, the yield of magnesium is increased rapidly at the initial leaching stage. After two hours the leaching rate declined gradually till the yield of magnesium had little increasing. It suggested that the reaction approaches equilibrium. The specific surface area of the residue particles after carbonate leaching is about twice as large as that of the initial residue, being $69.4m^2/g$ and $33.99m^2/g$, respectively. For porous particles, there are many active site on the outer and inner surface. As discussed before, the leaching process is controlled by mass transfer, so all ways of improving mass transfer is favorable to enhance the leaching process.

Effects of reaction temperature and CO_2 pressure on the yield of magnesium

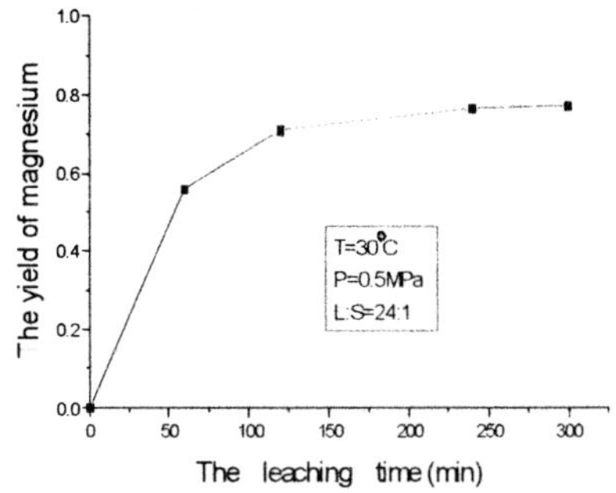

Fig. 2. The yield of magnesium as a function of reaction time

Effects of reaction temperature on the yield of magnesium is shown in Table 1. The yield is improved when reaction temperature was raised from 30 ℃ to 50 ℃. This situation is quite different from that of atmospheric carbonate leaching. In the pressurized system, the decomposition temperature of $Mg(HCO_3)_2$ is increased. Moreover, the carbonate leaching at higher temperature is beneficial to solid-liquid separation.

Table 2 indicates that the yield of magnesium is increased when the pressure is improved from 0.5MPa to 1.1MPa, but the increase is not as high as expected when the pressure is up to 4.0MPa. As discussed above, the process was controlled by mass transfer within the ash layer, so CO_2 pressure has slight effect on the yield of magnesium.

Table 1
Effect of reaction temperature on the yield of magnesium(P=0.55Mpa)

Batch number	Temperature(℃)	Liquid to the residue ratio	Yield of Magnesium
bg-4	30	24:1	71%
bg-6	50	24:1	74%
121-3	20	20:1	56%
121-2	50	20:1	58%

Table 2
Effect of CO_2 pressure on the yield of magnesium(t=2 hours)

Batch number	Pressure(MPa)	Temperature(℃)	Liquid to residue ratio	yield of magnesium
bg-4	0.5	30	24:1	71%
bg-2	4.0	30	24:1	80%
119-1	0.5	30	16:1	74%
119-2	1.1	30	16:1	78%

3.2 Preparation of magnesium oxide

The SEM photo of magnesium carbonate light(MCL) obtained by heating leaching solution illustrates that the particle size is in the range of 1-10μm. It is identified to be $Mg_5(CO_3)_4(OH)_2{\cdot}4H_2O$ from X-ray diffraction patterns. Because there is little other substance in the leaching solution except magnesium and MCL was precipitated as crystallites, it is easy to obtain high purity MCL. Magnesium oxide was prepared by roasting MCL at 900℃ for 5 hours. The SEM image of magnesium oxide is illustrated in Fig.3. It's purity is more than 99.5%. X-ray diffraction pattern of MgO is given in Fig.4.

3.3 Preparation of magnetic ferrite

Fe content in the residue is upgraded to 56% after carbonate leaching. Such ferric-basis substance is suitable to synthesis ferrite materials. Black powder with strong magnetism was acquired by roasting ferric-basis sludge in the atmosphere of hydrogen and steam at 350℃ for 1 hour. In the meantime, Cr^{6+} in residue was reduced to Cr^{3+} and fixed into the spinal lattices completely. Magnetic ferrite can be separated from nonmagnetic component through magnetic screening and used as hard magnetic materials. No chromium was detected in acid and alkali stripping test.

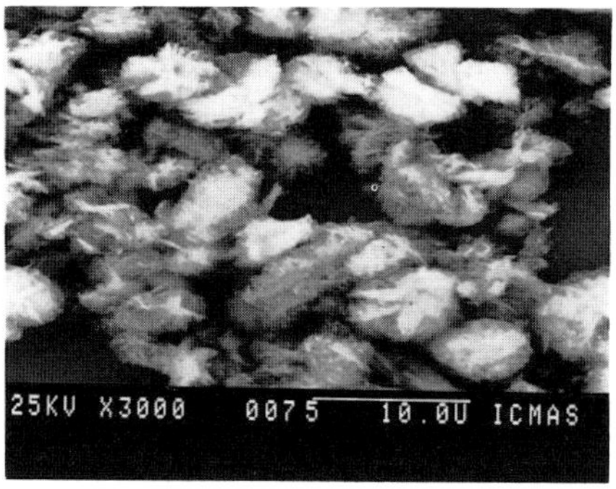

Fig.3 The SEM image of magnesium oxide

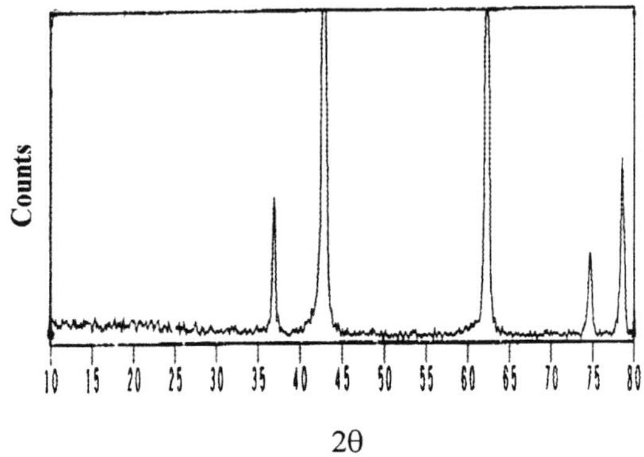

Fig.4 XRD pattern of MgO

4. CONCLUSION

The valuable magnesium compound in chromium containing residue can be extracted by the pressurized carbonate leaching process. The yield of magnesium is increased with the increasing of reaction time, reaction temperature, and CO_2 pressure. The leaching solution and the filtered residue can be used to prepare MgO and magnetic ferrite respectively. The toxic chromium containing residue can be utilized reasonably and comprehensively in this new process.

REFERENCE

1.Li Xiaohua, Environmental Science, 1995, 17(**2**), 49-52.(in Chinese)
2.Ren Qingyu, Environmental Engineering, 1989, 7(**3**), 50-55.(in Chinese)

ACKNOWLEGEMENTS

The authors wish to thank Professor Jiajun Ke and Professor Jiayong Chen of Institute of Chemical Metallurgy, Chinese Academy of Sciences for their helpful and insightful conversitions with appreciation.

Materials Science and Engineering Serving Society
S. Sōmiya, R.P.H. Chang, M. Doyama and R. Roy (Editors)

Concrete, the Gigaphase Ceramic: Its Key Role in Environmentally-Friendly Technology

Della M. Roy

The Pennsylvania State University
217 Intercollege Materials Research Laboratory
University Park, PA 16802, USA

Nanoscale structures which are ubiquitous in nature and have been utilized in industry for decades have received a great deal of attention in materials science. Yet the increasing impact of both the manufacture of some 1000 million (= giga) tons of cement, worldwide, and the creation of several hundred milion tons of fly ash (of cement-like composition) have been ignored. This paper discusses new science connected to the cement industry, including that which could point the way to non-CO_2 generating cementitious reactions.

1. INTRODUCTION

Cement and concrete constitute the largest volume manufactured materials in the world and therefore, it is a significant source of carbon dioxide. This is the result of:

1) decarbonation of limestone in the kiln during manufacturing,

$$5CaCO_3 + 2SiO_2 \rightarrow 3CaO{\cdot}SiO_2 + 2CaO{\cdot}SiO_2 + 5CO_2 \qquad (1)$$

2) use of fossil fuel in the kiln. The production of one ton Portland cement accounts for one ton of CO_2.

In addition, the annual output of fly ashes from power plants and other by-product materials is so enormous that there is a constant need to find new uses for them (1, 2). Approximately 49% of the utility wastes are simply land filled, 41% are contained in surface impoundments and about 10% are disposed of by discharging into old quarry operations. Increasingly, storage is conducted "on site" due to reduced avoided costs to the utilities.

Fly ash is also produced from fluidized bed combustion (FBC), an advanced coal-based energy process that produces energy at comparatively low cost and in an environmentally clean manner (3). But the fly ashes produced using FBC are highly enriched in sulfates, because of the need to remove SO_x from the flue gas. Their high sulfate content prevents their use in the production of ordinary portland cement (OPC). Recent work by Majling and Roy (4) describes the transformation of FBC fly ash into a cementitious material.

In the last two decades, there has been an increasing interest in the use of alkali activated materials as substitutes for Portland cement. Originally, it was limited to the activation of slag cements using alkali metal salts, but recently the alkali-activated cements have extended to include several industrial by-products and wastes. Developing means to utilize large volumes of such materials in blended cements for use in concrete construction for the infrastructure will provide significant economic and environmental benefits. The research reported here describes aluminosilicate materials as alternative cements with at least equivalent properties to Portland cement in order to reduce the CO_2 release. Because of the large volume in which they are used, they may be thought of as gigaphase ceramics which yet depend in their properties on fine tuning at the micro- and nano-structural levels.

At the other end of the spectrum, there is considerable interest in utilizing such by-product materials in the synthesis and processing of high-performance cementitious materials. Recent advances have led to the development of new families of high performance cementitious materials, including very high strength (for cements) materials. Some of these materials cross the boundaries of what have been defined as traditional cementitious materials, and the term chemically bonded ceramics (CBCs) has been used to classify these new materials. CBCs are defined as ceramic-like materials formed as the result of chemical reactions occurring at or near ambient temperatures. Additionally, these cements may be more durable than hydrated portland cement.

We shall discuss briefly some additional advances in processing of CBC materials, including microwave processing, and its application to formation of enhanced performance materials.

2. NEW RAPID SETTING ALKALI-ACTIVATED CEMENT COMPOSITIONS

The utilization of by-product granulated (glassy) blast furnace slag as a part of the cementitious component is now quite well known and used extensively in current commercial products. The commonly accepted mechanism for the hydration of slag is via an activation of the very finely ground glassy slag by portland cement that results in a rather slow initial hydration reaction, but the final products have a fine pore structure. Fly ashes may also be activated in a similar matter, and will be discussed also.

The advantage of utilizing blended cements incorporating by-product materials for many applications has been well documented. However, the use of these materials has been limited by the longer setting times associated with the use of these materials, and recent developments have formulated compositions employing alkali activation to shorten the setting times. The materials have been characterized using calorimetry, microscopic, X-ray diffraction, and mechanical property techniques.

Best known is the ability to activate blast-furnace-slags (BFS) by the addition of alkalis. The F-cements of Finland that utilize a finely ground, well-granulated BFS, together with alkaline activators and a superplasticizer to produce a cement suitable for the precast industry are one example of the utilization of slag activation. When used with thermal curing, these materials have strengths 20 - 30% higher than those found for ordinary portland cements (OPC). These materials typically require approximately 30% less water than conventional cements, which results in lower total porosities. Recent work suggests that it may be possible to utilize other, perhaps less alkaline, materials such as high alumina cements and aluminum sulfates (6) as activators for slags. Dosages in the range of 3 - 10% of NaOH, as Na_2O are utilized, with dosages in the range 3 - 6%, giving optimum strength development (7,8).

Combinations of conventional hydraulic cements with one or more of the class of latent hydraulic materials (i.e., fly ash, blast furnace slag, silica fume, etc.) plus a small amount of activator can lead to materials with high early strength and enhanced durability and/or strength. When the pH of the liquid phase is kept at approximately 12 and sulfate ions coexist, a well-crystallized calcium sulfo-aluminate can be produced (9) or zeolitic hydroaluminosilicates that have the general formula:

$$R_2O \cdot Al_2O_3 \cdot xSiO_2 \cdot nH_2O \qquad (2)$$

With normal short-term curing, x will be approximately 2, while with longer curing times or steam curing, x will be in the range of $2 \le x \le 3$ and with long-term steam curing x will approach 4. Under normal curing conditions, the reaction products will be mostly noncrystalline; however, the crystallinity increases with long curing times, especially at elevated temperatures. One type of composition was developed which is commercially available, the outgrowth of work on alkali-activated aluminosilicate materials originally inspired by studies of ancient building materials (10).

The strength development by accelerated hydration of alkali activated cements is shown in Table 1 (7). We and others have developed dense, low water/cement ratio formulations, employing alkali activation to shorten the setting time and increase the rate of strength development. Chemical compositions of the components used are given in Table 2.

Table 1
Compressive strengths (MPa) for alkali-activated cements and Portland cement cured at 25°C, at W/C = 0.25 (standard deviation in parenthesis); normally cast.

Time	Portland Type III	Alkali-activated P-14
5 hours	not set	23.3 (0.4)
1 day	83.5 (2.8)	64.7 (2.9)
3 days	91.0 (8.7)	60.1 (1.2)
7 days	109.9 (8.3)	77.3 (9.9)
14 days	122.4 (4.6)	92.4 (5.5)
21 days	118.3 (13.1)	93.0 (1.6)

Table 2
Chemical compositions of some materials used in Cements in Table 1 *.

	Class C Fly Ash	GGBF Slag	Class F Fly Ash
SiO_2	39.50	35.60	53.00
Al_2O_3	16.90	10.40	18.60
TiO_2			
Fe_2O_3	6.40	1.16	15.60
MgO	6.30	12.84	1.26
MnO			
CaO	24.80	37.59	6.37
Na_2O	1.44	0.07	0.94
K_2O	0.53	0.33	2.08
P_2O_5			
SO_3	1.99		2.02
S		0.67	
LOI	2.21	1.68	

* Type III cement, metakaolin, and silica fume also used (7).

The acceleration of hydration as evident from isothermal calorimetry by activation with an alkali, in this case, KOH, or a mixture of KOH and $Ca(OH)_2$, is additionally illustrated in Fig. 1 (11) in contrast with ordinary Portland cement.

Overall, the results of the studies describe a very promising new class of materials. They

* Usual cement chemistry notation is used in which C is CaO, S is SiO_2, A is Al_2O_3, F is Fe_2O_3, $\overline{S}$ is SO_3; belite is C_2S.

show that it is possible to combine the advantages of rapid strength development by alkali activation with high ultimate strength, by control of particle packing and chemistry in a single formulation of blended cements containing fly ash, slag and silica fume. More research will be required to fully optimize and characterize the long-term behavior of these materials. One unknown aspect is the effect of the somewhat higher alkali levels on the long-term behavior.

3. POTENTIAL USE OF FLY ASH AND FBC COMBUSTION PRODUCTS IN CEMENT MANUFACTURE

The fly ash from fluidized bed combustion (FBC) is normally not suitable as a raw material for cement manufacture, because of high sulfur content. Calcium sulfoaluminate belite cement clinkers may provide a way to make this fly ash usable as a cement raw material. A well-known computational technique can be used to identify mixture/phase composition relationships for the raw materials (4,12).

Portland cement, as is commonly known, consists mainly of four clinker minerals, C_3S, C_2S, C_3A, and C_4AF*, which are compatible in the four-component system CaO-Al_2O_3-Fe_2O_3-SiO_2. The major mineral component C_3S in particular, consumes much energy in its production because of the high calcium content (high $CaCO_3$ content of the usual source material) and the high temperature required in processing. Sulfoaluminate belite clinkers (SAB) contain phases that are inherently sulfates ($C_4A_3\overline{S}$, $C\overline{S}$)* so ashes high in sulfate content may be a valuable source material for their production.

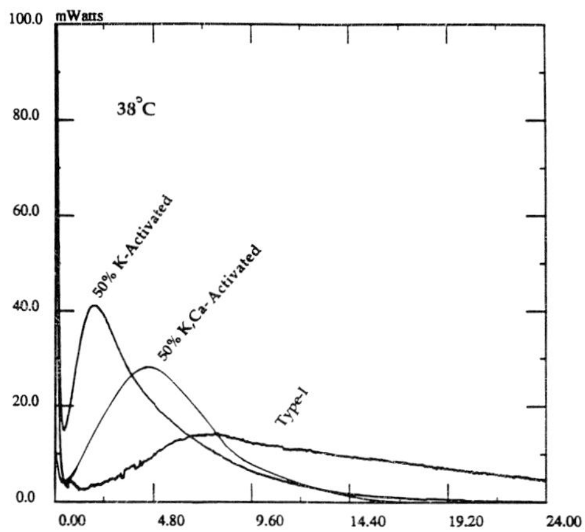

Fig. 1. Isothermal calorimetry data of 50% K-activated and 50% K, Ca-activated materials together with that of neat cement.

Development of cements containing the sulfoaluminate phase began in the middle 1970s and there is now an extensive literature in this field.

By "SAB clinkers" is meant primarily those containing the phases C_2S, C_4AF, $C_4A_3\overline{S}$, and $C\overline{S}$. In addition to utilizing sulfates in such clinkers, synthesis of the C_4AF phase makes possible the consumption of iron-containing byproducts from the environment. They offer the great advantage of low burning temperature, approximately in the range of 1200° - 1300°C, rather than the 1450°C range of portland cement clinkers (4).

This concept analyzes binary mixtures composed of high-sulfate, fluidized bed combustion ashes. Emphasis is given to creating suitable raw mixtures from byproducts (ashes) of the same fluidized bed reactor unit. If feasible, the mixture components would then be mutually corrective and there would be no need to transport either component for long distances when used in the manufacture of proposed cements.

The examples chosen also illustrate the use of the computation method, which may be applied broadly. Table 3 gives chemical composition data for some of the ashes used. Fig. 2 shows the calculated mineralogical compositions of clinkers resulting from particular raw mixture compositions. The figures show a number of (intersecting) lines composed of linear portions. Each line, however, is ascribed to a phase, for which it shows the change in content resulting from the raw mix composition. The sum of phase contents is always 100%.

Fig. 2 is relatively simple. The ashes are assumed to consist of five chemical components. The ashes used as raw constituents do not differ very much in chemical composition, so the join of the mixture end members passes only three phase assemblages: C_2S-C-$C\overline{S}$-C_4AF-C_2F, C_2S-$C\overline{S}$-C_4AF-C_2F-$C_4A_3\overline{S}$, and C_2S-$C\overline{S}$-C_2F-$C_4A_3\overline{S}$-CF. The free calcia content diminishes only slowly with an increasing content of the FA/CH raw component. The whole interval of mixtures shown is of interest for testing the properties of burnt mixtures. In addition, it is evident that the solid residues from fluidized bed coal combustion are prospective candidates for further processing by mixing with other suitable byproducts, or appropriate natural components, and by subsequent thermal treatment to produce the (calciumsulfoaluminate belite) cement clinkers. This procedure might be the first step to creating a nonwaste technology from the coal burning process.

Table 3
Comparison of FBC and PCFA ash bulk chemistries (4).

	Fly Ash	Fly Ash	FBC
SiO_2	52.5	36.9	24
Al_2O_3	22.8	17.6	6.05
Fe_2O_3	7.5	6.2	2.05
CaO	4.9	25.2	42
MgO	1.3	5.1	0.45
Na_2O	1.0	1.7	0.07
K_2O	1.3	0.6	0.51
SO_3	0.6	2.9	20.8
moisture	-0.11	-0.06	+0.25
LOI	2.6	0.33	2.03

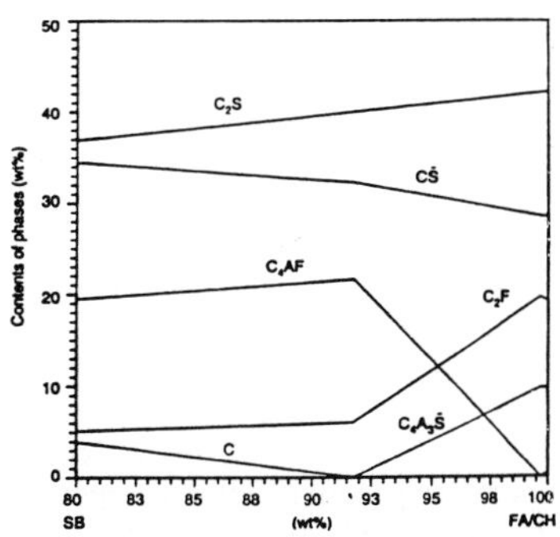

Fig. 2 Change in the mineralogical composition of clinkers for the SB-FA/CH raw mix, depending on the raw mix composition (4).

4. ZEOLITE-CEMENT COMPOSITES

Zeolites can be readily synthesized at elevated temperatures from laboratory prepared glasses and Class F fly ash, as well as Class F fly ash-blended cement pastes. Evidence suggests that the zeolites can coexist with a cementitious C-S-H (calcium silicate hydrate) phase. Furthermore, the results indicate that compressive strength and flexural strength are higher for samples in which zeolites have formed. Although it is not clear that the presence of zeolites is responsible for the observed increase in strength, it may be concluded that their presence does not degrade the mechanical properties of the composite when compared to equivalent samples made with water. Krivenko (14) has given more detail in general on alkali-activated cements and their classification.

5. HYDROTHERMAL PROCESSING OF CEMENT FROM FLY ASH TO PRODUCE REACTIVE CEMENT

Two types of hydrothermal processing of fly ash have been undertaken. The first is the synthesis of a reactive cement, and the second is an in-place formed cementitious composite. Both depend upon the activation of the fly ash by a combination of two methods (1) alkali (hydroxyl) and (2) hydrothermal.

In the case of activated fly ash cement, Fig. 3 shows a flow chart of the general procedure of hydrothermal processing of fly ash. The starting materials used are class F fly ash and CaO (analytical reagent grade). The samples were prepared with a Ca/Si molar ratio of 2 (stoichiometric ratio of dicalcium silicate). The starting materials in this stage were heated first up to 80°C in stirred suspensions for 4 h, then treated hydrothermally 4 h at 200°C under saturated steam pressure using a tubular pressure vessel. The main hydration products of fly ash-lime-water mixtures are CSH and C_3AH_6. After hydrothermal treatment, the reaction product was heated "dry" to produce the active phases.

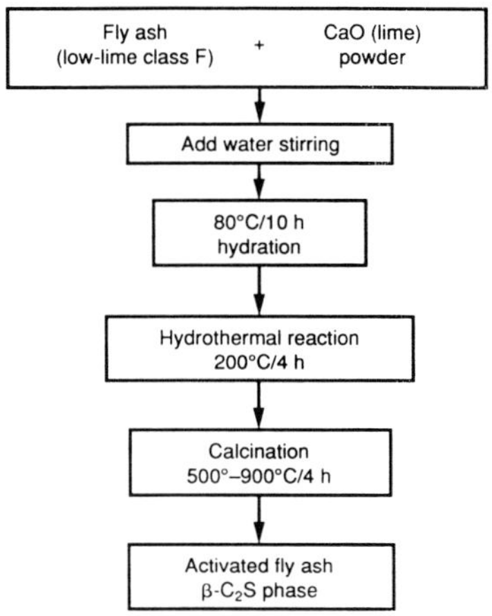

Fig. 3 Flow chart of the general procedure for the hydrothermal processing of fly ash to make cement (13).

The mineralogical compositions of reaction products generated were revealed by X-ray diffraction to contain predominantly a special active β-C_2S, and $C_{12}A_7$,. This new type of cementitious material generally involves the following two sequential steps in its formation.

The first step consisted of the hydrothermal synthesis reaction of fly ash and lime to form calcium silicate hydrates. Once formed, the fibrous calcium silicate hydrates are dewatered or filtered, then heated at 500° to 900°C to prepare the final product, a reacted cement. This kind of active β -C_2S synthesized by hydrothermal reaction is different from the conventional β -C_2S which is synthesized at high temperature (1350°C) with added stabilizer. The high surface area and twinned structure of the crystals are both consistent with the higher reactivity.

6. CEMENT CLINKER FORMATION FROM FLYASH BY MICROWAVE PROCESSING

Microwave processing is generally characterized by volumetric heating, high heating rate, and a short processing time; combined with substantial energy efficiency. Considerable potential for energy saving and reduction of emissions in cement production is possible through use of waste utilizing microwave processing to produce cement clinkers based on Class F flyash. These clinkers have been prepared by microwave processing at temperatures up to 1350°C (15). For appropriate chemical compositions, CaO or $CaCO_3$ was used as an additional component phase added to the raw material mixture. CaF_2 and $CaSO_4$ $2H_2O$ additives were used as fluxes or mineralizers to further lower the processing temperature. The clinkers thereby obtained have similar phases to those present in commercial portland cement.

7. SUMMARY

Major advances have been made in the synthesis of several cementitious materials and their composites of widely ranging compositions, critical to modern technology, which have, as a major parallel goal, reduced impact on the environment.

These may be thought of as gigaphase ceramics, which offer the opportunity for an environmentally friendly technology. A number of key properties have been investigated, and very high strengths have been attained. Such materials are already being used in novel applications (16), which include, but are not limited to, those depending on chemical properties. Perhaps most surprising are the facts that a substantial number of "advanced" materials may be developed utilizing large amounts of waste/by-product materials such as fly ash. The challenge remains to better understand the mechanisms involved in generating strength and toughness as well as chemical resistance (17). The effects of compositional and microstructural modifications, and the range of methods and parameters which control the properties of such composites are part of this challenge. This will open further horizons in the field of synthesis of new materials which will, at the same time, have low impact on the environment.

REFERENCES

1. D. M. Golden, 1, Sect. 8:1-24 in Proc., 8th Int'l Ash Utilization Symp., Electric Power Res. Inst., Palo Alto, CA, 1987.
2. O. E. Manz, B. A. Collings, J. S. Peri, and D. M. Golden, Electric Power Res. Inst., Palo Alto, CA, 1987.
3. C. L. Kilgour, Mater. Res. Soc. Symp. Proc., 245:81-87, 1992.
4. J. Majling and D. M. Roy, Am. Ceram. Soc. Bull. 72, 10, 1993.
5. B. Forss, Alkalis in Concrete, 101-104, Copenhagen, Denmark, 1983.
6. X. Wu, W. Jiang, and D. M. Roy, Cem. Concr. Res. 20:961-74, 1990.
7. D. M. Roy, M. R. Silsbee, and D. Wolfe-Confer, B. E. Scheetz et al., (Eds.), p.203 in Mat. Res. Soc., Vol. 179, 1991.

8. D. M. Roy and M. R. Silsbee, V. M. Malhotra, (Ed.), pp.700-715 in Proc. 2nd CANMET/ACI Int'l Symp. on Advances in Concrete Tech. Supp. Papers, Las Vegas, NV, 1995.
9. D. M. Roy and M. R. Silsbee, MRS Symp. Proc. 245:153-164, Materials Research Society, 1992.
10. J. Davidovits, Concr. Int'l. 23-25, 1987.
11. R. I. A. Malek and D. M. Roy, Xth Intl. Cong. on the Chemistry of Cement, Gothenburg, Sweden, Amarkai AB and Congrex Götenborg AB, Ed. Harald Justnes, V. I, li 024, 8 pp, 1997.
12. S. Sahu, J. Skolik, J. Havlica and J. Majling, Progmatech '91, Progressive Mat. and Tech.—Developing Clean Industry and Environment, Czechoslovakia, 1991.
13. W. Jiang and D. M. Roy, Am. Cer. Soc. Bull. 71:642-647, 1992.
14. P. Krivenko, Alkaline Cements: Terminology, Classification, Aspects of Durability, Xth Intl. Cong. on the Chemistry of Cement, Gothenburg, Sweden, Amarkai AB and Congrex Götenborg AB, Ed. Harald Justnes, V.4, iv 046, 6 pp, 1997.
15. Y. Fang, D. M. Roy, Y. Chen and M. R. Silsbee, Xth Intl. Cong. on the Chemistry of Cement, Gothenburg, Sweden, Amarkai AB and Congrex Götenborg AB, Ed. Harald Justnes,V. I, li 010, 4 pp, 1997.
16. D. M. Roy, 9th Intl. Congress on the Chemistry of Cement, Congress Reports, New Delhi, India, V.1, 357, 382; (NCB, New Delhi, India, 11049, 1992.
17. W. Jiang, Ph.D. Thesis, The Pennsylvania State University, 1997.

Materials Science and Engineering Serving Society
S. Sōmiya, R.P.H. Chang, M. Doyama and R. Roy (Editors)

Investigation on Diffusion Bonded Titanium-Alumina Dental Implants

G. Petzow, G. Soyez and G. Elssner

Max-Planck-Institut für Metallforschung, Seestr. 92, 70174 Stuttgart, Germany

1. INTRODUCTION

Titanium and alumina as parts of a dental implant are biocompatible, show corrosion resistance in physiological solutions and a large contact area to bone tissue [1-7]. For the passage through the gingiva high purity alumina ceramic is superior to titanium [8]. Deposition of plaque onto polished alumina is negligible [9]. The electric neutrality of alumina prevents irritations caused by galvanic effects. Its white colour complies with aesthetic demands. Titanium possesses a low modulus of elasticity, high strength and ductility. A design of a composite dental implant consisting of a conical titanium part situated in the spongiosa and an alumina ring penetrating the gingiva is shown in Fig. 1. To avoid gap corrosive processes at the metal-ceramic interface perfect joints between titanium and alumina were fabricated by diffusion bonding [10-12].

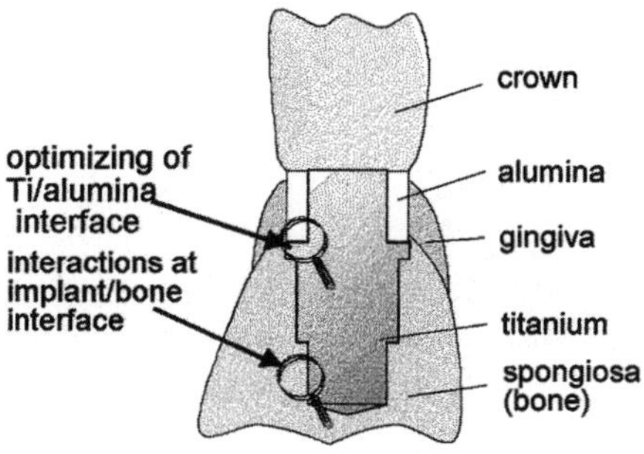

Fig. 1. Ti/Al_2O_3 composite implant.

2. DIFFUSION BONDING

Metallographic preparation techniques were applied to obtain flat surfaces of the starting materials [13]. Using an semi-automatic grinding and polishing machine the specimens were ground on diamond disks of grain size 20 µm for alumina or on SiC papers ending up with grit size P1000 for Ti, TiTa30 and Nb and polished in three steps with diamond suspensions of grain size 6, 3 and 1 µm on hard synthetic cloth. For final polishing of Ti, TiTa30 and Nb colloidal silica was applied.

The mean roughness R_a and the curvature h_{max} of the optical polished surfaces were measured with a stylus profilometer. The curvature was defined as the maximum elevation over a distance of 15 mm. The mean roughness is the mean peak-to-peak value over a distance of 2 mm. Table 1 shows the surface quality of the materials used for Ti/Al_2O_3 implants.

Table 1
Surface quality of materials used for Ti/Al_2O_3 implants

Property	Al_2O_3	Ti	TiTa30
R_a [nm]	21.4	10.9	31.4
h_{max} [μm]	1.8	3.3	3.0

The rather high value of surface roughness of the TiTa30 alloy is caused by its two-phase structure. The body-centered β-phase and the hexagonal close packed α-phase of the alloy differ in hardness. Titanium was of 99.4 wt.% purity. The alumina ceramic contained 99.7 wt.% Al_2O_3. Prior to diffusion bonding the polished materials were cleaned in pure ethanol and dried under nitrogen.

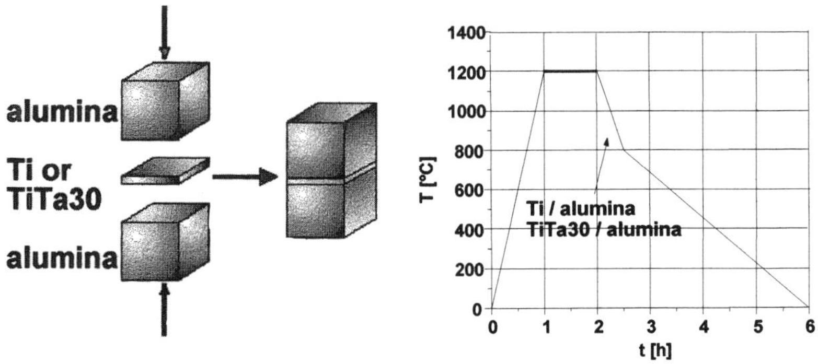

Fig. 2. Diffusion bonding of metal/ceramic joints.

Sandwich-like specimens consisting of Al_2O_3/Ti/ Al_2O_3 and Al_2O_3/TiTa30/ Al_2O_3 were manufactured by diffusion bonding to determine the bond strength and the microstructure of the metal-ceramic joints. The bonding process is illustrated in Fig. 2 together with time-temperature diagrams for the two types of joints. TiTa30 was chosen because of its low thermal mismatch to alumina. At the diffusion bonding temperature T_s the joints were uniaxially compressed under a mechanical pressure of 4 MPa. Heat was applied by an induction coil with a Nb susceptor. The temperature was measured by a thermocouple positioned directly at the metal-ceramic interface. During bonding the vacuum pressure was less than 10^{-5} mbar.

A fracture mechanical testing concept [14] was applied to determine the bond strength. With a high speed diamond saw one diffusion bonded sample was cut into 15 four point bending test specimens of 2x4x36 mm³. These test bars of width b and heigth w are notched at one of the metal-ceramic interfaces to a depth a with an a/w ratio between 0.3 and 0.4 and then loaded to the fracture load F_c. The width of the notch was 100 μm and the cross-head speed 96.8 μm/min. The value of half the difference between upper and lower span of the bend test device was e = 9 mm. As a measure of bond strength the interface fracture energy G_c and the interface fracture toughness K_c were calculated from the fracture load F_c, the sample geometry, the notch depth a and the elastic constants of the constituents. K_c values can be compared with fracture toughness data K_{Ic} of monolithic materials.

Table 2 and 3 show the elastic constants and the thermal expansion coefficient α_t of the materials and bond strength data.

Table 2
Elastic and thermal properties of the materials used

Property	Al_2O_3	Ti	TiTa30
E [GPa]	390	120	88
ν	0.27	0.36	0.41
α_t [10^{-6}/K]*	8.6	11.1	9.3

*20° to 1200°C

Table 3
Bond strength data of Ti/ Al_2O_3 and TiTa30/ Al_2O_3 joints

Bond strength	Ti/ Al_2O_3	TiTa30/Al_2O_3
G_c [J/m²]	15	38
K_c [MPa/m$^{1.5}$]	1.8	2.6

3. MANUFACTURING OF Ti/Al_2O_3 DENTAL IMPLANTS

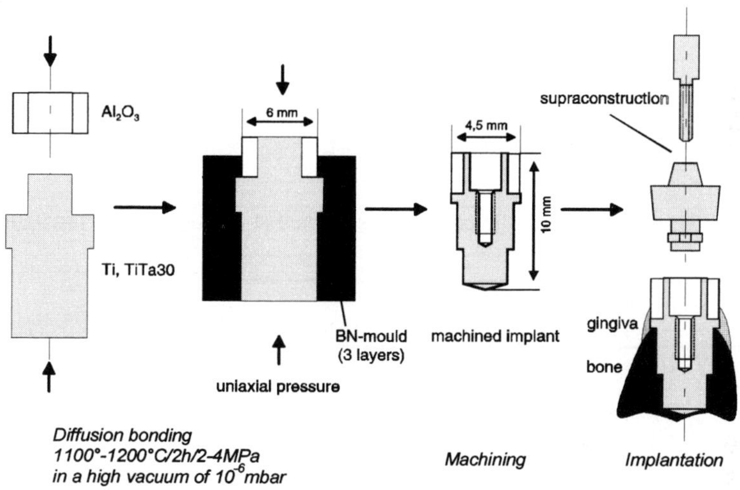

Fig. 3. Manufacturing of Ti/Al_2O_3 and TiTa30/Al_2O_3 dental implants.

Dental implants (Fig. 1) were produced by diffusion bonding of an alumina ring onto a Ti- or TiTa30-cylinder. They were used for biological and mechanical tests. For manufacturing a Ti/Al_2O_3 or TiTa30/Al_2O_3 implant (Fig. 3) an alumina ring of grain size 5 μm is pressed onto a machined titanium cylinder and fitted into a boron nitride mould. The composite is high vacuum diffusion bonded at 1100° to 1200°C under a load of 2 MPa. The boron nitride mould prevents a bulging of the metal cylinder and generates a three dimensional stress state.

Metallographic studies confirmed that this procedure leads to an intimate contact between metal and ceramic without any voids. After diffusion bonding the conical design of the implant is fabricated by machining. Finally, the supraconstruction is added which bears later on the artificial tooth. One example of such an implant used for the histological studies is shown on the left of Fig. 4 together with a titanium implant on the right.

Fig. 4. Dental implants made from a Ti/Al_2O_3 composite (left) and from titanium (right).

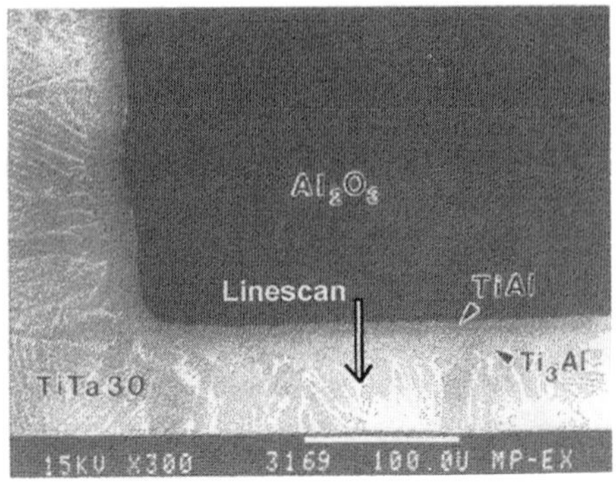

Fig. 5. Interface region between TiTa30 and alumina, SEM.

As predicted by the Ti-Al-Ta phase diagram SEM and microprobe measurements show that intermetallic reaction products are formed during diffusion bonding at 1200°C both at the Ti/Al_2O_3 and the $TiTa30/Al_2O_3$ interfaces. Reaction zones of about 10 µm thickness are visible in SEM micrographs of polished sections (Fig. 5). The reaction zone consists of two Ti-Al intermetallic compounds differing in contrast.

Table 4.
Microprobe analysis of intermetallic compounds formed at Ti/Al_2O_3 and $TiTa30/Al_2O_3$ interfaces

	Concentration in at.%			
	Ti/Al_2O_3		$TiTa30/Al_2O_3$	
Element	TiAl	Ti_3Al	TiAl	Ti_3Al
Ti	51.50	75.36	52.36	68.94
Al	48.50	24.64	47.12	28.10
Ta	0.00	0.00	0.52	2.96
Ti/Al	1.06	3.06	1.11	2.45

Microprobe measurements of Ti, Al, Ta and O in WDX mode showed that the dark phase adjacent to alumina is TiAl and the bright phase adjacent to titanium Ti_3Al. The intermetallic compounds formed at $TiTa30/Al_2O_3$ interfaces contain Ta. Table 4 gives the results of microprobe measurements within the reaction zones at Ti/Al_2O_3 and $TiTa30/Al_2O_3$ interfaces. The contrast between the lamellar phases of TiTa30 shown in the SEM micrograph is due to their difference in concentration of Ta and Al according to the phase diagramm Ti-Al-Ta. Ta stabilises the b.c.c β-phase of Ti, aluminium the h.d.p. α-phase of Ti. The thickness and the composition of the lamellae is dependent on the cooling rate used in the bonding experiment.

4. TESTING OF COMPOSITE DENTAL IMPLANTS

The functionality of the composite implants was determined using a special bending test device in which the implants together with differing supraconstructions were fractured under loading angles of 90° and 45°.

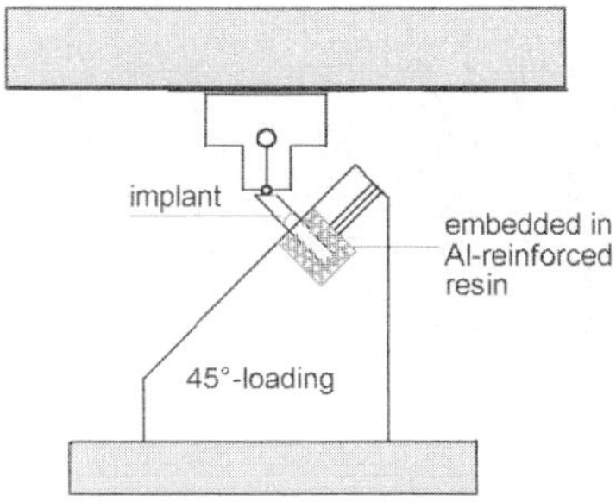

Fig. 6. Bending device for testing of composite implants.

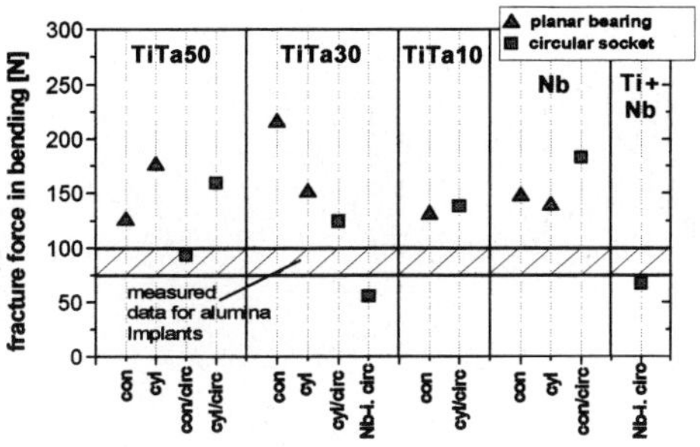

Fig. 7. Fracture forces of metal/ceramic composite implants in comparison to those of Tübingen alumina implants. Nb-i = Nb interlayer between metal and ceramic.

The implants were embedded in a mixture of resin and aluminum powder of a modulus of elasticity nearly equal to that of the bone and loaded in a distance of 10 mm to the metal/ceramic interface by means of a sphere. Load and deflection were recorded. Fig. 6 shows the bending test device set up for loading under 45°.

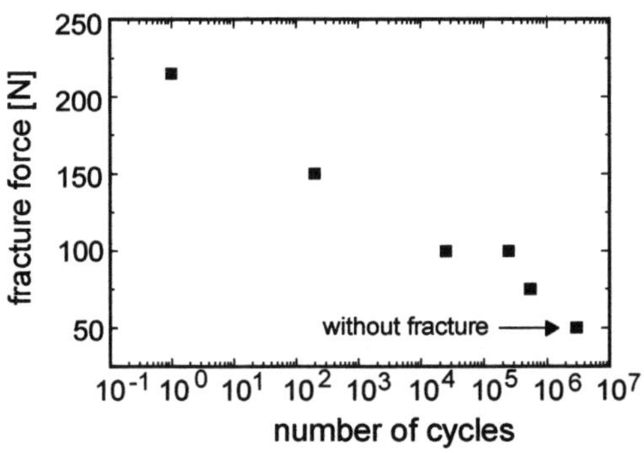

Fig. 8. Fracture force in bending of TiTa30/ Al_2O_3 implants as a function of the number of fatigue cycles.

Results of 90° bending test for different material combinations are given in Fig. 7. Composite implants with planar bearing surface or circular socket of the supraconstructions were used. The implants had a conical (con) or a cylindrical (cyl) interface geometry. Load drops in the load-deflection diagrams were due to crack formation in the alumina. A stabilization of the alumina ring by a circular socket of the supraconstruction was not observed as shown by the forces necessary to produce crack formation (see Fig. 7).

Fig. 7 demonstrates that the critical fracture forces of the composite implants are higher than those of the Tübingen alumina dental implants (hatched area). Furthermore, the TiTa30/Al_2O_3 implants showed the threefold of the fracture loads of the alumina implants. The results of fatigue tests performed with the special bending test device are given in Fig. 8.

5. THE IMPLANT/TISSUE INTERFACE

Interdisciplinary research is required to control interactions between the implant and the bone tissue [15,16]. Metallographic techniques uncommon in histology have to be applied to obtain sections through the hard metal and ceramic explants parts and the softer tissue. The Ti-30wt.%Ta/bone, Al_2O_3/bone interfaces and the Ti-30wt.%Ta/Al_2O_3/bone triple points were examined using SEM. Three different interface morphologies were observed: i) a direct contact; ii) an intermediate layer; iii) a gap at the implant/bone interface.

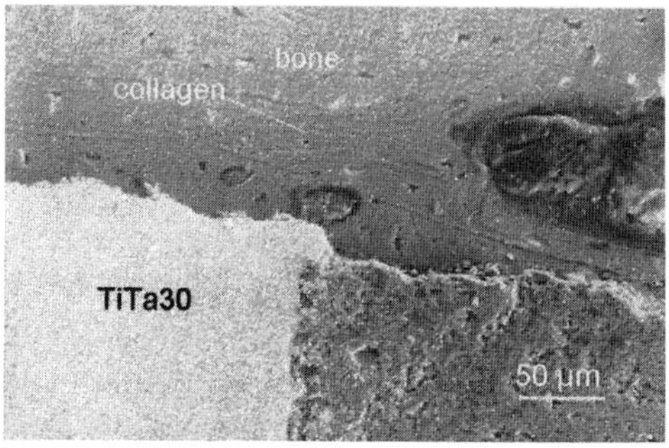

Fig. 9. The interface between bone and Ti-30wt.%Ta/alumina. SEM.

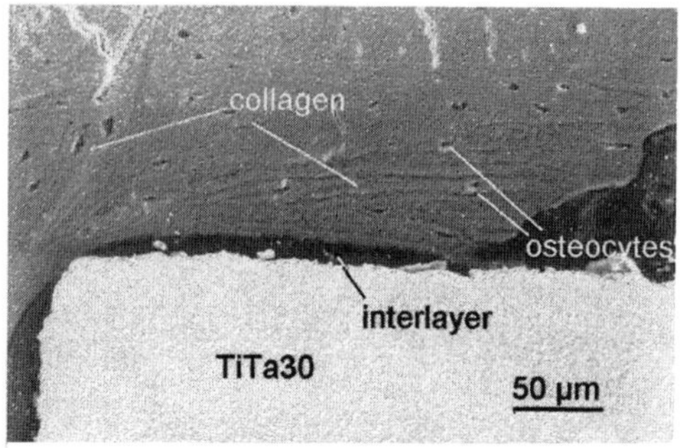

Fig.10. Ti-30wt.%Ta/bone interface of the the Ti-30wt.%Ta/Al_2O_3 implant separated by an intermediate layer of 100 µm maximum thickness mainly consisting of carbon. SEM.

A direct contact with bone is observed at the region where Ti-30wt.%Ta and Al_2O_3 are fitted (Fig. 9). The arrangement of cells inside the bone is well ordered. The osteocytes are included within a flux of collagen fibres surrounding the implant (Fig. 9). The bone exactly accommodates to the composite implant surface. Generally, the micrographical and analytical studies performed at the implant/bone interfaces of Ti, Ti-30wt.%Ta and Al_2O_3 show no significant differences in the attachment of bone to the different materials and in the concentration of osteocytes in the bone matrix near the implant surfaces. Collagen fibers and osteocytes were also detected at the Ti-30wt.%Ta side of the implant (Fig. 10). The welding zone between Ti-30wt.%Ta and alumina does not show any negative influence on bone, hence biocorrosive effects can be excluded at this location.

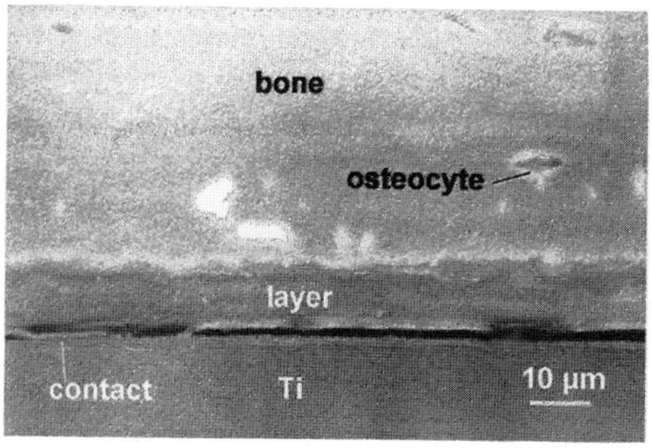

Fig. 11. SEM micrograph of the Ti/bone interface with a soft interfacial layer.

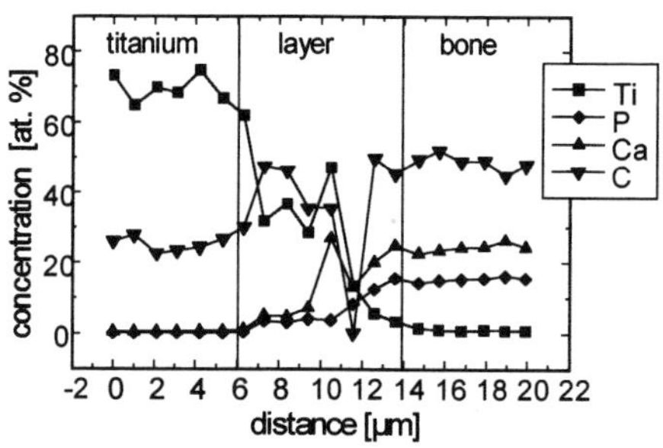

Fig. 12. EDX analysis of an interface region between titanium and bone tissue.

Fig. 11 shows the bone tissue at the top, then a soft tissue interlayer and at the bottom titanium. A microanalysis of this cross section (Fig. 12) reveals that the biocompatible titanium interacts with the tissue and is present in the tissue interlayer.

The results of the morphology and the chemical analysis of the implant/bone interface are in agreement with fracture surface studies performed by SEM and AES. The fracture surface of the composite implant shows a good contact between bone and both of the materials (Fig. 13). Bonding between implant and bone is achieved by contact points. The bone tissue adapts to the surface topology especially at the interface region. The presence of tissue at the top of the implant surface implies high fracture toughness of the implant/bone interface.

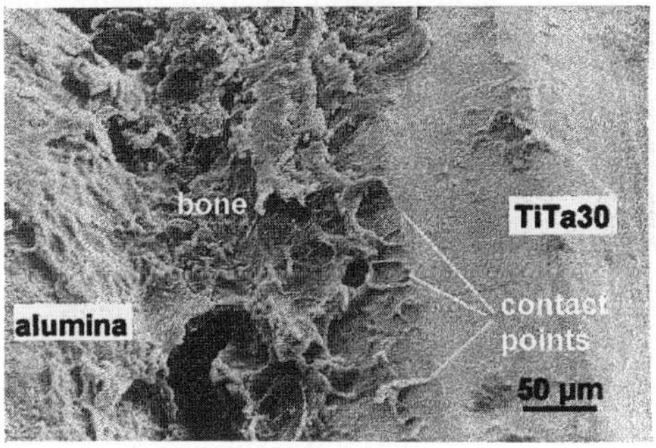

Fig. 13. Fracture surface of the TiTa30/alumina implant. SEM.

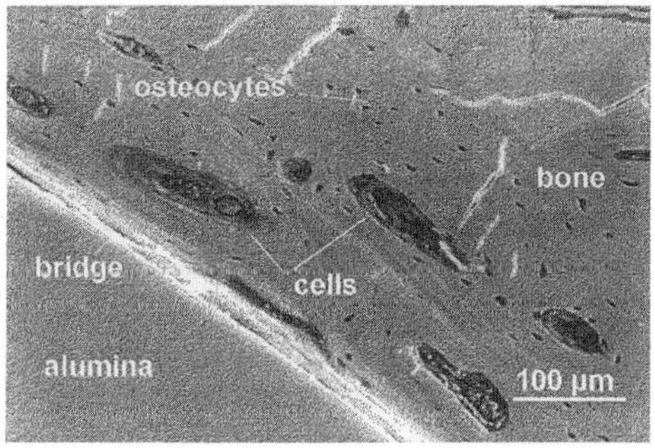

Fig. 14. Fibrous structure surrounding cell-like dark structures near the alumina/bone interface. SEM.

The results of the morphology and the chemical analysis of the implant/bone interface are in agreement with fracture surface studies performed by SEM and AES. The fracture surface of the composite implant shows a good contact between bone and both of the materials (Fig. 14). Bonding between implant and bone is achieved by contact points. The bone tissue adapts to the surface topology especially at the interface region. The presence of tissue at the top of the implant surface implies high fracture toughness of the implant/bone interface.

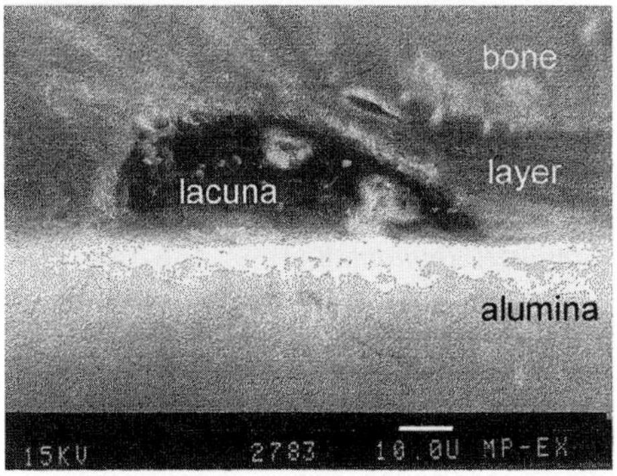

Fig. 15. Interface region between alumina and bone tissue, SEM.

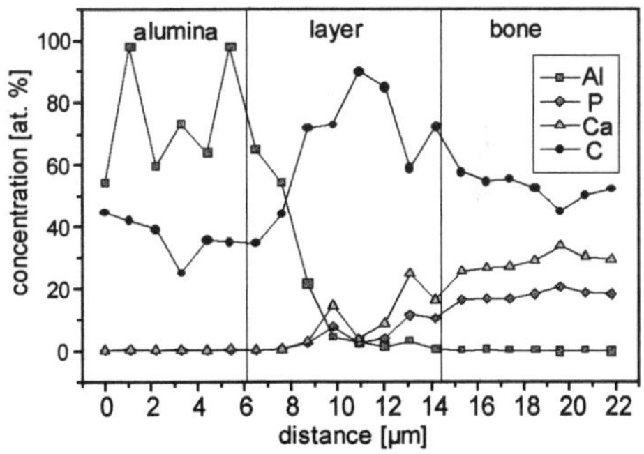

Fig. 16. EDX analysis of an interface between alumina and bone tissue.

The interface morphologies described above were also observed for an alumina implant. A fibrous structure surrounding cells in the adjacent bone is shown in Fig. 15. In some cases nuclei are observed in the cells. The fibres are mainly alligned parallel to the implant surface. Inside the fibre bundles osteocytes are visible. A thin gap with bridges between material and bone is observed. It cannot be excluded that this gap is in fact a crack produced during the preparation process. Dark areas within the bone tissue are surrounded by concentric bundles of collagen fibers. These areas are Haversian systems. Lacunas are sometimes visible at the interface between alumina and bone tissue and a tissue interlayer is formed (Fig. 15). As found by microanalysis measurements (Fig. 16) again a constituent of the implant, aluminum, penetrates from the implant into the tissue interlayer. Transmission electron microscopy studies were also performed to get a deeper metallographic and histological insight into the microstructure of this interface region.

6. CONCLUDING REMARKS

Composites manufactured by diffusion bonding from titanium alloys and pure alumina are due to their biocompatibility and their combination of favourable properties well suited for dental implants. This was shown in a ten years research project at the MPI comprising material science studies, evaluation of the microstructure and the properties of the implants, manufacturing of implants for in vivo experiments and microscopical and analytical studies of the implant/tissue interface. It was supported by the Deutsche Forschungsgemeinschaft within a program centered at the University of Tübingen under the guidance of Prof. W. Schulte. Part of the results are given in this paper.

REFERENCES

1. P. Tetsch, Enossale Implantationen in der Zahnheilkunde, Hanser Verlag, München-Wien (1983)
2. G. Heimke, Ceramics in Substitutive and Reconstructive Surgery, Editor P. Vincenzini, Elsevier Science Publishers B. V., (1991) 567-581
3. W. Schulte, B. d'Hoedt, D. Axmann, G. Gomez, *Z. für Zahnärztl. Implantologie* **8** (1992) 77-96
4. E. Dörre, Zahnärztliche Praxis **31** (1980) 343-346
5. P. Tengvall, I. Lundström, *Clinical Materials* **9** (1992) 115-134
6. T. Flechsig, Med. Diss. University Göttingen 1991
7. F. Henning, Med. Diss. University Hamburg 1991
8. G. Heimke, W. Schulte, B. d'Hoedt, *Clinical Materials* **1** (1986) 147-150
9. D. Korn, G. Elssner, G. Petzow, W. Schulte, *Z. Zahnärztl. Implant.* **8** (1992) 136-144
10. B. Gibbesch, G. Elssner, G. Petzow, *Int. J. of Oral & Maxillofacial Implants* **4** (1989) 131-137
11. B. Gibbesch, G. Elssner, E. Bischoff, G. Petzow, *Z. Zahnärztl. Implantol.* **5** (1989) 108-115
12. B. Gibbesch, G. Elssner, G. Petzow, *Biomaterials* **13** (1992) 455-461
13. G. Kiessler, D. Korn, G. Elssner, *Praktische Metallographie* **29** (1992) 597-610
14. T. Suga, G. Elssner, *J. de Physique* **46** (1985) C4 657-662
15. G. Soyez, G. Elssner, M. Rühle, G. Petzow, D. Korn, W. Schulte, in: Werkstoffe für die Medizintechnik, Herausg. J. Breme, DGM Informationsgesellschaft Verlag, Frankfurt 1997 29-34
16. D. Korn, G. Soyez, E. Bres, G. Elssner, G. Petzow, W. Schulte, will appear in *Journ. Mat. Science: Materials in Medicine* (1997)

Materials Science and Engineering Serving Society
S. Sōmiya, R.P.H. Chang, M. Doyama and R. Roy (Editors)

Advanced metallization in ULSI devices

L. J. Chen[a], J. Y. Yew[a], S. L. Cheng[a], B.Y. Tsui[b], and W.F. Wu[c]

[a]Department of Materials Science and Engineering, National Tsing Hua University, Hsinchu, Taiwan, Republic of China

[b]Electronics Research and Service Organization, Industrial Technology and Research Institute, Hsinchu, Taiwan, Republic of China

[c]National Nano Devices Laboratory, Hsinchu, Taiwan, Republic of China

Low resistivity $TiSi_2$ and $CoSi_2$ are two primary candidates for metal contacts in ULSI devices. In the present paper, we report the results of investigations on 1. the formation of $TiSi_2$ on N^+ implanted ultrashallow contacts, 2. self-aligned formation of $CoSi_2$ on the selective epitaxial growth (SEG) silicon layer on (001)Si inside 0.1-2 μm oxide openings prepared by electron beam lithography.

1. INTRODUCTION

As the microelectronics device dimensions scale down to the deep submicron size, silicides in the source/drain area are needed to reduce the contact resistance to an acceptable level. For self-aligned silicidation devices, silicides are formed on source, drain and gate regions simultaneously to ease the lithography requirements and to lower the contact resistance. Low-resistivity $TiSi_2$ is currently the most common silicide for this application [1]. On the other hand, $CoSi_2$ is also being considered as the metal contacts on shallow junctions [2]. For the deep submicron devices, it is essential to understand the interactions of metal contacts with highly doped shallow junctions [3].

Shallow junction formation has long been one of the major challenges in the fabrication of ultralarge scale integrated circuits (ULSI) devices [4]. N^+ implantation was introduced in an attempt to suppress both channeling and TED [5].

$CoSi_2$ is the only silicide that offers properties and reliability for continued use in the sub-0.25 μm devices [6]. Elevated source/drain structures are a potential solution to many of the problems by the deposition of an ultra-shallow and defect-free junction [7]. The common approach to fabricating elevated source and drain has been the selective epitaxial growth of silicon (SEG-Si) on the source and drain areas. To achieve low-temperature SEG-Si, a promising technique is ultrahigh vacuum chemical vapor deposition (UHV-CVD). High quality epitaxial silicon layer has been grown on exposed Si by applying UHV-CVD with disilane (Si_2H_6) as the source on silicon wafers with oxide patterns. In this paper, we report the successful integration of the self-aligned $CoSi_2$ and SEG-Si processes inside 0.1-2 μm oxide openings prepared by electron beam lithography.

2. EXPERIMENTAL PROCEDURES

2.1 The formation of $TiSi_2$ on N^+ implanted ultrashallow contacts

Single crystal, 3-5 Ω-cm, (001)Si wafers were used in the present study. Following a standard cleaning procedure, (001)Si wafers were implanted by 30 keV N^+ to doses of 5×10^{14} to 1×10^{16} /cm^2 and 20 keV N_2^+ to doses of 2.5×10^{14} or 1×10^{15} /cm^2. Some of the N^+- or N_2^+-implanted samples were then implanted by 30 keV BF_2^+ to a dose of 2×10^{15} /cm^2 or 30 keV As^+ to a dose of 3×10^{15} /cm^2. The implantation of N_2^+ was carried out to achieve an effective N^+ implantation of half energy and double dose.

Some of the implanted samples were then cleaned by a standard RCA process and then dipped in a dilute HF solution ($HF:H_2O$=1:50) before loading into an electron gun evaporation chamber. 30-nm-thick Ti thin films were deposited onto the samples. For Ti thin films on ion implanted silicon, unless otherwise mentioned, a 30-nm-thick amorphous silicon (a-silicon) was deposited as a capping layer in the electron beam evaporation chamber. The capping layers were used to protect the films from oxidation during the subsequent heat treatments.

Heat treatments were carried out in a three-zone diffusion furnace at 600-1000 °C for 1 h or in a rapid thermal annealing (RTA) apparatus at 600-1000 °C for 30 s in N_2 ambient unless otherwise mentioned.

2.2 Self-aligned formation of $CoSi_2$ on the selective epitaxial growth (SEG) silicon layer on (001)Si inside 0.1-2 μm oxide openings

The patterned oxide wafers were prepared by using the electron beam lithography (EBL) and reactive ion etching (RIE). 6-in.-diam, boron doped (100) Si wafers, 1-20 ohm-cm in resistivity, were first cleaned by standard RCA process followed by a low pressure chemical vapor deposition (LPCVD) to form 240-nm-thick blanket SiO_2 film on both sides of the wafer at 700 °C by decomposing tetraethoxysilane (TEOS). After the SiO_2 film was deposited, the oxide film on the back side of the wafer was subsequently removed. The oxide film stress was determined by a thin film stress measurement system through measurements of wafer curvature using a scanning laser beam reflection technique. The radius of curvature of each of the silicon wafers was measured prior to the deposition of the oxide film and following the deposition.

In order to measure resistivity of metal contacts on narrow linewidth openings, the patterned wafer was implanted by 10 keV As^+ to a dose of 5×10^{15} /cm^2. Prior to loading into the ultra-high vacuum chemical vapor deposition (UHV-CVD) chamber, the wafers were chemically cleaned in H_2SO_4/H_2O_2 and $NH_4OH/H_2O_2/H_2O$ and dipped in a dilute HF solution. The base pressure of the UHV-CVD system was 1.5×10^{-9} Torr. Prior to the SEG Si deposition, The in-situ thermal process was done at 900 °C for 10 min. In this process, the dopant activation and the residual native oxide on the Si surface removal were achieved. The selective epitaxial growth (SEG) Si deposition temperature was 560 °C and the deposition time was 10 or 30 min. The Si_2H_6 gas flow rate was 0.5 sccm.

For the metal deposition in E-Gun evaporator, 30-nm-thick Co thin films were then electron beam deposited onto the patterned substrates. Before the metal deposition in the sputtering system, the residual native oxide was removed by argon ions sputter cleaning. Thin Co and TiN thin films, 30-nm-thick, were deposited by sputtering, consequently. The silicidation reaction was carried out in a RTA system in N_2 ambient. For the one-step RTA process, The samples were annealed at 400 to 900 °C with a heating rate of 100 °C/s and annealing time of 60 s.

Selective non-reacted Co removal was carried out in an etchant solution ($HCl:H_2O_2$=3:1). For the two-step RTA process, The first RTA step was done at 480 °C with a heating rate of 100 °C/s and time of 60 s. After a selective etch of $H_2SO_4:H_2O_2$ (3:1) to remove the TiN and unreacted Co films, a second RTA step followed at 700 °C with a heating rate of 100 °C/s and annealing time of 30 s.

3. RESULTS AND DISCUSSION

3.1 30 keV N^+, 30 keV N^+ + 30 keV BF_2^+, 30 keV N^+ + 30 keV As^+ Implanted Samples

Sheet resistance data of BF_2^+-N^+ and As^+-N^+ implanted samples, from standpoint of electrical activation, showed that the N^+ doses of 5 x 10^{15} /cm^2 and 1×10^{16} /cm^2, respectively, are too high to be useful in forming the shallow junctions. In addition, to activate the BF_2^+-2×10^{15} /cm^2 N^+ samples, annealing at a temperature of 800 °C was not sufficient.

From SIMS data, the suppression of boron and As diffusion is seen to increase with the N^+ dose.

3.2 Titanium on 20 keV N_2^+, 20 keV N_2^+ + 30 keV BF_2^+ and 20 keV N_2^+ + 30 keV As^+ Implanted Samples

A sharp drop in sheet resistance was found for all implanted samples at 700 °C. The sheet resistances maintained the same low level in implanted samples annealed at 800-900 °C. The sheet resistance data are consistent with those obtained by glancing angle XRD and TEM analysis.

For Ti on 2.5×10^{14} and 1 x 10^{15} /cm^2 N_2^+ implanted samples. no amorphous surface layer was observed in samples implanted to a dose of 2.5×10^{14} /cm^2. On the other hand, as the N_2^+ dose was increased to 1×10^{15} /cm^2, a 33-nm-thick amorphous layer was found to form in as-N_2^+ implanted samples. In both sets of samples annealed at 600 °C, C49-$TiSi_2$ was found to be the only silicide phase present. A continuous C54-$TiSi_2$ layer was found to form in all samples annealed at 700-950 °C. An example is shown in Fig. 1. Severe island formation of C54-$TiSi_2$ was observed in 1000 °C annealed samples. A simple calculation indicated that he formation of

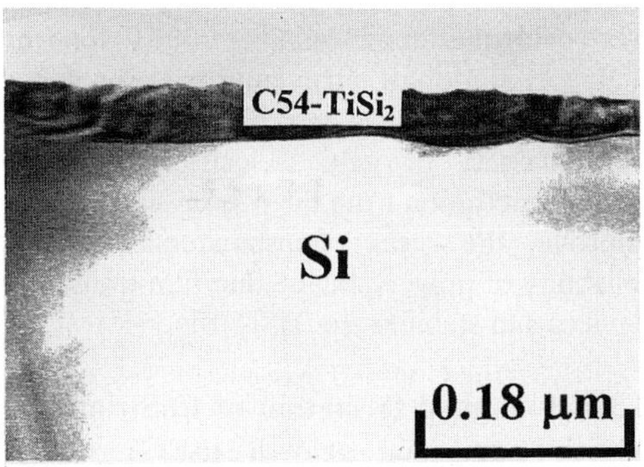

Fig. 1 Ti on 1×10^{15} /cm^2, 20 keV N_2^+ implanted (001)Si, 900 °C, 30 s, c-s.

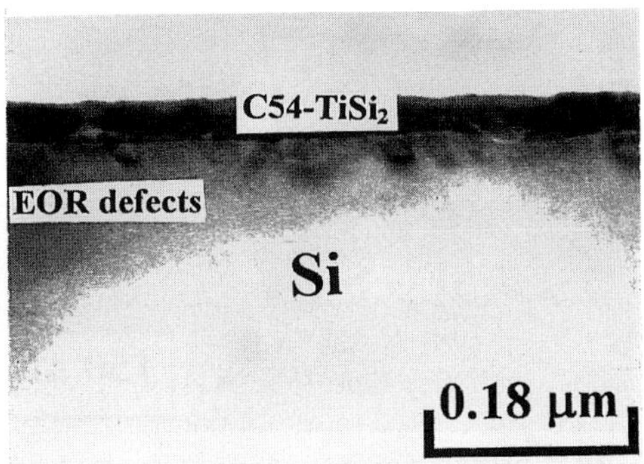

Fig. 2 Ti on 1×10^{15} /cm^2, 20 keV N_2^+ + 30 keV BF_2^+ implanted (001)Si, 900 °C, 30 s, c-s.

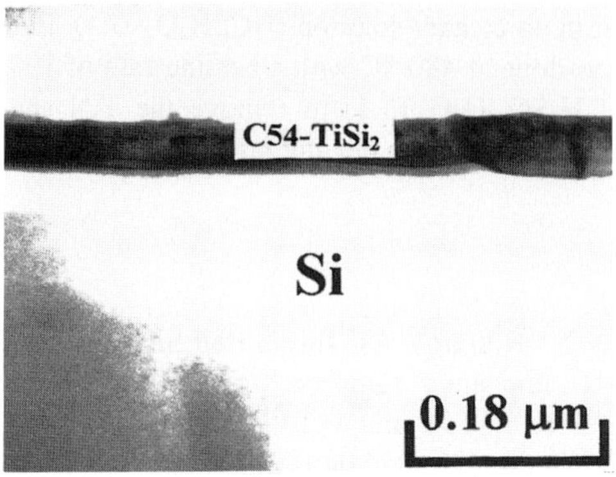

Fig. 3 Ti on 1×10^{15} /cm^2, 20 keV N_2^+ + 30 keV As^+ implanted (001)Si, 900 °C, 30 s, c-s.

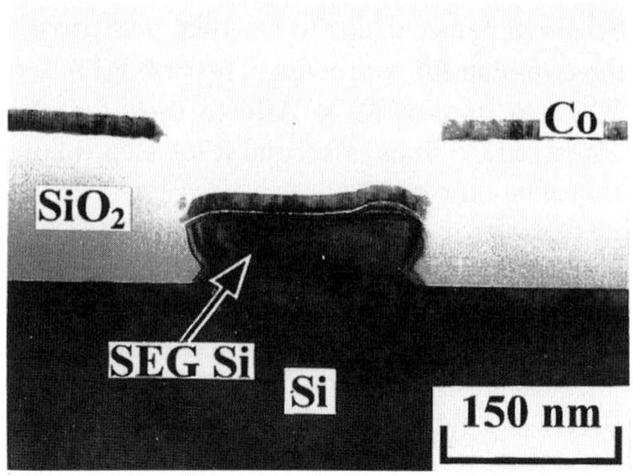

Fig. 4 XTEM image of the as-deposited Co film on SEG Si layer inside 0.2 μm linear oxide opening.

the $TiSi_2$ layer consumed the regions with EOR defects in silicon.

For Ti on 2.5 x 10^{14} and 1×10^{15} /cm^2 N_2^+ and BF_2^+ implanted samples, a 45-nm-thick amorphous layer was found to form in as-N_2^+- and BF_2^+ implanted samples. Continuous C54-$TiSi_2$ was the only silicide phase formed in 700-950 °C annealed samples. In 1000 °C annealed samples, severe island formation of C54-$TiSi_2$ was found. EOR defects were found to be prevalent in 600-950 °C annealed samples. An example is shown in Fig. 2.

For Ti on 2.5×10^{14} and 1×10^{15} /cm^2 N_2^+ and As^+ implanted samples, a 51-nm-thick amorphous layer was found to form in as-N_2^+- and As^+ implanted samples. Continuous C54-$TiSi_2$ was the only silicide phase formed in 700-900 °C annealed samples. In 950-1000 °C annealed samples, severe island formation of C54-$TiSi_2$ was found.

For 1×10^{15} /cm^2 N_2^+ and As^+ implanted samples, no EOR defects were observed in all samples annealed at 700-950 °C. An example is shown in Fig. 3. In 2.5×10^{14} /cm^2 N_2^+ and As^+ implanted samples without Ti overlayer, the rodlike defects located at the surface region can be observed. However, no EOR defects were evident after annealing at 900 °C for 1 h.

Incorporation of nitrogen in the processing of silicon devices has been shown to improve the reliability of gate dielectrics, suppress boron penetration from the p^+-polysilicon gate, improve the peak transconductance and reliability of polycrystalline thin film transistor devices and stabilize the $TiSi_2$ [8].

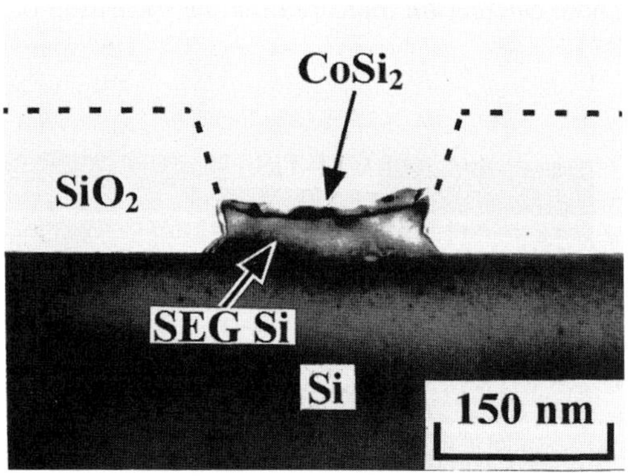

Fig. 5 XTEM image of the $CoSi_2$ film on the SEG Si layer inside a 0.23 μm linear opening after one-step RTA at 700 °C and selective etching treatment.

3.3 Self-aligned formation of $CoSi_2$ on the selective epitaxial growth (SEG) silicon layer on (001)Si inside 0.1-2 μm oxide openings

Fig. 4 shows a XTEM image of the E-gun evaporator as-deposited Co film on SEG Si layer inside a 0.2 μm linear opening. Cross-section TEM image of the $CoSi_2$ film on the SEG Si layer inside a 0.23 μm contact hole after one-step RTA treatment at 700 °C is shown in Fig. 5. The 15-nm-thick polycrystalline $CoSi_2$ was formed on the SEG Si layer without lateral growth of the silicide.

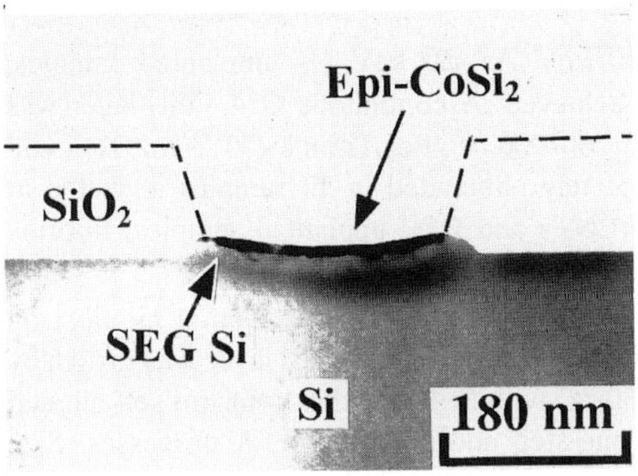

Fig. 6 (a) XTEM image showing the epitaxial $CoSi_2$ film on the SEG Si layer on (001)Si inside a linear opening of 0.25 μm in size.

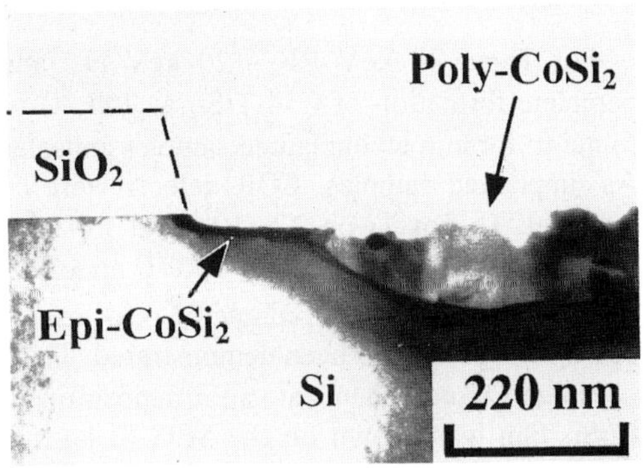

Fig. 6 (b) XTEM image showing the epitaxial and polycrystalline $CoSi_2$ films on oxide edge on (001)Si inside a linear opening of 0.18 μm in size.

Figs. 6 (a) and (b) shows XTEM images of epitaxial and polycrystalline $CoSi_2$ on 0.25 μm and 1.8 μm linear openings, respectively, by a two-step RTA process. The morphology of $CoSi_2$ is consistent with that of planview TEM images. A thin and uniform $CoSi_2$ epitaxy was grown inside 0.25 μm linear openings. Inside the 1.8 μm linear opening, the thickness of epitaxial $CoSi_2$ at the oxide edge was thinner than the polycrystalline $CoSi_2$ from the center to a distance away from the oxide edge. The epitaxial growth of $CoSi_2$ was found to form along the oxide edge.

Fig. 7 shows the XTEM image of a sample annealed at 480 °C. Two distinct regions with different amounts of Co consumed are seen. Analysis of selective area electron diffraction data of the region inside 1.8 μm linear opening, the polycrystalline CoSi phase was identified be present. Similar results were obtained for the blank and patterned wafers. The thickness of CoSi in the blank wafer was the same as that of the completely consumed area in the patterned wafer. In the partially consumed area near the oxide sidewall, the thickness of CoSi was increased with the distance from the edge of the oxide sidewall. The growth of CoSi is known to be diffusion limited. Maker experiments show that Si is the dominant diffusing species. For the diffusion limited case, the growth of CoSi is governed principally by the diffusion of dominant diffusing species. For the first step RTA process (480 °C), the local compressive stress induced from the oxide edge was

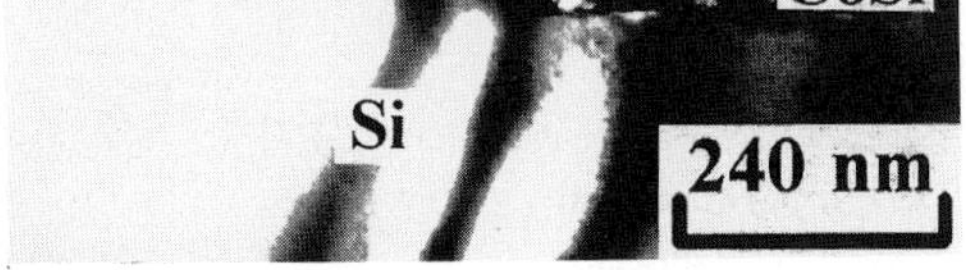

Fig. 7 XTEM image showing the $CoSi_2$ films on oxide edge on (001)Si inside a linear opening of 0.18 μm in size, 480 °C.

thought to hinder the migration of Si through the Co/Si interface, so that the formation of CoSi film was retarded at the oxide edge.

4. CONCLUSIONS

For Ti on 30 keV BF_2^+- 20 keV N_2^+ and 30 keV As^+-20 keV N_2^+ implanted samples, complete formation of C54-$TiSi_2$ at 700 °C was achieved. A continuous C54-$TiSi_2$ layer was found to form in all implanted samples annealed at 700-900 °C. For Ti on 1×10^{15} /cm^2 N_2^+- and As^+ implanted samples, EOR defects were completely eliminated in all samples annealed at 700-900 °C. For Ti on 2.5×10^{14} and 1×10^{15} /cm^2 N_2^+- and BF_2^+ implanted samples, fluorine bubbles were observed in 600-900 °C annealed samples.

The formation of self-aligned $CoSi_2$ inside 0.2-2 μm oxide openings prepared by electron beam lithography has been demonstrated. The uniform, high quality SEG Si layer was grown by ultrahigh vacuum chemical vapor deposition at 560 °C with Si_2H_6. A thin, uniform self-aligned $CoSi_2$ film was grown on the SEG Si layer by one-step and two-step RTA processes. Size effect of $CoSi_2$ formation is correlated to the stress induced at the oxide wall edge.

Acknowledgments

The research was supported by the Republic of China National Science Council and ERSO, IRTI. Helps from many NDL technical personnel are gratefully acknowledged.

REFERENCES

1. 1. C.Y. Wong, L.K. Wang, P.A. McFarland, and C.Y. Ting, J. Appl. Phys. 60, 243 (1983).
2. C.Y. Wong, F.S. Lai, P.A. McFarland, F.M. d'Heurle, and C.Y. Ting, J. Appl. Phys. 59(1986)2773.
3. W. Lur and L.J. Chen, J. Appl. Phys. 66, 3604 (1989).
4. M. Ono, M. Saito, T. Yoshitomi, C. Fiegna, T. Ohguro, and H. Iwai, IEEE IEDM 1995 Tech. Digest, 119 (1995).
5. T. Murakami, T. Kuroi, Y. Kawasaki, M. Inuishi, and Y. Matsui, Nucl. Instrim. Methods 121, 257 (1997).
6. S.P. Murarka, Mater. Res. Soc. Symp. Proc. 320, 3 (1994).
7. S. S. Wong, D. R. Bradbury, D. C. Chen, and K. Y. Chiu, IEDM Tech. Dig. 634 (1984).
8. T. Kuroi, Y. Yamaguchi, M. Shirahata, Y. Okumura, Y. Kawasaki, M. Inuishi, and T. Tsubouchi, IEEE IEDM 1993 Digest, 325 (1993).

Materials Science and Engineering Serving Society
S. Sōmiya, R.P.H. Chang, M. Doyama and R. Roy (Editors)

Multilayered optical memory using silica glasses as data-recording media

Jianrong Qiu [a], K. Miura [a] and K. Hirao [a,b]

[a]Hirao Active Glass Project, ERATO, Keihanna Plaza, Seika-cho, Kyoto 619-02, Japan
[b]Faculty of Engineering, Kyoto University, Sakyo-ku, Kyoto 606-01, Japan

Multilayered optical data storage was examined in SiO_2 and Ge-SiO_2 glasses by using a focused 800 nm, 120 fs, 200 kHz pulsed laser. The glasses are transparent in the wavelengths ranging from 300 to 1000 nm. After the irradiation of the 800 nm tightly focused femtosecond laser, a refractive-index bit was formed in the glasses. The relations between the size of refractive-index bit versus average power and irradiation duration of the laser have been examined. ESR spectra showed a significant increase in the concentration of SiE' and GeE' color centers in the laser irradiated glasses. Writing 3-dimensional refractive-index bit inside the transparent glasses based on a multi-photon absorption process is expected to become a promising method to fabricate optical memory with both an ultra high storage density and an ultra high recording speed.

1. INTRODUCTION

3-dimensional optical memories have been investigated over the last years. In these memories, data are recorded through the volume of the media. If we record 10 layers of data, we can achieve 10 times higher storage than that with current optical memories such as compact disk.

Usually photopolymers and photorefractive crystals were used as the data-recording media of 3-dimensional optical memory [1, 2]. As photopolymer is concerned, there are problems of distortion owing to shrinkage and flow or of isomerization due to ultraviolet light. In the case of photorefractive crystals, it is difficult for one to fabricate a crystal in large size. If 3-dimensional data storage can be achieved inside transparent glasses, it would become a very useful method since glasses are mechanically, chemically and thermally more stable than photopolymer and can be easily fabricated into various forms such as a thick and flat plate in large size. However, few studies have been performed on the 3-dimensional data storage inside glasses. Glasses are usually used only as substrates to support the recording medium in optical memory. In this paper, we report on the 3-dimensional optical storage inside transparent glasses by using a focused femtosecond pulsed laser. The structural change in the glasses before and after the laser irradiation has also been discussed.

2. EXPERIMENTAL

Glass samples used in the present experiment were SiO_2 and Ge-SiO_2 ($3GeO_2 \cdot 97SiO_2$ (mol%)) glasses fabricated by the VAD process. A regeneratively amplified 800 nm Ti:sapphire laser that emits 120 fs, 200 kHz, mode-locked pulse was focused through 10-100X microscope objectives and injected into inside the samples. By using an XYZ stage, it is possible to control the focused position inside the glasses. For the ESR measurement of the samples after laser irradiation, we wrote many damage lines inside the sample at room temperature. Then, ESR spectrum measurement was performed at liquid N_2 temperature for the samples before and after laser irradiation. Absorption spectrum was measured using a JASCO-570 spectrophotometer at room temperature.

3. RESULTS AND DISCUSSION

Figure 1 shows the absorption spectra of the SiO_2 and Ge-SiO_2 glasses. The glasses are transparent in the wavelengths ranging from 300 to 1000 nm. An absorption band peaking at 240 nm is observed. This band is due to the deficit of oxygen in the glasses. When irradiated by the 800 nm focused femtosecond pulsed laser, white light due to the effect of self-phase-modulation and an emission at 400 nm for the Ge-SiO_2 glass due to the second harmonic generation were observed. After the irradiation of the focused femtosecond laser, refractive index bit was formed in the samples. The shape of the bit was read out by use of transmitted light in a microscope. We measured the refractive index profiles across the cross sections of the line, a continuous assembly of the refractive-index bits, that was formed perpendicular to the laser beam. A remarkable refractive-index increase was observed at the center of the irradiated region [3].

Figure 2 shows a result of the read out of the data stored inside the SiO_2 glass. The average laser power was 40 mW and a 20X lens with an NA of 0.46 was used to focus the laser light. The irradiation duration of laser was 1s on one dot and the distance between the adjacent dots was about 10 μ m.

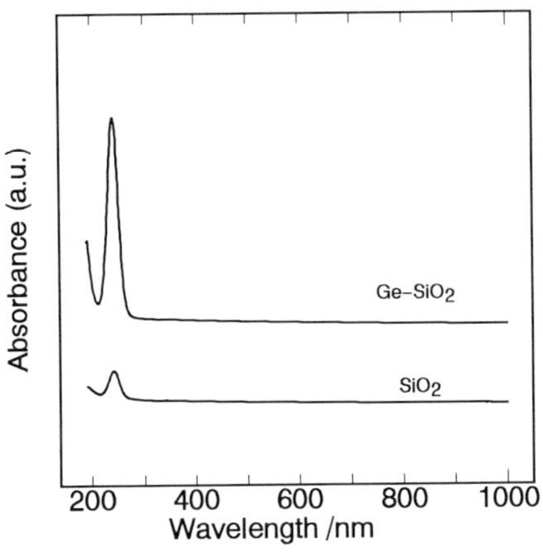

Figure 1 Absorption spectra of the SiO_2 and Ge-SiO_2 glasses.

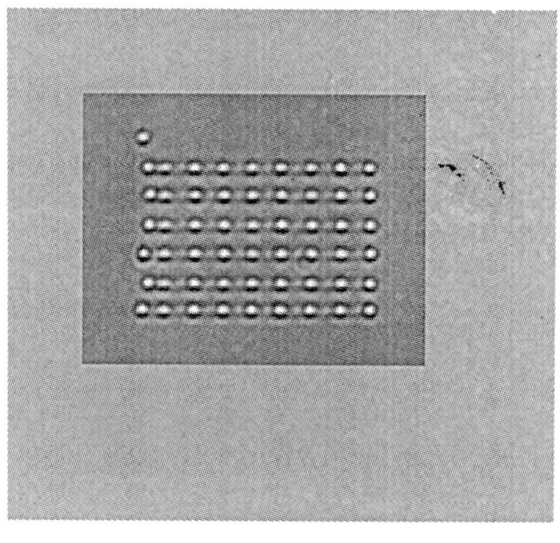

Figure 2 A result of the read out of the data information stored inside the SiO_2 glass.

Figure 3 shows the relation between the size of refractive-index bit formed in SiO_2 glass and average power of the laser. Irradiation duration of the laser was 1 s. The size of refractive-index bit clearly increases with increasing the average power of the laser. Moreover, contrast between the irradiated and unirradiated parts increase apparently with an increase in the average power of the laser. In addition, the contrast between the irradiated and unirradiated parts increases with an increase in irradiation duration as shown in Fig. 4 for SiO_2 glass, while the size of refractive-index bit seems to remain constant. The average power of the laser was 24 mW. Therefore, it is possible to adjust the size of refractive-index bit and the contrast between the irradiated and unirradiated parts by controlling the average power and irradiation duration of laser.

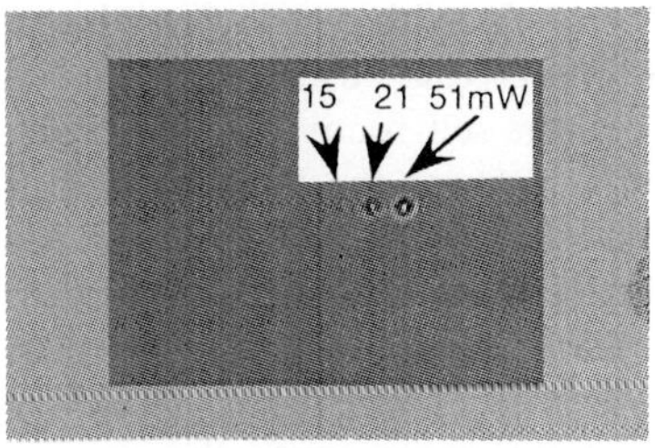

Figure 3 Relationship between the size of refractive-index bit formed in SiO_2 glass and average power of the laser .

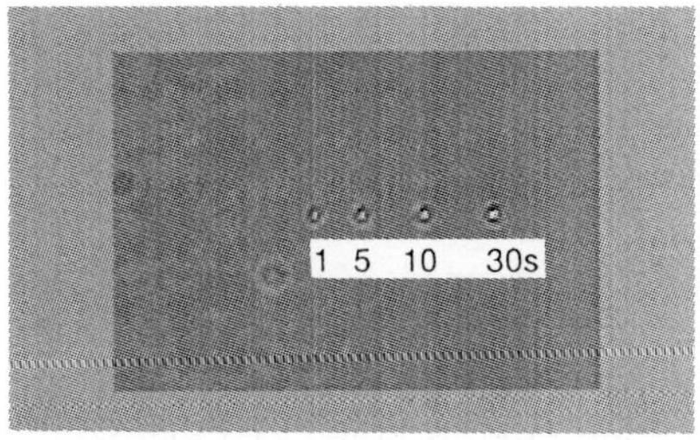

Figure 4 Relationship between the size of refractive -index bit formed in SiO_2 glass and irradiation duration of the laser .

Figure 5 shows the ESR spectra of the SiO_2 and Ge-SiO_2 glasses before and after the irradiation of the laser. The average power of the laser was 65 mW for SiO_2 and 30 mW for Ge-SiO_2, respectively. It is clear that the concentration of SiE' and GeE' color centers increases significantly in the laser irradiated glass. The formation of peroxy -radicals and non-bridge oxygen hole centers were also observed in the laser irradiated samples.

The formation of SiE' and GeE' defect centers was mainly observed in UV and X-ray irradiated SiO_2 and Ge-SiO_2 glasses [4-6]. However, there are few reports on the radiation effect on SiO_2 and Ge-SiO_2 glasses by infrared laser. From ESR spectra, apparently permanent atomic displacements were created and electron transfer between atoms occurred during the irradiation duration of the focused pulsed laser with a very high peak power of 10^7 W. Since the glass samples are transparent in 800nm, therefore, we suggest that a multiphoton absorption process occurred in the glasses when irradiated by the femtosecond pulsed laser. The emission at 400 nm may be due to the formation of periodic electronic structure in the Ge-doped SiO_2 glass after the laser irradiation.

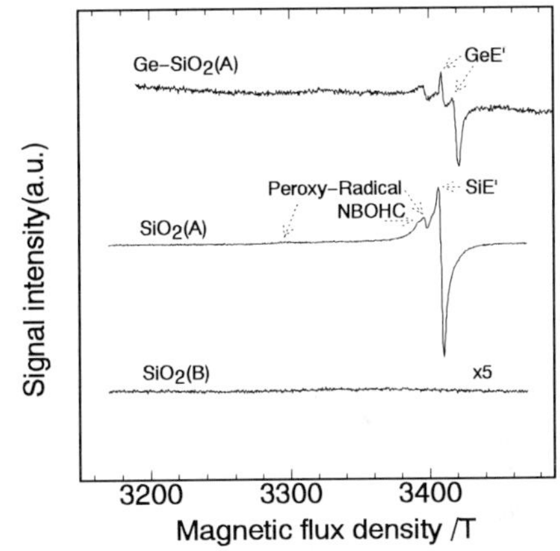

Figure 5 ESR spectra of the SiO_2 and Ge-SiO_2 glasses before (B) and after (A) the laser irradiation.

We have also observed a densification phenomenon in the glass irradiated by the 800 nm focused femtosecond pulsed laser from the observation of AFM [7].Therefore, we suggest that the formation of defect centers such as SiE' and GeE', and densification may be the reasons to cause the increase in refractive-index at the center of the laser irradiated part. In addition, the ultrafast energy deposition may also create high temperatures and pressures inside the glasses.

The remelting and rapid quenching may occur during the laser irradiation process, resulting in densification of a domain in the glasses. However, the detailed mechanism of the structural change induced by the focused femtosecond laser still remain unclear at present. A related investigation program is in progress in our group.

We have also observed that refractive index bit can be induced by a single pulse of the femtosecond laser [8]. The size of refractive-index bit was observed to be about 400 nm, which is smaller than the wavelength of laser. This may be due to the self-focusing of the laser and non-linear effect of the glass, resulted from the interaction between the laser and the glass. Therefore, by using a focused femtosecond laser, it is possible to record a data within several hundred femtoseconds. Moreover, with an objective having a long working distance, it is possible to record more than 100 layers spaced by less than 10 μ m. A recording density of about 10^{13} bits/cm^3 can be achieved.

Therefore, we are convinced that writing three-dimensional spots inside the glasses will become a promising method to fabricate optical memory with both an ultra high storage density and an ultra high recording speed.

4. CONCLUSION

We examined the feasibility of multilayered optical data storage inside SiO_2 and Ge-SiO_2 glasses using a focused 800 nm, 120 fs, 200 kHz pulsed laser. The glasses are transparent in the wavelengths ranging from 300 to 1000 nm. A refractive index bit was formed in the samples after the irradiation of the 800 nm focused femtosecond laser. The size of refractive-index increased with increasing the average power of the laser. The contrast between the irradiated and unirradiated parts increases with an increase in average power and irradiation duration of the laser. ESR spectra showed a significant increase in the concentration of SiE' and GeE' color centers in the laser irradiated glass samples. A multi-photon absorption process is considered to have occurred in the glasses when irradiated by the pulsed laser. Writing three-dimensional dots inside the glasses will become a promising method to fabricate optical memory with both an ultra high storage density and an ultra high recording speed.

ACKNOWLEDGEMENT

J. Qiu thanks Dr. J. Nishii of Osaka National Research Institute, Japan for his kind help in the measurement of ESR spectra. J. Qiu is also grateful to Drs. H. Inouye, S. Fujiwara, T. Suzuki and Ms. T. Murase of Hirao Active Glass Project, ERATO, JST, Japan for their corporation in the experiments.

REFERENCES

1. J. H. Stricker and W. W. Webb, Opt. Lett., 16 (1991) 1780.
2. Y. Kawata, T. Tanaka and S. Kawata, Appl. Opt., 35 (1996) 5308.
3. K. Davis, K. Miura, N. Sugimoto and K. Hirao, Opt. Lett., 21 (1996) 1729.
4. H. Hosono, H. Kawazoe and J. Nishii, Jpn. J. Appl. Phys., 35 (1996) 2234 .
5. J. Nishii, N. Kitamura, H. Yamanaka, H. Hosono and H. Kawazoe, Opt. Lett., 20 (1995) 1184.
6. D. L. Griscom, Phys. Rev., B 40 (1989) 4224.
7. K. Miura, J. Qiu, H. Inouye and K. Hirao, submitted to Appl. Phys. Lett.
8. J. Qiu, K. Miura, H. Inouye and K. Hirao, in preparation.

Materials Science and Engineering Serving Society
S. Sōmiya, R.P.H. Chang, M. Doyama and R. Roy (Editors)
1998 Elsevier Science B.V.

Possibility of Material Recycling using Hydrothermal Process

Nakamichi Yamasaki

Research Laboratory of Hydrothermal Chemistry,
Faculty of Science of Kochi University
Akebono-cho, Kochi 780, Japan

Hydrothermal solution has a wide possibility for reaction media from the range of ionic to radical reactions. The hydrothermal solution around at critical point have a wide variations of dielectric constant from 80 to 2 according to P-T conditions. And so, the choice of suitable reaction media can be easily controlled by only P-T-conditions.

In this paper, we have aimed to evaluate the employment of the mini-autoclave to the practical continuous tube reactor and its application for decomposition of halogenated organic compounds, degradation of coal, liquefaction of polymer and rubber are described.

1. Characteristics of hydrothermal solution containing supercritical water.

The characteristics of hydrothermal reaction is in a high reaction velocity in the solution, and accordingly the continuous flow system autoclave[1)] with using the tube digester has been expected to wide application with developments of censers and computer control. Using the continuous tube type autoclave, the P-T conditions can be independently controlled by an external pressure pump system compared to ordinary batch type autoclave only using under saturated vapor pressure. The characteristics of hydrothermal reaction is usually mentioned in the large hydrolysis ability. Hydrolysis is generally presented on acidic and alkaline catalytic reaction corresponding to objective materials, however the magnitude of hydrolysis capacity can be explained by the Kw value[2)], which is largely affected by temperature. On the other hand, the selectivity of reaction media is affected by the dielectric constant. Large values of 80

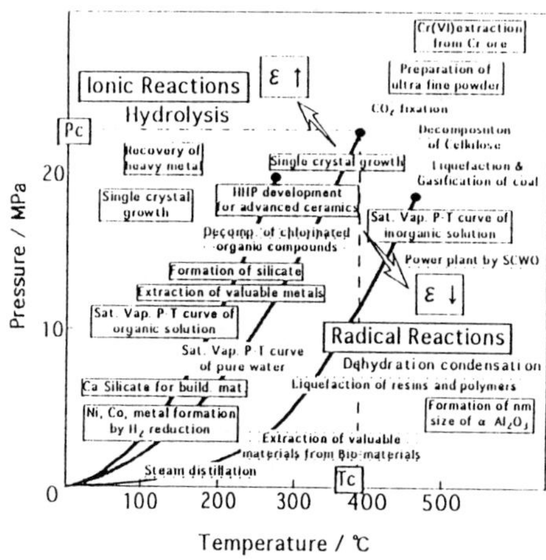

Figure 1 The suitable P-T regions for various reactions

are suitable media to enhance an ionic reaction such as complex decomposition of ionic compounds, nevertheless the Kw values under 10 are good media as organic reactions. These dielectric constants(ε) are largely affected by density[3]. The dielectric constant increases with pressure at the range of high density region and at contrast decreases with pressure at high temperature region. The P-T relation of water and a typical reaction example at each suitable reaction media, as well as, the variations of saturated vapor pressure curves of organic and inorganic solute solutions at the P-T region are shown in Figure 1. The non-polar organic compounds contained polymer and resin easily dissolve in water with low dielectric constant under high temperature(above Tc) and low pressure, and so these low dielectric fields are suitable reaction media to enhance organic reactions. However, at the region of high temperature above 500℃ pyrolysis of organic substance is easily complicated and to form a pitch and/or char by radical condensation. *Arai* and *Adshiri* had studied[4] the selective decomposition of in order to form glucose using a very high speed reaction velocity.

4. Degradation of halogenated organic waste under subcritical conditions[5)6)]

Figure 2 shows the results of hydrothermal degradation of CCl_3F. Various halogenated organic compounds can not dissolve in water whereby the first step of reaction may be occur at the inter surface. The addition of alcohol show a large effects, it decrease the decomposition temperature for about 100℃ of the common. This figure shows the initial stage of the decomposition for tests conducted at several temperatures and environmental conditions.

In the case of the chloroalkanes, those are chloromethane and ethane, trichloroethylene, etc. easily decompose under the same conditions of the case of CCl_3F. The decomposition of stable aromatic compounds such as PCBs, DIOXINEs, chlorobenzenes require higher temperature approximately 350℃ and the use of same solvent[7)8)9)]. The reaction products are halogenide ion in water phase among other organic products such as alco-

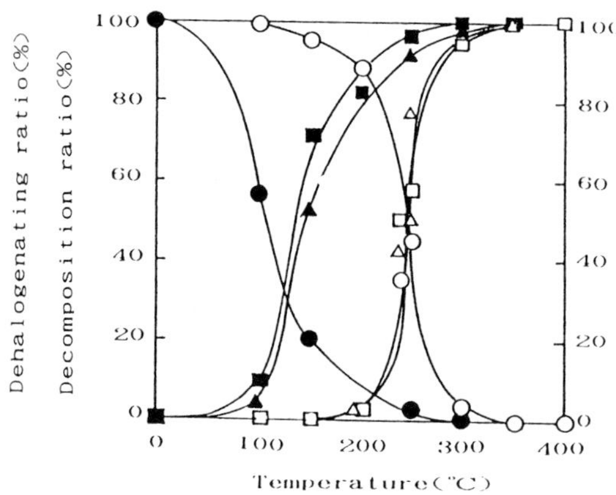

Figure 2 Decomposition of CFC-11(Comparison between alkaline only and methanolic alkaline solvent, $20cm^3$ mini-autoclave, degree of filling;58%, reaction time;30min, rising temperature;60 ℃/min, CFC-11;0.5g)
○:residual CFC-11,4M-NaOH,
●: residual CFC11,methanol/4M-NaOH;1/1,
△: defluorine ratio,4M-NaOH,
▲: defluorine ratio,4M-NaOH/methanol;1/1,
□: dechlorine ratio,4M-NaOH,
■: dechlorine ratio,4M-NaOH/methanol;1/1,

hols, phenols and their derivatives. Waste of these halogenated organic materials are commonly by the environmental contaminant materials because of its stability and toxicity. The toxicity is based on halogen, and so the perfect release of halogen from organic molecule to ion in water might be expected. A successful decomposition of chloride easily dissolved in water under alkaline conditions and accordingly the perfect decomposition is expected.

On the other hand, the decomposition under acidic conditions such as supercritical and pyrolitic conditions easily form hydrogen chloride and/or chlorine molecule. These are easily act to organic molecules often forms other chlorinated organic compounds such as DIOXINEs.

The main products from alkane groups are alcohol and its derivatives and from aromatic groups are phenol and its derivatives. Byproducts from alkanes are olefins and from aromatics are biphenyl compounds, and therefore dehydrogen chloride may be occurred inner and intra molecules. When trichloroethane was used, the n-hexane was formed under hydrothermal conditions with only pure water. It is expected that the n-haxane may be formed by hydration substitution of chlorine and condensation.

The reaction mechanism of decomposition of halogenated organic compounds under hydrothermal conditions can not be explained using the simple model of S_N2 under ambient conditions. Moreover, the chlorinated organic compound containing same numbers of chlorine and hydrogen in molecule decomposed with hydrogen chloride to form amorphous carbon, and in some cases graphite[10].

3. Oilification of polyethylene and rubber

The liquefaction of polyethylene has been reported by Enomoto et al[11]. They demonstrated that the oilification occurred around at 400-450℃ with supercritical water. Figure 3 shows a relation between viscosity and degree of filling of water in vessel(corresponded to density and pressure). Viscosity exhibits a maximum value near to the critical volume(30-40%), however, it has extremely lower values at 20% of lower pressure. The GC-MS spectra for several conditions of both degree of filling and temperature are shown in Figure 4. The products are consisted of paraffin with C_8- C_{20} and the lower filling ratio only the formation of low molecular weight was observed in comparison with these spectra(that of 20% filling ratio lowest molecular weight). The spectra of various experiments conducted by using 30% degree of filling, those also showed the formation of low

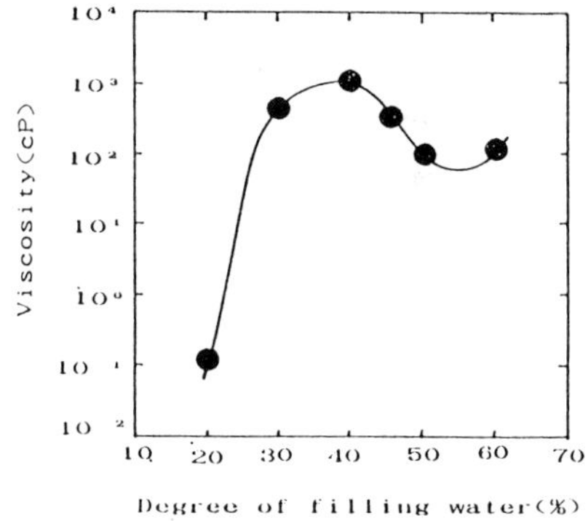

Figure 3 Viscosity of product oil from polyethylene as a function of filling ratio temperature; 415℃, reaction time;3hrs.

molecular weight compounds at high temperature. This results under supercritical region reveal that the decomposition of low molecule products may be occurred by the lower pressure and higher temperature zone, that is, at lower dielectric constant values. As mentioned above, high density water(high pressure and degree of filling) shows a high dielectric constant and act as a suitable reaction media for ionic reaction. However low density water(lower pressure and degree of filling) is considered for organic reaction to be a successful media for reaction. The characteristics of hydrothermal solution as reaction media can be continuously changed by the variation of P-T conditions, in accordance with these processing conditions may be a possibility to treat successfully every reaction media from ionic to radical reaction is expected.

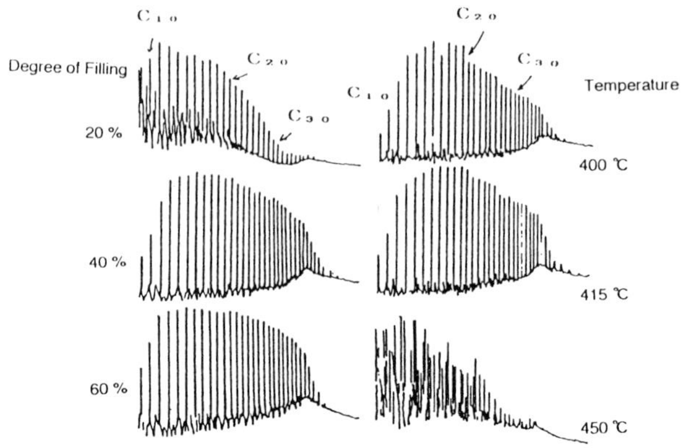

Figure 4 GC-MS Chromatograms of products from polyethylene(comparisons among those of degree of filling;20,40 and 60% , and of reaction temperatures; 400, 415 and 450 ℃)

Table 1 Composition of vulcanization rubber

Ingredient	phr*1
EPDM*2	100
Zinc oxide	5
Stearic acid	2
Carbon black	80
Paraffinic oil	60
Sulfur	2
Dibenzothiazyl disulfide	2

*1: parts per hundred parts of rubber
*2: ethylene-propylene-diene methylene linkage polymer

Liquefaction of vulcanization rubber[11] (Table 1) can be easily liquefied using high temperature and lower pressure side(420℃, up to 20MPa) as shown in Figure 5. As it can be seen from this Figure, the results show a similar reaction behavior as in the case of polyethylene, on other words, same oily products under same conditions. In addition, the oily products have not showed

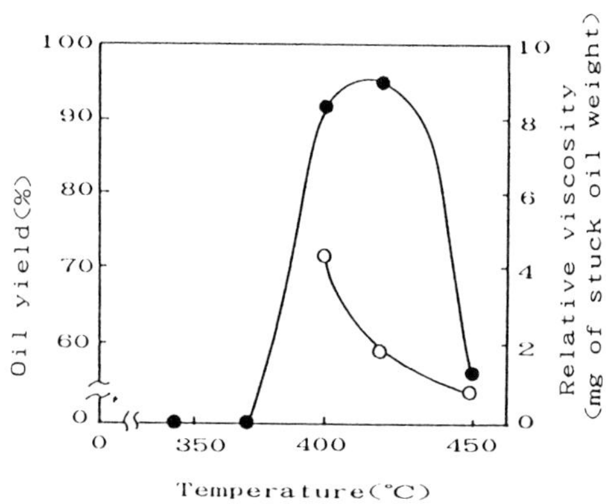

Figure 5 Oil yields and relative viscosity of resulted oil from vulcanic rubber as a function of temperature

the formation of sulfur or its solid such as zinc sulfide precipitation. These results represent the possibility to use this process for desulfurization of petroleum.

4. Continuously flow system tube reactor[1)12)]

Figure 6 shows a continuously flow system tube reactor for hydrothermal treatment. The main characteristics of hydrothermal process is its extreme large reaction velocity. As well as, the wide selectivity as a reaction media of hydrothermal solution. This is commonly difficult to obtain by using batch type autoclave under the conditions of saturated vapor pressure. The continuous pipe line system with pressure liquid has been developed from this century. The first studies were conduced on synthesis of ammonia by Harber-Bosch process, liquefaction of coal by Bergius process before world war II. The pipeline system for oil industry products has been recently developed. The hot water pipeline system has been developed to be used at atomic power plants, however the system water solution containing some ions and slurry does not developed yet for the treatment of organic compounds such as oil and coal fired, this system has not been developed yet, because of high corrosion limitation and defect of basic data.

On the other hand, large possibilities of tube reactor hydrothermal process using the P-T control will be expected for the frontier fields of advanced material format ion, metallurgy, new energy transformation technology, recycle system from waste to products, treatment of toxic and hazardous materials. Recently, new sensor and tough alloys such as Hastelloy, Inconel, Stiletto, etc. in combination with computer control system have been under evaluation. The heating small diameter tube system has a large advantage for low process cost, safety, closed system, not requires large space, many problems of corrosion, control system with pressure valve make it not useful for treatment of organic compounds.

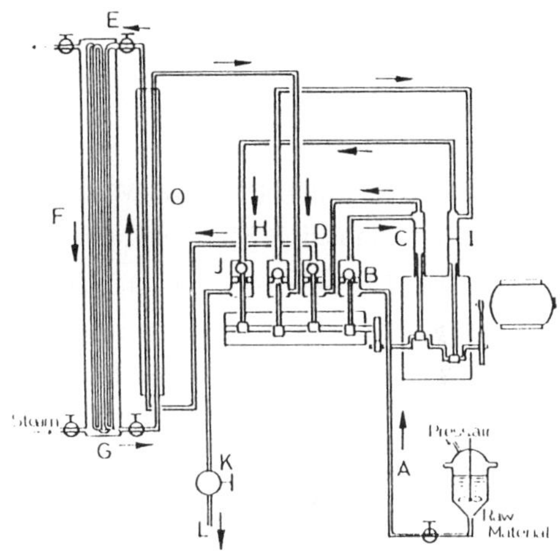

Figure 6 Continuous flow system tube reactor consist of pressure exchanger
A: flow direction from slurry entrance, B and J: ball valve for pressure exchanger, C and I: plunger pump of inlet and exhaust pressure, O: heat exchanger, F: tube reactor, K: pressure control valve

Conclusion

The high temperature and pressure water are a suitable reaction media to conduce every reaction from ionic to radical, controlling the P-T conditions, species and concentration of solute.

A bond with maldistributed existence of electron, that is, hetero-organic compounds are easily decomposed in the liquid phase by using saturated vapor pressure at the high pressure region under supercritical conditions. The condition for decomposition of organic polymer with covalent bond exist in the vapor phase under saturated vapor pressure and in the low pressure site under supercritical region.

Considering a slurry transformation using heating small diameter tube, new process which combines the reaction and separation can be expected to be used as a recycle system containing waste treatment and new material formation.

References

1) N.Yamasaki, M.Tsubouchi, M.Doi, S.Kanahara, K.Matsuoka, J.Yamasaki, *Rep. Res. Lab.* Hydrotherm. Chem., 3,1 (1979).

2)drawing by data of IAPWS report (1994).

3) R.W.Shaw, T.B. Brill, A.A.Clifford, C.A.Eckert, E.U.Franck, *C & En.*, December 23, 26 (1991).

4) K.Arai, M.Adjiri, *Rep. Focused Res.* by Min. Ed., 211 (1994).

5) N.Yamasaki, ALPHA, (5), (1991).

6) N.Yamasaki, M.Fujita, T.Moriya, M.Kanazawa, *Shigenkankyotaisaku*, **34**, (4), 39 (1995).

7) N.Yamasaki, T.Yasui, K.Matsuoka, *Env. Sci. Tech.*, **14**, 550(1980).

8) N.Yamasaki, J.Katada, S.Kanahara, K.Maysuoka, *Rep. Res. Lab. Hydro. Therm. Chem.*, **2**, 33(1978).

9) H.Yamaguchi, N.Yamasaki, *Chem. Equipment*, **37**, (7) 55(1995).

10) N.Yamasaki, K.Hosoi, K.Yanagisawa, *J. Chem.Soc.Jpn,* (11), 1909 (1988).

11) S.Sato, A.Hatakeyama, T.Moriya, H.Enomoto, N.Yamasaki, *Proc. 58th Conf.* (3), 90(1993).

12) M.Nishioka, N.Yamasaki, K.Yanagisawa, S.Yamasaki, J. Soc. Agricul. Struct. Jpn. 20(2), 5 (1989).

Materials Science and Engineering Serving Society
S. Sōmiya, R.P.H. Chang, M. Doyama and R. Roy (Editors)

Crystal growth of CeO_2 under hydrothermal conditions

Michiyo Kamiya, Eriko Shimada and Yasuro Ikuma

Department of Applied Chemistry, Kanagawa Institute of Technology,
1030 Shimoogino, Atsugi, Kanagawa, 243-02, Japan

CeO_2 powder precipitated by NH_4OH was hydrothermally treated with LiOH as a mineralizer at the conditions of 300~800°C, 100 MPa, for 0~6 h. After hydrothermal treatment, the samples were examined by X-ray diffraction, and observed by SEM and TEM to measure the particle sizes. The particle sizes of CeO_2 powder after treatment grew bigger except for the conditions of 300~400°C, 100 MPa, for 0 h. All of the X-ray diffraction patterns obtained in this study were those of CeO_2. As treatment temperature increased in the range of 500 to 800°C, the (111) peaks became gradually strong. This was closely related to the results obtained by electron microscopy.

1. INTRODUCTION

Ceria (CeO_2) possesses the fluorite structure and is known to show both electric and ionic conduction mechanisms. The presence of wide range of non-stoichiometry [1] is largely responsible for the nature of electrical conductivity of this oxide. To reach better understanding of the property of CeO_2 it is very important to make stoichiometric CeO_2. However, the formation of dense stoichiometric CeO_2 is very difficult because CeO_2 deviates from stoichiometry at high temperature and low oxygen partial pressure [2]. In this sense, hydrothermal processing is a good technique for the formation of stoichiometric CeO_2 because materials are treated at relatively low temperature and at high pressure.

In the past, CeO_2 powder was prepared by Tani *et al.* [3,4] using hydrothermal synthesis technique. Tani *et al.* studied the effect of mineralizer on the hydrothermal processing of CeO_2 and concluded that KF and $LiSO_4$ solutions considerably enhanced the growth of CeO_2 crystallites up to 100 μm at 500 and 600°C. However, the kinetics of hydrothermal treatment of CeO_2 was not studied in detail. In the present work,

LiOH has been used as a mineralizer and the effects of treatment temperature and time on the size of CeO_2 powder during the hydrothermal treatment have been investigated.

2. EXPERIMENTAL PROCEDURE

Cerium nitrate was used as starting material. NH_4OH was added to the solution of 0.2 mol·dm^{-3} cerium nitrate. The mixed solution was stirred for 30 min. The precipitate was separated through a filter paper by suction filter, washed with distilled water and dried at 85°C overnight in an oven in air. Cerium oxide powder obtained in this way was slightly ground and then about 0.2 g of the powder and about 1.1 g of 5 mol·dm^{-3} LiOH solution were put into a Au tube (inner diameter:4.7 mm, length:75 mm, thickness:0.1 mm). LiOH was a mineralizer. Mouth of the Au tube was closed and the tube was put into the test tube (inner diameter:6 mm, thickness:10 mm, length:200 mm, depth:190 mm).

Hydrothermal treatment was carried out at 300~800°C, 100 MPa for 0~6 h. After the treatment, the test tube was quenched into tap water. It took about 5 min to cool down to 30°C. The sample was taken out of the Au tube, filtered and washed with distilled water. Then, the sample was dried at 85°C overnight in an oven in air.

The dried samples were examined by X-ray powder diffraction to verify the chemical composition and crystal structure of the sample. To measure the particle sizes of samples, they were observed under scanning electron microscope (SEM) and transmission electron microscope (TEM).

3. RESULT AND DISCUSSION

The powder precipitated by NH_4OH was identified to be CeO_2 by X-ray diffraction. However, the peaks of X-ray diffraction were small, indicating low degree of crystallinity and fine particle size. The small particle size was confirmed by TEM observation. Average particle size of precipitated CeO_2 was about 11 nm. The results of X-ray diffraction of CeO_2 powder treated by hydrothermal processing at 300~800°C, 100 MPa for 3 h are shown in Fig. 1. All of the X-ray diffraction patterns obtained in this study were those of CeO_2. As treatment temperature increased, the X-ray peaks changed from broad to sharp. This indicates that crystallinity and particle size increase with temperature. If we observe Figure 1 more carefully, we notice that as treatment temperature increased in the range of 500 to 800°C, the (111) peak become gradually strong, and the other peaks became slightly weak. This may be related to the results observed by SEM and TEM.

The samples were observed by TEM and SEM after treatment at 300~800°C, 100 MPa for 3 h (Figure 2). At low temperatures (300~400°C), the shape of samples seemed to be spherical or cubic with (100) plane at the faces of cube. At 500°C, the particle size of sample was considerably large and corner of cube was slightly removed to bring (111) planes to surface. At higher temperatures (>500°C), the particle size of samples grew bigger and the (111) planes were clearly at the surface. This corresponds to strong X-ray diffraction from (111) planes in Figure 1. These results are in agreement with the results of crystal growth by hydrothermal treatment using the different mineralizer [3] and by flux method [5].

At each temperature, hydrothermal treatment time was changed from 0 h to 6 h and the samples were observed under SEM and TEM. The particle sizes of samples were plotted as a function of treatment time (Figure 3). The particle sizes of all samples were larger than original particle (11 nm) except for the samples treated at 300°C and 400°C, 100 MPa for 0 h. In the latter conditions, the particles were dissolved in the beginning of hydrothermal treatment. Consequently, these particle sizes were smaller than the particle size of sample without treatment. With increasing time, the particle

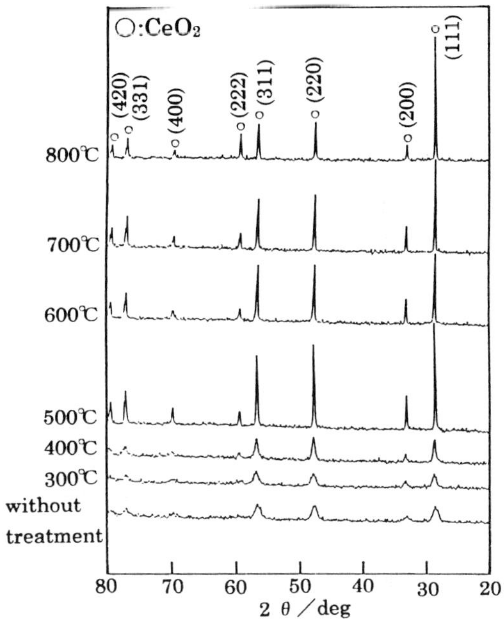

Figure 1. X-ray diffraction patterns after hydrothermal treatment at 300~800℃, 100 MPa, for 3 h

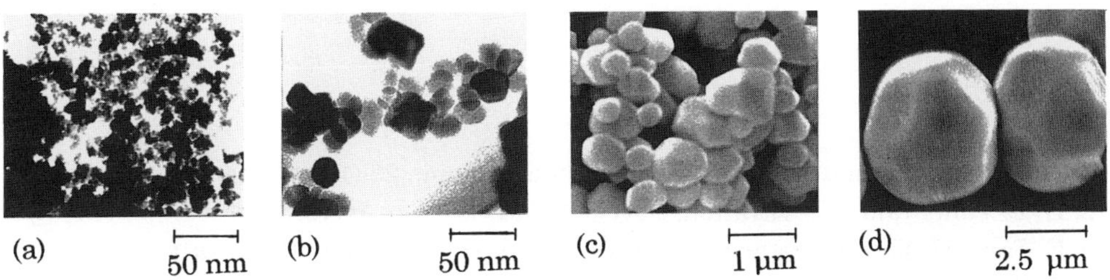

Figure 2. The micrographs of CeO_2 powders after hydrothermal treatment ; (a) without treatment, (b) at 400°C, (c) at 500°C, and (d) at 700°C, 100 MPa, for 3 h

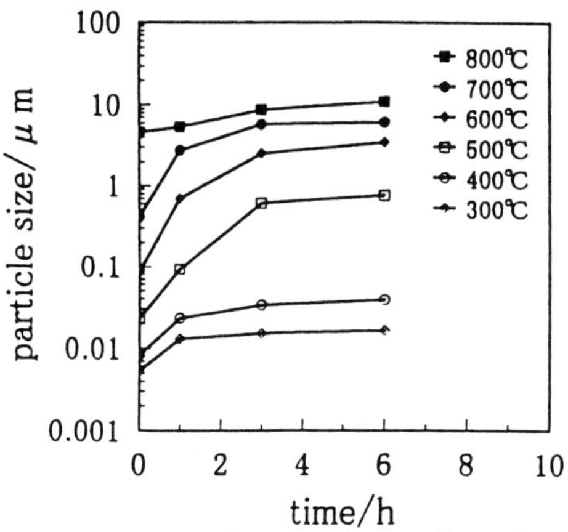

Figure 3. Particle size of samples after hydrothermal treatment at 300~800°C, 100 MPa, for 0~6 h

grew big but the growth rate slowed down at long treatment time. We should apply a rate equation to these results, but no rate equation is available at present. As high treatment temperatures, the grain growth has started before sample reached the treatment temperature. In addition the time to reach the constant temperature was different in these treatment temperatures. It is difficult, therefore, to give quantitative explanation to the data in Figure 3.

4. CONCLUSIONS

The CeO_2 powder was treated hydrothermally at 300~800°C, 100 MPa, for 0~6 h with LiOH as a mineralizer. The samples after treatment were characterized by X-RD, TEM and SEM. From the results it was found :

(1) As hydrothermal treatment temperature was increased and treatment duration was extended the particle size grew bigger.

(2) At low temperatures, the particle sizes in the beginning of treatment were smaller than the particle size without treatment, indicating that the particle dissolved into solution before reaching treatment temperature.

(3) From the results of X-RD, it was suggested that during the hydrothermal treatment of CeO_2, (111) planes developed preferentially. This observation was consistent with the results in the literature.

REFERENCES

1. D. J. M. Bevan and J. Kordis, J. Inorg. Nucl. Chem., 26 (1964) 1509.
2. H. L. Tuller and T. S. Stratton, NATO ASI Series. Ser. B (1985) 87.
3. E. Tani, M. Yoshimura and S. Somiya, J. Mater. Sci. Lett., 1 (1982) 461.
4. E. Tani, M. Yoshimura and S. Somiya, Report of The Research Laboratory of Engineering Materials, Tokyo Inst. Tech. 8 (1983) 47.
5. I. V. Vinokurov, Z. N. Zonn and V. A. Ioffe, Soviet Phys.-Solid State, 7 (1965) 814.

Materials Science and Engineering Serving Society
S. Sōmiya, R.P.H. Chang, M. Doyama and R. Roy (Editors)

Hydrothermal–electrochemical synthesis of $ATiO_3$ (A = Ba, Sr) thin films

Koji Kajiyoshi, Yukio Sakabe and Kikuo Wakino

Functional Materials Research Department, Murata Manufacturing Co., Ltd.,
2288 Oshinohara, Yasu–cho, Yasu, Shiga 520–23, Japan

The hydrothermal–electrochemical method has a feature that ABO_3 complex oxide thin film can potentially be synthesized on B–site metal substrate in any form and shape. Growth of $ATiO_3$ (A = Ba, Sr) thin films of controlled thickness and microstructure has been investigated by this method on plates and porous bodies of titanium. Coulometric film growth was obtained on Ti plates using a single–step electrolysis based on the three–electrode technique. Similar film growth was realized on porous Ti bodies using a cyclic electrolysis.

1. INTRODUCTION

The hydrothermal–electrochemical method is an attractive technique for synthesis of complex oxide thin films [1,2]. This method permits a perovskite–type compound, ABO_3, to be synthesized on a substrate of B–site metal being oxidized anodically in an aqueous alkaline solution containing A–site ions, under hydrothermal conditions at low temperatures, generally below ca. 200°C. Since this synthesis technique utilizes a solution process, it has an advantage over sputtering and sol–gel techniques in that the ABO_3 thin film can potentially be grown on the B–site metal substrate in any form and shape such as plate, flexible foil, wire and porous body. This method is hence expected to provide applications in many fields of electronic devices. Although $ATiO_3$ (A = Ba, Sr) thin films have been synthesized on several forms of titanium thus far, not a few room for improvement still remains with controllability and uniformity of film thickness and microstructure [2,3]. The present paper describes our recent developments on the hydrothermal–electrochemical synthesis of the $ATiO_3$ thin films on plate and porous body of titanium.

2. EXPERIMENTAL

2.1. Preparation of titanium substrates

Titanium plates and porous titanium bodies were used as substrates. The titanium plates (Sumitomo Sitix, Amagasaki, Japan) with ≥ 99.5% purity were cut into dimensions of 50 x 20 x 0.5 mm^3, and their surfaces were mechanically polished to a mirror finish. The porous titanium bodies were fabricated from a spherical titanium powder with ≤ 45 μm particle size and ≥ 99.0% purity (TILOP–45, Sumitomo Sitix, Amagasaki, Japan). The Ti powder was pressed into cylindrical pellets having a diameter of 5.0 mm and a thickness of 1.0 mm, using an uniaxial hydraulic press under pressure of 100 MPa. The green pellets were lightly sintered at 650°C for 2 h in vacuum to establish electrical connection among every Ti particles. The porosity of the sintered pellets was estimated to be 32.7 vol.% from their density.

2.2. Hydrothermal–electrochemical treatment

Guaranteed reagents with ≥ 98% assay of $Sr(OH)_2 \cdot 8H_2O$ (Kojundo Chemical Laboratory, Osaka, Japan) and $Ba(OH)_2 \cdot 8H_2O$ (Nacalai Tesque, Kyoto, Japan) were used to prepare aqueous solutions of $Ba(OH)_2$ or $Sr(OH)_2$. The concentration of Ba or Sr was adjusted to 0.2–0.5 mol/L and the pH value was controlled in the range from 13.3 to 14.2 at 25°C using a NaOH solution. The electrolytic cell assembled in an autoclave is illustrated in Fig. 1. It accommodates a Ag/AgCl external reference electrode (Toshin Kogyo, Tokyo, Japan) to perform electrolysis potentiostatically by the three–electrode technique. The Ti plate or the porous Ti body was suspended as the working electrode with a Ti wire in the electrolytic cell containing 500 mL of the $Ba(OH)_2$ or $Sr(OH)_2$ solution. A platinum plate having ≥ 99.0% purity and dimensions of 50 x 20 x 0.5 mm^3 was suspended as the counterelectrode with a Pt wire, maintaining a separation of 30 mm between the two electrodes in the solution. The temperature profile of the hydrothermal treatment was controlled so that it would follow a heating process up to a synthesis temperature of 150°C with an approximate heating rate of 1.5°C/min, and a subsequent isothermal process followed by a cooling process. The electrolysis was performed using an automatic polarization system (Model HZ–1A, Hokuto Denko, Tokyo, Japan). Two types of electrolysis profiles were investigated to grow $ATiO_3$ thin films on the two different Ti substrates of plate and porous body, as described below. After the hydrothermal–electrochemical treatment, grown films were investigated using scanning electron microscopy (SEM, Model S–4000, Hitachi, Tokyo, Japan) at an acceleration voltage of 10–20 kV.

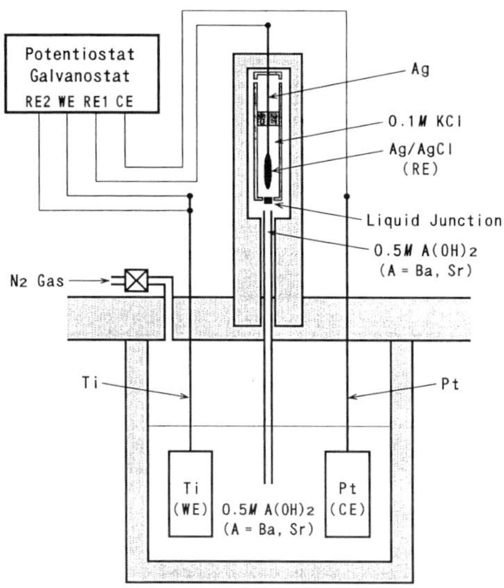

Figure 1. Schematic diagram of electrolytic cell used for hydrothermal–electrochemical synthesis. The electrolytic cell is assembled in an autoclave on the basis of the three–electrode arrangement.

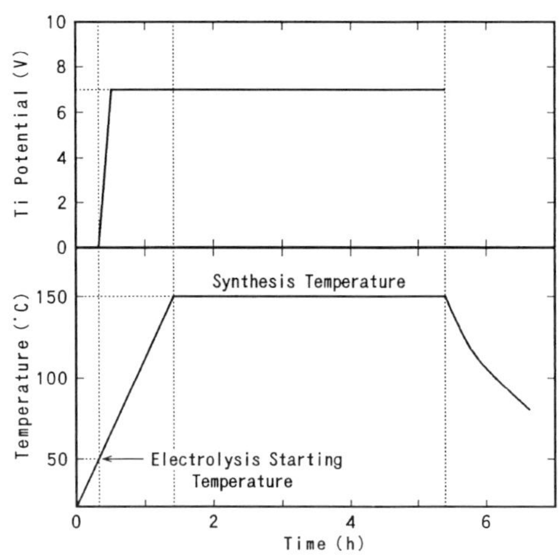

Figure 2. Single–step electrolysis profile (upper) and temperature profile (lower) adopted to grow $ATiO_3$ thin films on Ti plates. The potential of working electrode is represented in units of volts vs. Ag/AgCl.

3. RESULTS AND DISCUSSION

3.1. Film growth on titanium plate

A single–step electrolysis in which a constant anodic potential was applied to the working electrode of Ti plate during the hydrothermal treatment was tested so as to grow $ATiO_3$ thin films of controlled thickness. Such electrolysis treatments in previous studies [1–3] were performed exclusively from the beginning to the end of the isothermal process at a synthesis temperature. The electrolysis in the present study was performed from various temperatures in the heating process to the end of the isothermal process. It was found that densely packed $ATiO_3$ thin films can be grown up to several micrometers in thickness by starting the electrolysis at a temperature below about 90°C in the heating process (the electrolysis started above 90°C results in fragile porous films) [4]. A typical profile of the single–step electrolysis is depicted in Fig. 2. The electrolysis was started at 50°C by polarizing the working electrode from its natural potential to an anodic potential ranging from +7.0 to +12.0 V vs. Ag/AgCl, and was continued at the anodic potential. Thickness of $SrTiO_3$ thin film grown by this electrolysis increased with an increase in the quantity of electricity passed, and hence the film thickness can be accurately controlled by this factor [5]. Figure 3 demonstrates such a coulometric growth process of the $SrTiO_3$ thin film obtained at the potential of +8.0 V vs. Ag/AgCl. The current efficiency for the film growth varied in a range from 0.7% to 3.8%, depending on the synthesis conditions [4].

3.2. Film growth on porous titanium body

The electrolysis shown in Fig. 2 succeeded in growing $ATiO_3$ thin films uniformly on the Ti plate. However, $ATiO_3$ thin films could be grown by this electrolysis only on Ti particles located in the surface region to a depth of 100–200 μm into the porous Ti body. It was found that $ATiO_3$ thin films can be grown on all Ti particles in the porous Ti body by performing the electrolysis cyclically. A typical potential profile of the cyclic electrolysis used for the porous Ti body is demonstrated in Fig. 4. One cycle consisted of an anodic potential of +7.0 V vs. Ag/AgCl and a rest potential of −1.0 V vs. Ag/AgCl with respective soak times of 1 and 2 min. Prescribed number of cycles were applied to the working electrode of porous Ti body.

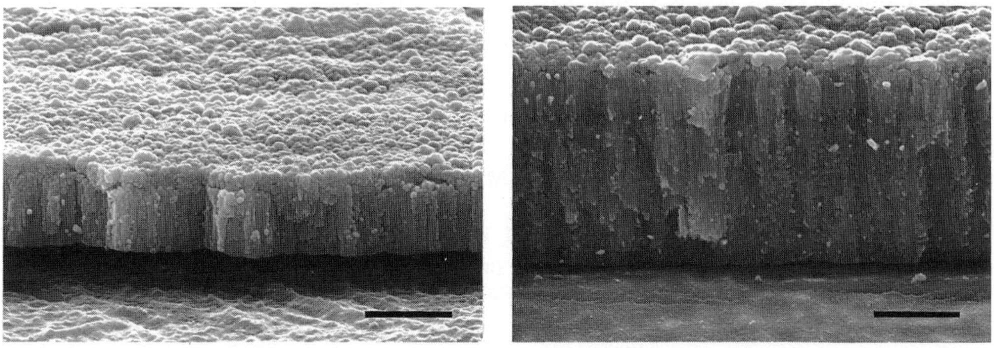

(a) (b)

Figure 3. SEM micrographs of fractured cross sections of $SrTiO_3$ thin films grown on Ti plates by single–step electrolysis (Fig. 2) at +8.0 V vs. Ag/AgCl with quantities of electricity passed: (a) 50 and (b) 100 C/cm^2 (bar = 1 μm).

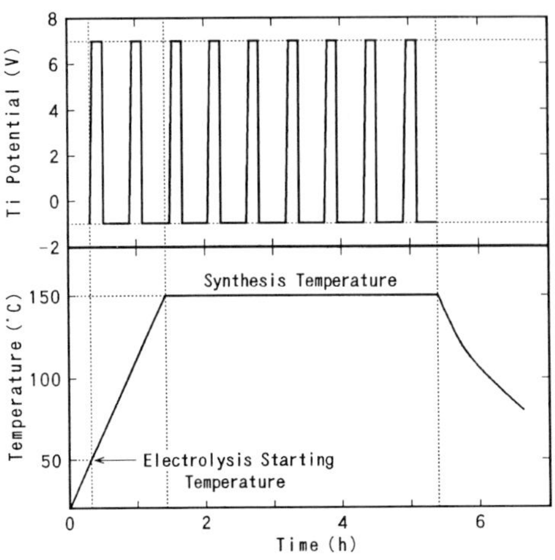

Figure 4. Cyclic electrolysis profile (upper) and temperature profile (lower) adopted to grow $ATiO_3$ thin films on porous Ti bodies. The potential of working electrode is represented in units of volts vs. Ag/AgCl.

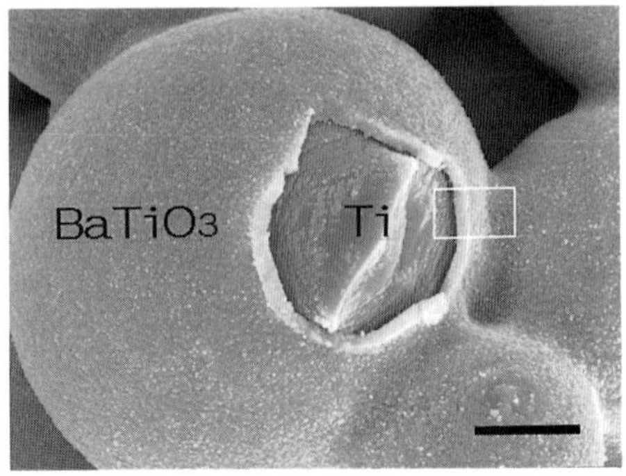

Figure 5. SEM micrograph of $BaTiO_3$ thin film grown on porous Ti body by cyclic electrolysis (Fig. 4) with 120 cycles. The rectangular marker indicates a region of fractured cross section of the $BaTiO_3$ thin film surrounding a necking of spherical Ti particles (bar = 2 μm).

Similarly as Fig. 2, the electrolysis was performed from 50°C in the heating process to the end of the isothermal process at 150°C. Figure 5 shows a SEM micrograph of $BaTiO_3$ thin film grown by the cyclic electrolysis in Fig. 4 on Ti particles at the inner region of the porous Ti body. Since the porous Ti body was lightly sintered to form neckings among Ti particles, it could be fractured at these neckings. Cross section of the $BaTiO_3$ thin film are observable around such a fractured surface of necking of Ti particles, as indicated by the marker in Fig. 5. The $BaTiO_3$ thin film displays a sufficiently uniform film thickness. It was also confirmed that the film thickness increases with increasing number of cycles of electrolysis performed.

REFERENCES

1. M. Yoshimura, S. E. Yoo, M. Hayashi and N. Ishizawa, Jpn. J. Appl. Phys., 28 (1989) L2007.
2. S. E. Yoo, M. Hayashi, N. Ishizawa and M. Yoshimura, J. Am. Ceram. Soc., 73 (1990) 2561.
3. S. Venigalla, P. Bendale, R. E. Chodelka, J. H. Adair, and S. A. Costantino, Proceedings of the 8th IEEE International Symposium on Application of Ferroelectrics, M. Liu et al. (eds.), IEEE Service Center, Piscataway, 1992, pp. 89–93.
4. K. Kajiyoshi, K. Tomono, Y. Hamaji, T. Kasanami and M. Yoshimura, J. Am. Ceram. Soc., 77 (1994) 2889.
5. K. Kajiyoshi, K. Tomono, Y. Hamaji, T. Kasanami and M. Yoshimura, J. Mater. Res., 9 (1994) 2109.

Materials Science and Engineering Serving Society
S. Sōmiya, R.P.H. Chang, M. Doyama and R. Roy (Editors)

Synthesis of Lithiophorite by Hydrothermal Soft Chemical Process

Qi Feng, Chiaki Honbu, Kazumichi Yanagisawa and Nakamichi Yamasaki

Research Laboratory of Hydrothermal Chemistry, Faculty of Science, Kochi University, 2-5-1 Akebono-cho, Kochi-shi 780, JAPAN

Lithiophorite which has a sandwich layered structure was prepared from birnessite-type manganese oxide by using hydrothermal soft chemical process. A birnessite-type sodium manganese oxide with layered structure was prepared as a precursor. The Na^+ ions in the interlayer space of the birnessite-type manganese oxide were ion-exchanged with $Li_xAl_n(OH)_m^{z+}$ complex ions. The ion-exchanged birnessite was treated under hydrothermal conditions to transform the birnessite structure to the sandwich layered structure. The transformation reactions from the birnessite strcuture to the sandwich layered structures were investigated by X-ray, DTA-TG, and chemical analyses, and IR spectroscopy. The lithiophorite can be obtained by hydrothermally treating the ion-exchanged birnessite at 150℃. A reaction model was proposed for the formation reaction of the sandwich layered structure.

1. INTRODUCTION

Manganese oxides with layered and tunnel structures show similar properties as zeolites and clay minerals. Since metal ions and organic molecules can be topotactically inserted into and extracted from the interlayer space and the tunnel of the manganese oxides, these oxides can be used as an ion-sieve,[1-6] molecule-sieve,[7,8] catalysts,[9,10] and cathode material of lithium battery.[11-13] Lithiophorite is a naturally occurring manganese oxide mineral with a sandwich layered structure, as shown in Fig. 1(a).[14] The structure is constituted by stacking $LiAl_2(OH)_6$ octahedral cation sheets and MnO_6 octahedral anion sheets. Wadsley has reported that the lithiophorite can be obtained by hydrothermally treating birnessite-type manganese oxide with Al_2O_3 in LiOH solution.[15] However, the reaction mechanism is not clear.

Recently, we have proposed a hydrothermal soft chemical process for synthesis tunnel manganese oxides from a layer manganese oxide.[16,17] This process comprises two steps: the first step being preparation of a framework precursor with layered structure and insertion of template ions or molecules (structure-directing agents) into its interlayer space by a soft chemical reaction, and the second step transformation of the template-inserted precursor into a tunnel structure by hydrothermal treatment. The dimension of the resulting tunnel can be controlled by the size of the used template. We think that this process is useful not only for tunnel structure but also for other structure. The present paper describes the synthesis of lithiophorite from a layer manganese oxide by using the hydrothermal soft chemical process.

2. EXPERIMENTAL

A Na^+-form birnessite (NaBIR) which was used as a precursor was prepared by pouring a mixed solution (100 mL) of 3% H_2O_2 and 0.6 M NaOH into a solution (50 mL) of 0.3 M $Mn(NO_3)_2$ with stirring. A solution of lithium-aluminium hydroxide complex ion ($Li_xAl_n(OH)_m^{z+}$) was prepared by slowly dropping a solution of 0.5 M LiOH (1000 mL) into a solution of 0.2 M $Al(NO_3)_3$ (800 mL) (Li/Al mole ratio = 2.0) at 60°C with stirring. The solution was aged at 60°C for 2 hr, and filtered to remove precipitate. NaBIR was treated with the solution of lithium-aluminium hydroxide complex ion for 1 day to exchange Na^+ in the interlayer space of NaBIR with $Li_xAl_n(OH)_m^{z+}$ complex ions. The ion-exchange treatment was repeated for two-times to complete the ion-exchange reaction. The ion-exchanged sample was hydrothermally treated in distilled water at 100 and 150°C, respectively, for 1 day under autogenous pressure.

An XRD analysis was carried out using an X-ray diffractometer (Rigaku RAD-RC). Infrared spectrum measurement was performed by KBr method on an infrared spectrometer (Perkin Elmer 1600). DTA-TG curves were obtained on a thermal analyzer (Seiko TG-DTA 320) at a heating rate of 10°C/min. The sodium, lithium, alminium, and manganese contents in the prepared samples were determined by atomic absorption spectrometry and ICP after dissolving the samples in a mixed solution of HCl and H_2O_2.

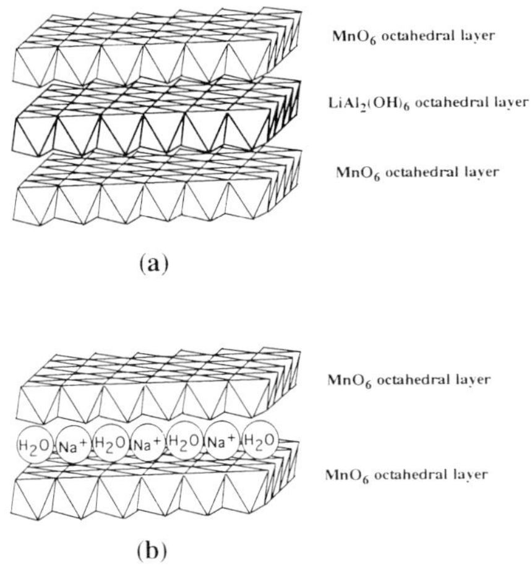

Fig. 1. Structures of (a) lithiophorite and (b) birnessite.

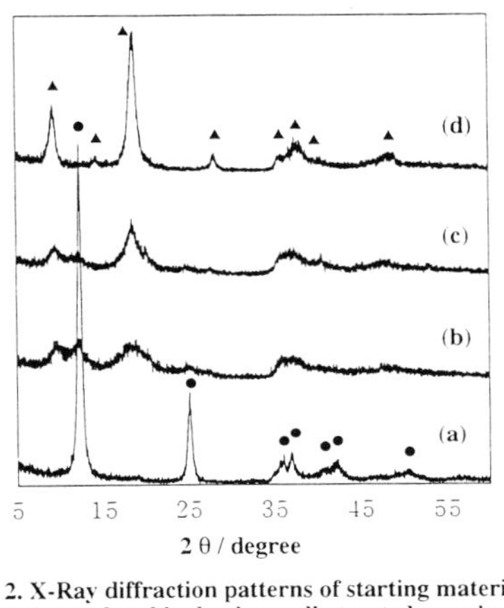

Fig. 2. X-Ray diffraction patterns of starting material, ion-exchanged and hydrothermally treated samples. (a) NaBIR (birnessite); (b) ion-exchanged sample; (c) and (d) hydrothermally treated samples at 100 and 150oC. ●: birnessite phase. ▲: lithiophorite phase.

3. RESULTS AND DISCUSION

X-Ray diffraction analysis (Fig. 2(a)) indicated that NaBIR has a layered structure with a basal spacing of 0.72 nm, which contains two-dimensional sheets of edge-shared MnO_6 octahedra, with single-crystal water sheets and Na^+ between the sheets of the MnO_6 octahedra (Fig. 1(b)).[5,6] Two layer phases with basal spacings of about 0.72 and 0.95 nm were observed in the ion-exchanged sample (Fig. 2(b)). The phase with 0.72 nm basal spacing has a single-crystal water sheet between the sheets of the MnO_6 octahedra, and the phase with 0.95 nm basal spacing has a double-crystal water sheet between the sheets of the MnO_6 octahedra.[6,18]

The hydrothermally treated samples were investigated by X-ray diffraction, FT-IR, DTA-TG, and chemical analyses. The X-ray diffraction studies on the ion-exchanged and hydrothermally treated samples indicated that the amount of 0.72 nm layer phase decrease and that of the 0.95 nm layer phase increase after the hydrothermal treatment (Fig.2). Only the 0.95 nm layer phase was observed after the hydrothermal treatment at 150°C, and the X-ray diffraction pattern of the 0.95 nm layer phase corresponds to that of natural lithiophorite, except a small diffraction peak with a d value of 0.62 nm is observed in the hydrothermally treated sample.[19] The IR spectrum of the ion-exchanged sample varies after the hydrothermal treatment, and the sample hydrothermally treated at 150°C showed same IR spectrum as the natural lithiophorite.[20] These facts suggest that a single phase of lithiophorite can be obtained by the hydrothermal treatment at 150°C. The sandwich layered structure of the lithiophorite is stable until heating at 400°C in air.

The compositional analysis indicated that most of Na^+ in the NaBIR were exchanged with the $Li_xAl_n(OH)_m^{z+}$ complex ions by the ion-exchange treatment. The Li/Mn and Al/Mn mole ratios (0.29 and 0.84) were kept almost constant before and after the hydrothermal treatments. The hydrothermal treatment causes a large decrease in water content of the sample. There is almost not crystal water molecule in the hydrothermally treated sample at 150°C.

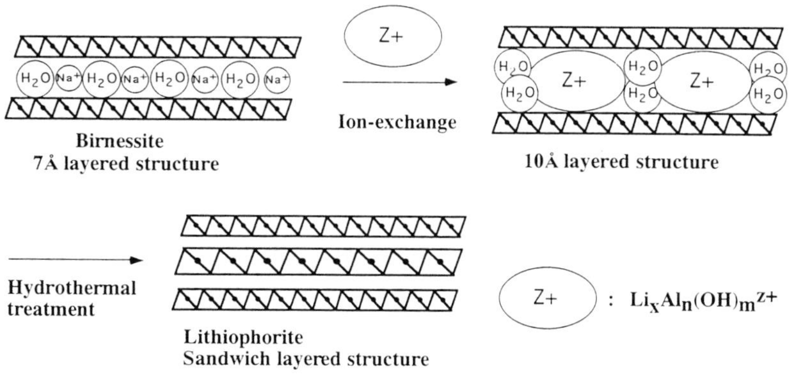

Fig. 3. A model of hydrothermal soft chemical reaction for synthesis of lithiophorite

The reaction of transformation from the ion-exchanged birnessite to the lithiophorite can be described by a reaction model of Fig. 3. The $Li_xAl_n(OH)_m^{z+}$ complex ions in the interlayer space are unstable, and polymerize together to form large complex ions under the hydrothermal conditions. Since the layered structure of birnessite acts as a template (structure-directing agent) in the polymerization reaction, the polymerization occurs in two-dimension, and forms the $LiAl_2(OH)_6$ octahedral sheet between the MnO_6 octahedral sheets of the birnessite structure. The polymerization accompanies dehydration of the crystal water from the interlayer space of the birnessite. Since there is almost not crystal water molecule in the lithiophorite prepared here, most of space between MnO_6 sheets is occupied by $LiAl_2(OH)_6$ sheet. The smaller mole ratio of Li/Al (=0.4) than the theoretical value (=0.5) suggests that there are some Li^+ vacancies in the $LiAl_2(OH)_6$ octahedral site.

The above facts suggest that the hydrothermal soft chemical process is useful not only for the synthesis of tunnel oxides but also for other oxides, such as the sandwich layer oxides, at low temperature conditions.

REFERENCES

1. Q. Feng, Y. Miyai, H. Kanoh, and K. Ooi, Langmuir 8 (1992) 1861.
2. Q. Feng, Y. Miyai, H. Kanoh, and K. Ooi, Chem. Mater. 5 (1993) 311.
3. M. Tsuji and M. Abe, Solv. Extr. Ion Exch. 2 (1984) 253.
4. Q. Feng, H. Kanoh, Y. Miyai and K. Ooi, Chem. Mater. 7 (1995) 148.
5. Q. Feng, H. Kanoh, Y. Miyai and K. Ooi, Chem. Mater. 7 (1995) 1226.
6. Q. Feng, H. Kanoh, Y. Miyai and K. Ooi, Chem. Mater. 7 (1995) 1722.
7. Y. F. Shen, R. P. Zerger, S. L. Suib, L. McCurdy, D. I. Potter and C. L. O'Young, J. Chem. Soc. Commun. 17 (1992) 1213.
8. Y. F. Shen, R. P. Zerger, R. N. DeGuzman, S. L. Suib, L. McCurdy, D. I. Potter and C. L. O'Young, Science 260 (1993) 511.
9. K. Matsuo, M. Nitta and K. Aomura, J. JJpn. Pet. Inst. 22 (1979) 212.
10. S. T. Wong and S. Cheng, Inorg. Chem. 31 (1992) 1165.
11. J. B. Goodenough, M. M. Thackeray, W. I. F. David and P. G. Bruce, Rev. Chim. Miner., 21 (1984) 435.
12. Q. Feng, H. Kanoh, K. Ooi, M. Tani and Y. Nakacho, J. Electrochem. Soc. 141 (1994) L135.
13. P. Strobel and C. Mouget, Mat. Res. Bull. 28 (1993) 93.
14. R. Giovanoli, Chem. Erde 44 (1985) 227.
15. A. D. Wadsley, Amer. Miner. 35 (1950) 485.
16. Q. Feng, K. Yanagisawa and N. Yamasaki, Chem. Commun. (1996) 1607.
17. Q. Feng, K. Yanagisawa and N. Yamasaki, 2nd International Conference on Solvothermal Reaction, Proceeding (1996) 235.
18. D. C. Golden, C. C. Chen and J. B. Dixon, Science 231 (1986) 717.
19. JCPDS 41-1378.
20. R. M. Poter and G. R. Rossman, Amer. Miner. 64 (1979) 1199.

Materials Science and Engineering Serving Society
S. Sōmiya, R.P.H. Chang, M. Doyama and R. Roy (Editors)

Synthesis of Birnessite-Type Sodium Manganese Oxide and the Behavior under Hydrothermal Conditions

Qi Feng, En-Hai Sun, Kazumichi Yanagisawa and Nakamichi Yamasaki

Research Laboratory of Hydrothermal Chemistry, Faculty of Science, Kochi University, 2-5-1 Akebono-cho, Kochi-shi 780, JAPAN

A birnessite-type sodium manganese oxide with a layered structure was synthesized by reacting $Mn(NO_3)_2$ solution with a mixed solution of H_2O_2 and NaOH at room temperature. This sodium manganese oxide was hydrothermally treated in neutral, acidic, and alkaline solutions at 150°C. The original samples which were synthesized at room temperature and hydrothermally treated samples were characterized by XRD and chemical analysis. The formation of the birnessite-type sodium manganese oxide is dependent on the mixed mole ratio of $NaOH/Mn(NO_3)_2$. The fraction of the birnessite phase increased with increase in the mixed mole ratio of $NaOH/Mn(NO_3)_2$. The birnessite-type manganese oxide has a basal spacing of 1.004 nm before air-drying, but it changes to 0.724 nm after the air-drying at room temperature. A good crystalline birnessite phase with a basal spacing of 0.724 nm was obtained by hydrothermally treating the birnessite-type sodium manganese oxide prepared at room temperature with a NaOH solution.

1. INTRODUCTION

Birnessite-type manganese oxides have a layered structure which contains two-dimensional sheets of edge-shared MnO_6 octahedra and water molecules and alkali metal ions between the sheets of edge-shared MnO_6 octahedra.[1,2] Since metal ions can be topotactically extracted/inserted from/into the interlayer space in aqueous phase and organic phase by chemical and electrochemical reactions, the birnessite-type manganese oxides can be used as a metal ion adsorbent and a cathode for lithium rechargeable battery. The manganese oxides can be used also as precursors for synthesis of manganese oxides with tunnel structure by using a hydrothermal soft chemical process.

Recently we have developed a new method for the preparation of layer manganese oxides with Li^+ or K^+ in the interlayer space by oxidation Mn(II) with H_2O_2 in alkaline solutions.[3,4] In the present study, we describe preparation of birnessite-type sodium manganese oxide with layered structure by using this method in detail, and characterization of behavior of the manganese oxide under hydrothermal conditions by XRD and chemical analysis.

2. EXPERIMENTAL

The original samples of manganese oxides were prepared by quickly pouring a solution (400 mL) containing 1 M H_2O_2 and NaOH of varied concentrations into a solution (200 mL) of 0.3 M $Mn(NO_3)_2$, under rapidly stirring at room temperature. After aging for 1 h at room temperature, the suspension was filtered, washed with water and air dried at room temperature for 2 days.

The original sample (1 g) prepared at room temperature was placed in a Teflon-lined, stainless steel vessel (30 mL) and hydrothermally treated with 1 M NaOH, 1 M NaCl, and 1 M HCl solutions (15 mL), respectively, at 150°C and autogenous pressure for 24 h. The hydrothermally treated sample was filtered, washed with water and dried in air at room temperature for 2 days.

An XRD analysis was carried out using an X-ray diffractometer (Rigaku RAD-RC). The sodium and manganese contents in the prepared samples were determined by atomic absorption spectrometry after dissolving the samples in a mixed solution of HCl and H_2O_2. The available oxygen of the manganese oxide was determined by the standard oxalic acid method as described previously.[5] The mean oxidation number of manganese in the oxide was evaluated from the value of the available oxygen.

3. RESULTS AND DISCUSSION

3.1 Preparation of original sample

XRD patterns of the original samples before air-drying, which were prepared at various mixed mole ratios of NaOH/$Mn(NO_3)_2$, are shown in Fig. 1. A β-MnO(OH) phase and a phase corresponding to layered structure with major diffraction peaks at d values of 1.004, 0.503, 0.336, and 0.248 nm were observed in the XRD patterns. The fraction of β-MnO(OH) phase decreases, and that of the layered structure phase increases with the increase in the mixed mole ratio of NaOH/$Mn(NO_3)_2$. After air-drying at room temperature, the layered structure phase shows diffraction peaks at d values of 0.724, 0.358, 0.249, and 0.241 nm (Fig. 2). These phases correspond to a birnessite-type sodium manganese oxide which has layered structure with crystal water and Na^+ between the MnO_6 octahedral sheets, as shown in Fig. 3. The sodium manganese oxide has a basal spacing of 1.004 nm along the **c**-axis with a double sheet of crystal water between the MnO_6 octahedral sheets before air-drying,[6] and a basal spacing of 0.724 nm along the **c**-axis with a single sheet of crystal water between the MnO_6 octahedral sheets after air-drying.[1,2] The transformation from the 1.0 nm birnessite to the 0.7 nm birnessite is irreversible. This property is different from birnessite-type lithium and potassium manganese oxides prepared by similar method,[3,4] where the basal spacings (0.712 and 0.726 nm for the lithium and potassium manganese oxides, respectively) keep almost constant before and after air-drying. This suggests that only 0.7 nm birnessite phases with the single sheet of crystal water are formed in the lithium and potassium manganese oxides. The basal spacings of dried birnessites prepared by this method increase in an order of potassium manganese oxide > sodium manganese oxide > lithium manganese oxide, corresponding to the increasing ionic radii of alkali metal ions in the interlayer space.

When the mole ratio of NaOH/$Mn(NO_3)_2$ is around 4.0, the intensity of the diffraction peak at 0.724 nm shows a maximum, indicating that the good crystalline birnessite phase can be obtained at this condition.

3.2 Behavior of manganese oxides under hydrothermal conditions

XRD patterns of the hydrothermally treated samples in 1 M NaOH solution at 150°C are given in Fig. 4. The β-MnO(OH) phase in the original sample was transformed to Mn_2O_3 phase and birnessite phase after the hydrothermal treatment in the NaOH solution. The fraction of the birnessite phase increased, and the diffraction peaks at d values of 0.724, 0.358, 0.249, and 0.241 nm, corresponding to the birnessite phase, became high and sharp after the hydrothermal treatment in NaOH solution. The diffraction peak intensity of the hydrothermally treated birnessite phase is about four times as that of the original sample. The basal spacings (0.724 nm) for the hydrothermally treated birnessite phase keep almost constant before and after drying, indicating it has a single sheet of crystal water between the MnO_6 octahedral sheets.

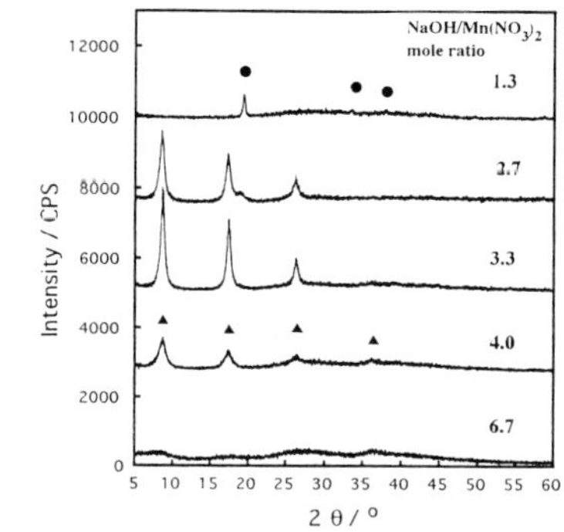

Fig.1. X-ray diffraction patterns of original samples before air-drying which are prepared at different NaOH / $Mn(NO_3)_2$ mixed mole ratio conditions. ▲ birnessite phase, ● β-MnO(OH) phase

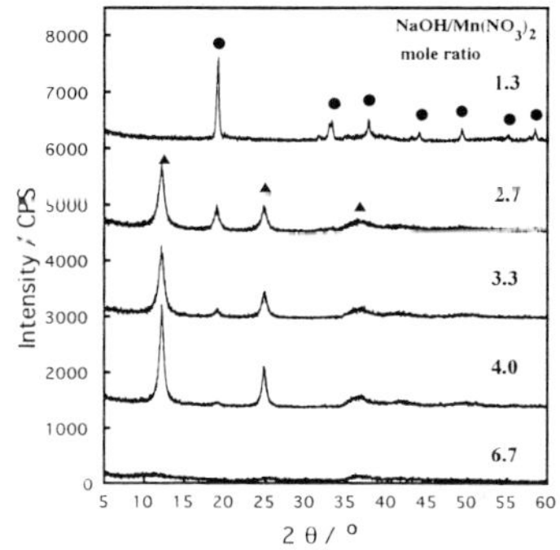

Fig.2. X-ray patterns of origanal samples after air-drying. ▲ birnessite phase, ● β-MnO(OH) phase

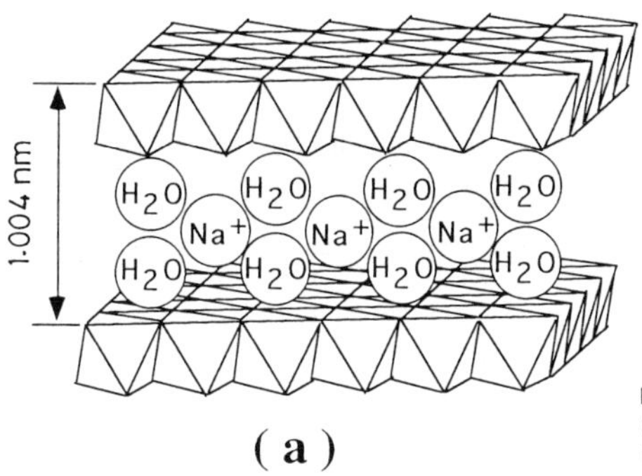

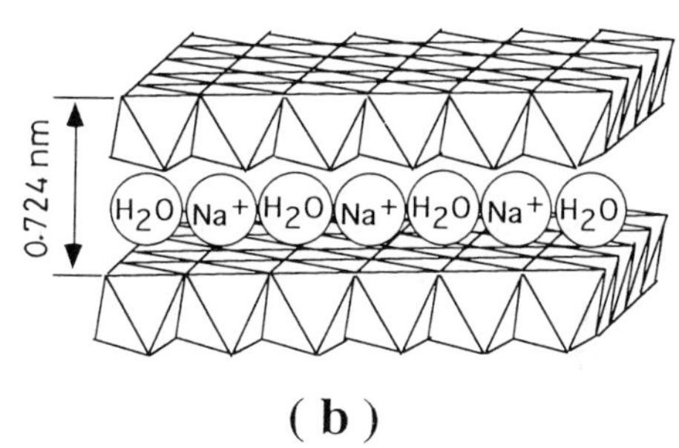

Fig. 3. Structures of birnessite-type sodium manganese oxides. (a) birnessite before air-drying with double crystal water sheet, (b) birnessite after air-drying with single crystal water sheet.

The β-MnO(OH) phase was transformed to Mn_2O_3 phase, and the diffraction peaks of the birnessite phase became broad after the hydrothermal treatment in NaCl solution. Only a pyrolusite-type manganese oxide with a one-dimensional (1×1) tunnel was formed by the hydrothermal treatment in HCl solution. Above facts suggest that the birnessite phase is stable in the alkaline solution, and a high crystalline birnessite phase can be obtained by hydrothermal treatment under the alkaline condition. The hydrothermal treatments in the neutral and acidic solutions cause a damage to the birnessite structure.

3.3 Composition of manganese oxide

The plots of sodium contents (Na/Mn mole ratio) in the original samples and hydrothermally treated samples versus the mixed mole ratio of NaOH/$Mn(NO_3)_2$ are shown in Fig. 5. The

Na/Mn mole ratios in the original samples and hydrothermally treated samples in NaOH and NaCl solutions increase with increasing $NaOH/Mn(NO_3)_2$ rapidly in the range of $NaOH/Mn(NO_3)_2 < 3.33$, respectively, but almost keep constant in the range of $NaOH/Mn(NO_3)_2 > 3.33$. The increases in the Na/Mn mole ratios in the range of $NaOH/Mn(NO_3)_2 < 3.33$ are due to the increase in the fraction of the birnessite phase. The Na/Mn mole ratio increased after hydrothermally treating in NaOH solution, but it was almost constant after hydrothermally treating in NaCl solution. The Na^+ increase in the interlayer space of birnessite structure results an increase in electrostatic force attracting between the negatively charged sheets of MnO_6 octahedra and the Na^+ ions in the interlayer space, which stabilizes the birnessite structure. Since there are almost not Na^+ in the HCl treated samples, they change to pyrolusite structure.

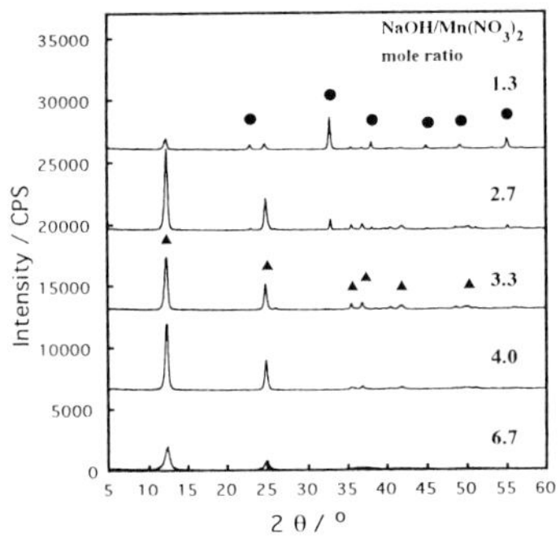

Fig.4. X-ray diffraction patterns of hydrothermally treated samples in 1M NaOH solution at 150°C for 24h. ▲ birnessite phase, ● Mn_2O_3 phase

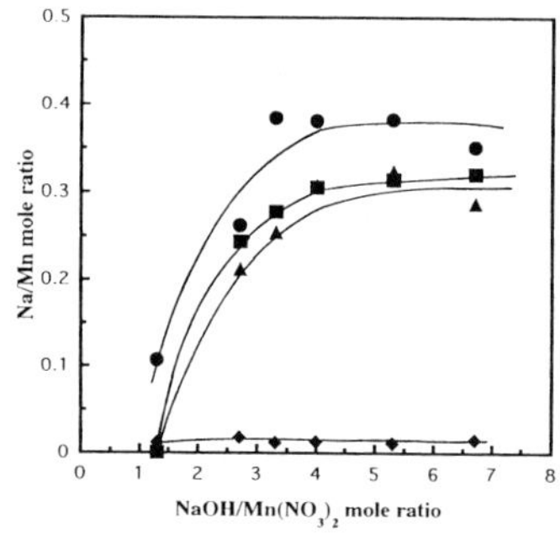

Fig. 7. Plots of sodium contents (Na/Mn mole ratio) in the original and hydrothermally treated samples versus the mixed mole ratio of $NaOH/Mn(NO_3)_2$. ■ original sample, ● hydrothermally treated sample in 1 M NaOH solution, ▲ hydrothermally treated sample in 1 M NaCl solution, ◆ hydrothermally treated sample in 1 M HCl solution.

The mean oxidation number of manganese for the original sample (BirMO) prepared at $NaOH/Mn(NO_3)_2 = 4.0$ and samples (BirMO-OH, BirMO-Cl, and BirMO-H) which were obtained by hydrothermally treating BirMO in NaOH, NaCl, and HCl solutions, respectively, was measured. The mean oxidation number increases after the hydrothermal treatment in HCl solution, but it is almost same after the hydrothermal treatments in NaOH and NaCl solutions. The increase of the mean oxidation number by hydrothermal treatment in the acidic solution is due to a disproportionation reaction of Mn(III) to Mn(IV) and Mn(II) in acidic solution. Mn(IV) remains in solid phase and Mn(II) dissolves into solution phase.[3]

The compositional analysis results indicate that the compositional formulas of birnessite-type sodium manganese oxides BirMO and BirMO-OH which have good crystallinities can be written as: $Na_{0.31}MnO_{1.91}{\bullet}0.7H_2O$ and $Na_{0.38}MnO_{1.95}{\bullet}0.7H_2O$.

RERENCES

1. R. G. Burns and V. M. Burns, "Manganese Dioxide Symposium", Vol. 2, I. C. MnO_2 Sample Office, Cleveland, Ohio (1980) 97.
2. S. Hirano, R. Narita and S. Naka, Mater. Res. Bull. 19 (1984) 1229.
3. Q. Feng, K. Yanagisawa, N. Yamasaki, J. Ceramic Soc. Japan, 104 (1996) 897.
4 Q. Feng, K. Yanagisawa, N. Yamasaki, J. Mater. Science 16 (1997) 110.
5. Q. Feng, H. Kanoh, Y. Miyai and K. Ooi, Chem. Mater. 7 (1995) 1226.
6 S. Turner, P. Buseck, Science 212 (1981) 1024.

Materials Science and Engineering Serving Society
S. Sōmiya, R.P.H. Chang, M. Doyama and R. Roy (Editors)

Processing science for engineering ceramics

Mamoru Mitomo

National Institute for Research in Inorganic Materials, Ibaraki, 305, Japan.

The mechanical properties are determined by a small fraction in a ceramic specimen. The present paper emphasizes that the quantitative evaluation and reasoning of the minor part is crucial for finding a scientific basis for improving mechanical properties of engineering ceramics. The examples of strengthening and toughening silicon nitride are presented and discussed.

1. INTRODUCTION

Ceramics science has been developed by relating average characters with atomistic processes. For example, the densification of a compact is related to the diffusion of atoms or ions. It has been successful as far as the properties of materials are directly determined by an average character. The strength decrease with grain size of ceramics, i.e. Petch equation, is a typical example. Transmittance of light in alumina ceramics is inversely proportional to the porosity and is another example.

The mechanical properties of ceramics, i.e. strength and fracture toughness, are not simply explained by an average microstructural character. For example, the strength is determined by the largest flaw in a material. Various types of flaws, which have their own distribution, formed in a specimen [1]. It is generally observed that test specimens from a single batch fracture from different types of flaws. The petch equation is only valid when the fracture originates from a large grain. Fracture toughness is much more dependent on the average microstructure, but recent tough ceramics are based on development of elongated grains in a fine matrix , i.e. in-situ reinforced ceramics [2-4]. This means that fracture toughness depends on the amount and size or shape of reinforcing grains due to crack deflection or crack bridging mechanisms [5,6].

Thus the mechanical properties of ceramics largely depend on the minor phase. To understand this kind of problem, we must focus on evaluating and controlling the minor phase. The purpose of the present paper is to show how this problem can be solved scientifically. Thus we propose "minority science" for engineering ceramics instead of "average science" for traditional ceramics.

2. DEFECT CONTROL IN SILICON NITRIDE

2.1. Preparation of fine-grained ceramics

In-situ reinforced microstructures have been developed during α to β phase transformation [7], and also by bimodal distribution of grains [8]. The sintering of fine and uniform β powder at low temperature resulted in fine-grained silicon nitride ceramics [9,10]. The characteristics of β powder are shown in Table 1. The powder was liquid phase sintered by hot-pressing with oxide additives at 1700℃. The fine microstructure shown in Figure 1 has an average diameter of 0.18 μm and an aspect ratio of 2.3 [11].

The purpose of minimizing grain size is to maximize the strength by minimizing flaw size. However, the strength is much lower than calculated from the Griffith equation,

$$\text{strength} = K_{1C}/(\pi c)^{1/2}. \quad (1)$$

The low strength, i.e. 676 MPa, is partly due to low fracture toughness of 2.8

Table 1
The characteristics of starting powder

Average grain size (μm)	β ratio (%)	Metal impurities (ppm) Al	Cr	Fe	Ca	Cu	Ti
0.3	95	100	80	30	30	20	10

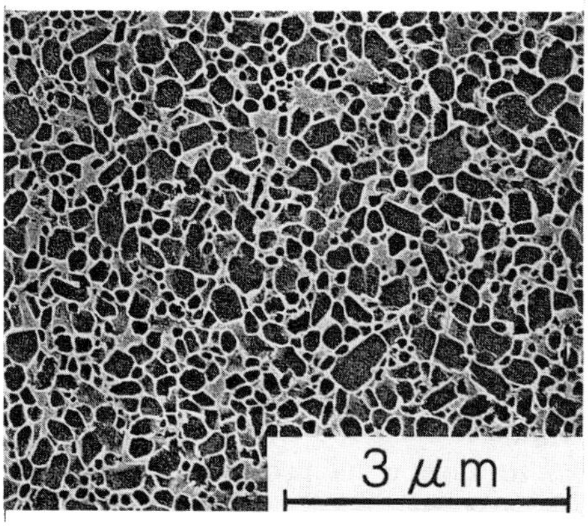

Figure 1. Microstructure of fine-grained silicon nitride.

Figure 2. Silicide particle as a fracture origin.

MPa· $m^{1/2}$ and mostly to the formation of large silicide particles as flaws. The silicides are formed by the reaction between impurity metals, Fe and Cr, with matrix silicon nitride. Silicon nitride powder was formed by Si nitridation. Fe metal was added for a catalyzer and remained after chemical leaching. Cr metal contamination was generated from pulverizing media. The size of the flaw is about 5 to 20 μm , which is fairly larger than that of grain size (Figure 2). It should be pointed out that even a small amount of metallic impurities of the order of 100 ppm or less sometimes seriously affects strength of silicon nitride ceramics. This suggests that the distribution of impurities is not uniform. In fact, the number of large silicide particles is less than $1/mm^2$.

2.2. Effect of homogenization by annealing

The formation of silicides is thought to be due to the reactions,

$Si_3N_4 + 3Fe = 3SiFe + 2N_2$ (2)

$Si_3N_4 + 3Cr = 3SiCr + 2N_2$ (3)

Themodynamic calculation revealed the effect of Cr is larger than that of Fe due to higher equilibrium pressure of N_2. It was shown that the application of 1.0 MPa N_2 is enough to prevent silicide formation during the annealing. We actually observed that fractures originate from residual pores, which are smaller than those of silicide particles (Figure 3). We have clarified that fine-grained ceramics grow a little with an appreciable increase in fracture toughness by annealing at 1800℃ for 1 h in 1.0 MPa N_2 [9,11]. The fracture toughness increased to 4.3

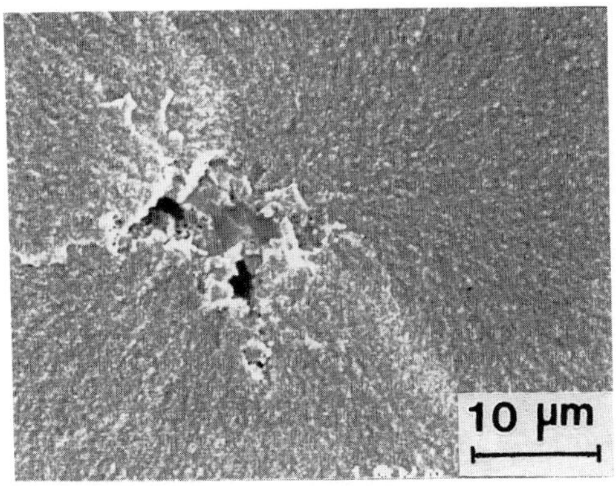

Figure 3. Fracture initiation form a pore.

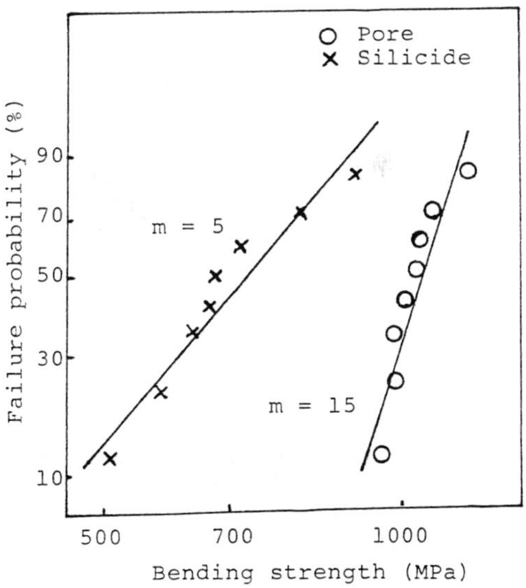

Figure 4. The reration between strength distribution and fracture origin.

$MPa \cdot m^{1/2}$. The average strength increased to 1.05 GPa.The narrower distribution of strength as compared to that before annealing is shown by Weibull plot (Figure 4).

Present work demonstrates that the mechanical properties could be improved by homogenizing grain size distribution and minimizing flaw size.

3. TOUGHENING BY MICROSTRUCTURAL CONTROL

The development of bimodal microstructures for toughening of engineering ceramics has been investigated for about 10 years. This approach seeks to develop a small number of elongated grains in a fine matrix grain during sintering. The microstructures are similar to whisker-reinforced ceramics (WRC). Because the WRC microstructures develop during fabrication, the materials are referred to as self-reinforced ceramics or in-situ composites. The process of in-situ toughening is particularly successful in silicon nitride and silicon carbide ceramics. A typical microstructure of a silicon nitride in-situ composite is shown in Figure 5.

Large grains increase fracture toughness by crack deflection or bridging. The large contribution of energy dissipation at the crack wake to the R-curve effect suggests that the toughening is directly related to the presence of elongated grains. The control of the size and amount of large grains is therefore very important. It is clear that fine β silicon nitride powder grows normally and produces fine-grained ceramics. The addition of large β grains as nuclei was investigated as a model experiment. The effect of nuclei addition on the average grain size of the matrix and abnormally elongated grains is shown in Figure 6.

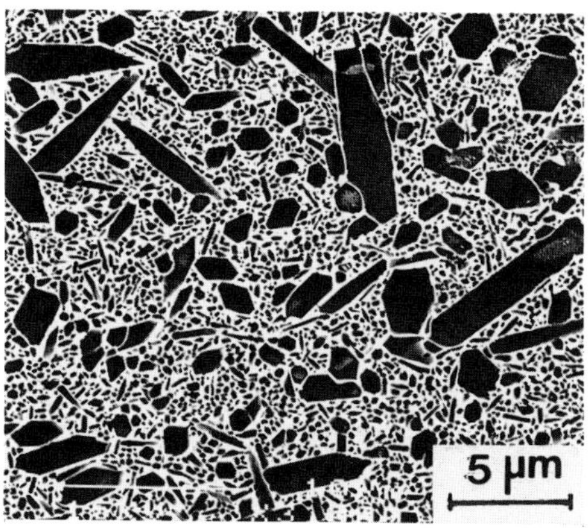

Figure 5. Typical microstructure of in-situ composite.

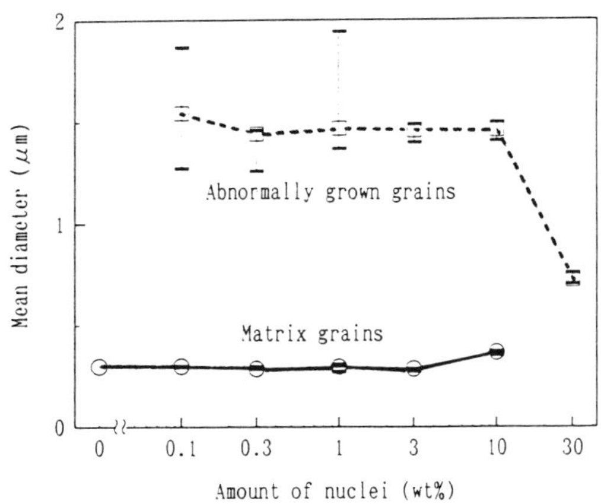

Figure 6. The effect of nuclei on average diameter.

Only large seed grains grow abnormally, so the area fraction, which corresponds to the volume fraction, of large grains is directly related to the amount of seed grains as shown in Figure 7. The rate of grain growth during liquid phase sintering is shown by the grain size difference [12],

$$dr/dt = Kr^{-1}(r_0^{-1} - r^{-1}) \quad (4)$$

$$K = 2DC\Omega\gamma/kT \quad (5)$$

where r_0 and r are the diameter of the matrix and seed grain, and D, C, Ω and γ is the diffusion coefficient of the solid in the liquid, the concentration of solid in the liquid, the atomic volume of the solid, and the solid/liquid interfacial energy, respectively. The rate of grain growth of a seed is proportional to the amount of diffusion (K) and grain size difference. This means that the size of large grains should be controlled by the size of seed and heating conditions.

The addition of a large amount of seed grains, however, resulted in unimodal microstructure. An important conclusion obtained in the present work is that there are maximum and minimum seed additions for developing in-situ composite microstructures (Figure 8). It should be noted that strength-limiting-flaws are always elongated grains in these materials [13]. The strength, strength distribution and fracture toughness should therefore be optimized scientifically based on grain growth theory.

4. CONCLUSIONS

Two examples show that mechanical properties can be improved by analyzing minor parts in specimens. The quantitative evaluation of the type and size of minor part was very important. By evaluating important microstructural parameters, we can provide a scientific approach to solving the problem. The

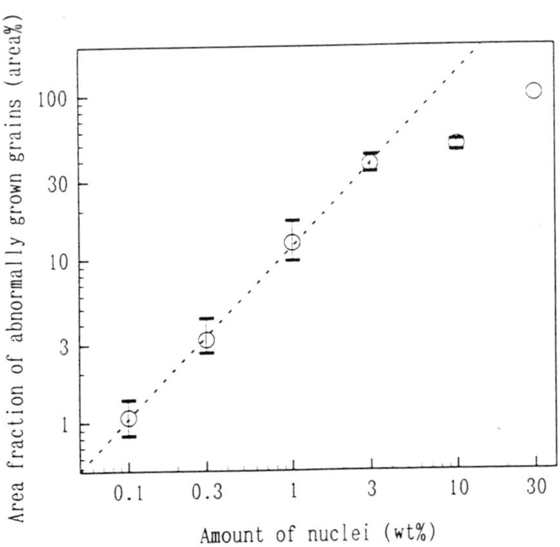

Figure 7. The relation between the amount of nuclei and abnormally grown grains.

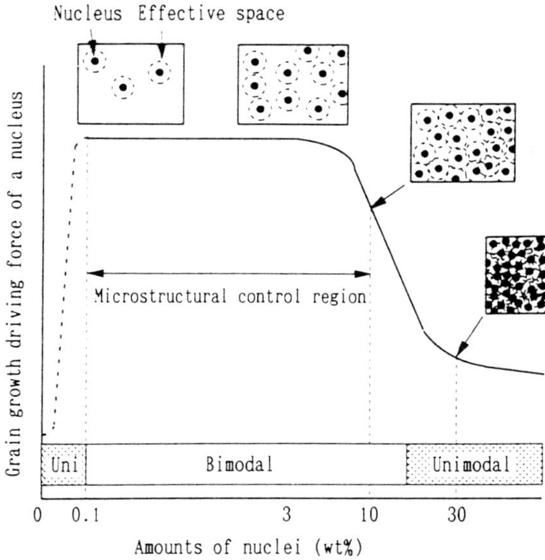

Figure 8. The region for developing in-situ composite microstructure.

relation between formation process of a small fraction and bulk mechanical properties should always be considered. Thus "heterogeneity science" or "minority science" should be a key approach for further investigation of engineering ceramics.

REFERENCES

1. A.G.Evans, J. Am. Ceram. Soc. 65 (1982) 127.
2. M.Mitomo and S.Uenosono, J. Am. Ceram. Soc. 75 (1992) 103.
3. C.W.Li and J.Yamanis, Ceram. Eng. Sci. Proc. 10 (1989) 632.
4. N.P.Padture, J. Am. Ceram. Soc. 77 (1994) 519.
5. K.T.Faber and A.G.Evans, Acta Metall. 13 (1983) 577.
6. P.Becher, J. Am. Ceram. Soc. 74 (1991) 255.
7. F.F.Lange, J. Am. Ceram. Soc. 62 (1979) 428.
8. M.Mitomo, T.Nishimura and H.Hirotsuru, Eur. J. Solid State Inorg.Chem. 32 (1995) 693.
9. M.Mitomo, H.Hirotsuru, H.Suematsu and T.Nishimura, J. Am. Ceram.Soc. 78 (1995) 211.
10. H.Emoto and M.Mitomo, J. Eur. Ceram. Soc. 17 (1997) 797.
11. H.Sato, M.Mitomo, T.Nishimura and H.Emoto, J. Ceram. Soc. Jpn. submitted.
12. R.M.German, Liquid Phase Sintering Plenum Pub., N.Y. (1985).
13. M.Mitomo, N.Hirosaki and H.Hirotsuru, MRS Bull. 20(2) (1995) 38.

Materials Science and Engineering Serving Society
S. Sōmiya, R.P.H. Chang, M. Doyama and R. Roy (Editors)

THICKNESS DEPENDENCE OF IMPACT DAMAGE BEHAVIOR IN SIALON CERAMIC

Yoshio Akimune*, Toru Akiba, Naoto Hirosaki, and Takao Izumi

Nissan Research Center, Nissan Motor Co., Ltd.
1, Natsushima-cho, Yokosuka, 237 Japan

1.INTRODUCTION

Impact damage behavior of ceramics for turbine use has been studied and analyzed by various researchers using the Hertzian cone-fracture theory[1-3]. Although those recent studies[1-5] have analyzed stress caused by foreign particle impact, the thickness dependence of impact damage and fracture is still not fully understood. In this study, the thickness dependence of impact damage shown by turbine ceramics was examined experimentally and computer simulations were carried out to analyze the resulting stress level.

2.EXPERIMENTAL PROCEDURES

2.1 Impact tests

Commercially available sialon (NGK Spark Plug Co., Ltd), having the material properties shown in Table 1, was used in this study. Partially stabilized zirconia (PSZ) spheres 1.0 mm in diameter were used in the spherical impact test. Sialon specimens, measuring 50 mm x 8 mm and with varying thickness of 1.0, 1.5, 2.0, and 3.0mm, were polished with diamond paste to eliminate machining damage. A He gas gun was used to carry PSZ spheres and the impact velocity was calculated by the time-of-flight principle at room temperature.

2.2 Dynamic stress analysis using DYNA3D

A dynamic stress analysis was conducted to investigate the thickness dependence of the stress level[6]. The FEM (Finite Element Method) model of target was structured with a geometry of 8mm by 20mm by 1, 1.5, and 2mm, with the target composed of 2,900-5,400 hexagonal solid elements and sphere diameter of 1mm, with the sphere composed of 400 hexagonal elements. The impact velocity was set at 300 m/s, where crack initiation behavior was seen to differ according to the specimen thickness.

3. RESULTS AND DISCUSSION

3.1 Impact tests

Specimens 1.0 mm in thickness showed craters below an impact velocity of 200m/s, and failed at a impact velocity above 247 m/s. Specimens 1.5 to 3.0 mm in thickness displayed craters at low velocity (<220 m/s), ring cracks at intermediate velocity (220-250 m/s), and ring cracks and radial cracks at high impact velocity (>350 m/s). These surface damage morphologies suggest that the specimen response behavior to impact varied according to the impact velocity and specimen thickness.

Residual strength is plotted in Fig. 1. For specimens 1.0 mm in thickness, sudden fracture occurred at an impact velocity of 247 m/s. For 1.5-3.0 mm specimens, strength degradation occurred above a 400 m/s impact velocity. Impact damage and strength degradation were found to be related to specimen thickness and may have an effect on crack initiation behavior at impact[1,5].

Hertzian cone cracks under the impact site were detected in almost all of the strength degraded specimens (Fig. 2 and 3) [7] and fracture caused by bending stress was also observed in the 1.0mm specimens [5](Fig. 4). It is thought that the stress wave was reflected at the back face of the specimen, which affected the stress level and resulted in sudden crack propagation near the back face[8]. On the other hand, Hertzian cone cracks were arrested in specimens having thickness of 1.5-3.0 mm.

3.2 Dynamic stress analysis using DYNA3D

Figures 5 and 6 show the maximum principal stress on the surface at 0.1-0.6 μs after impact for specimens having six thicknesses. The maximum stress level generated at 0.3 μs were in the same range and equivalent to the material strength level. As time elapsed, the stress level stayed in the same range but the location of peak stress shifted in the radial direction. The stress level generated at the back face was thickness dependent and increased with decreasing thickness. For the 1mm specimen, the stress level generated at the back face became higher than that at the surface in the early stage of contact as shown in Figure 5. Fracture morphologies in 1.0 mm thick specimen were supposed to depend on the local materials strength.

Two types of fracture behavior (Hertzian fracture and bending fracture) were observed for the 1mm thick specimen. With decreasing specimen thickness the stress level at the back face of the specimen increased, resulting in bending fracture, although the stress level near impact site became the maximum level for 2.0-3.0mm thickness resulting in Hertzian cone crack.

4. SUMMARY

The thickness dependence of impact damage shown in sialon ceramic was examined. The development of Hertzian cone cracks was arrested in specimens having a thickness of 1.5-3.0 mm, although the bending fracture was also observed in the specimen with1.0mm thickness.

Computer simulation results indicated that the stress level generated at the target surface became the strength level of the materials and it coincided with surface ring crack formation at impact. This calculation also exhibited the stress level at the back face became higher than the materials strength. The stress calculations support the explanation of fracture behavior.

REFERENCES

1. J. Cuccio and H. Fang, Proceedings 26th Auto.Tech. Development Coordination Meeting(1988).
2. K. C. Dao et al.,Office of Naval Research Contract, No N00014-76-057(1976).
3. H. R. Hertz, Hertzs Miscellaneous paperts, McMillan, London, 1886.
4. Y. Akimune et al., J. Am. Ceram. Soc., 73[12]3607-10(1990).
5. H. Tsuruta, et al.,J. Am. Ceram. Soc. 73[6]1714-18(1990).
6.T. Teramae, et al. ,Transactions of the Japan Soc. of Mech. Engineers,55-520(12)2423-26(1989)
7.A. Suzuki et al., ibid,56-532(12)2431-36(1992)
8.J. E. Field et al., Proc. 4th int. Conf. Mech. Prop. Materials at High Strain Rate(1889).

Table 1 Materials properties

Density(g/cm^3)	3.26
Bending Strength(MPa)	850
Young's Modulus(GPa)	305
Poisson's Ratio	0.26
Fracture Toughness(MPa $\sqrt{\ }$ m)	7.0

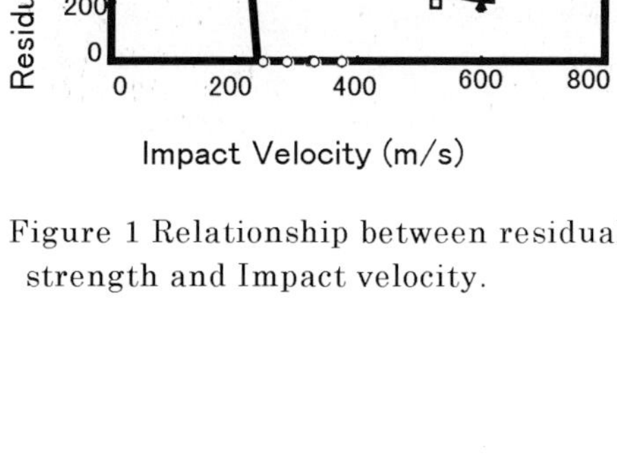

Figure 1 Relationship between residual strength and Impact velocity.

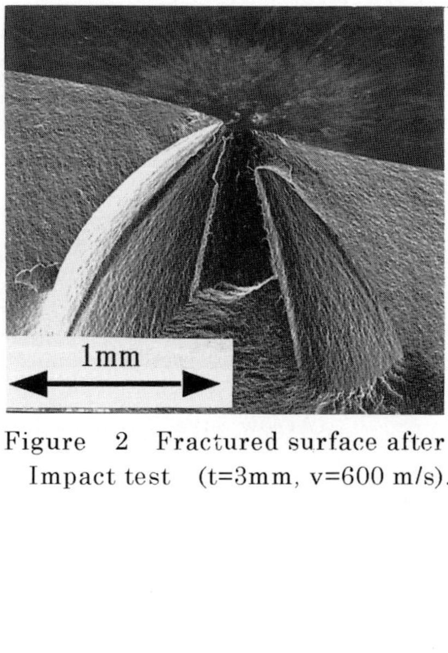

Figure 2 Fractured surface after Impact test (t=3mm, v=600 m/s).

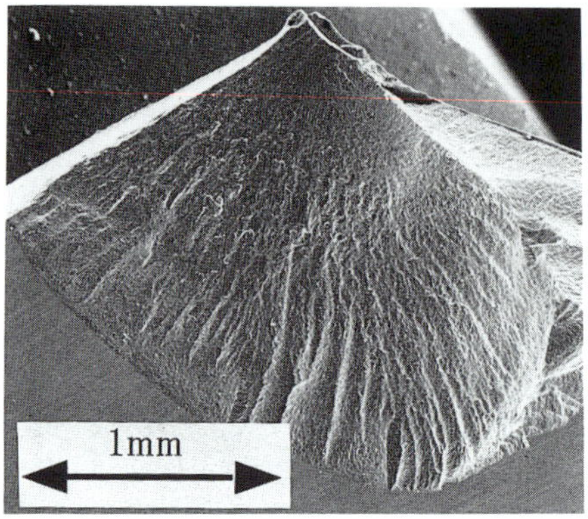

Figure 3 Fractured surface after Impact test. (t=1mm, v=300m/s((cone crack))

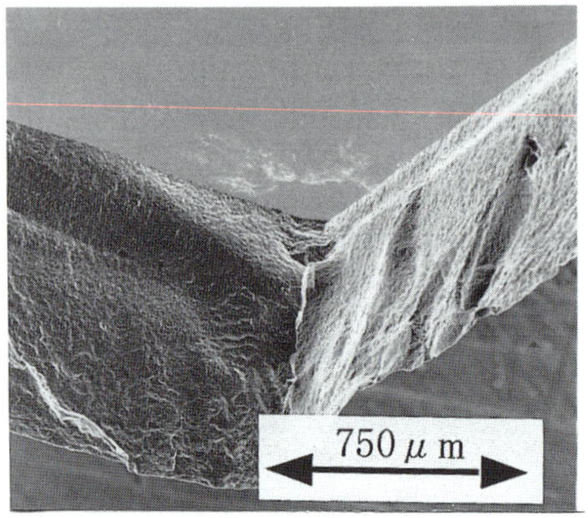

Figure 4 Fractured surface after Impact test by bending force (t=1mm, v=320m/s)

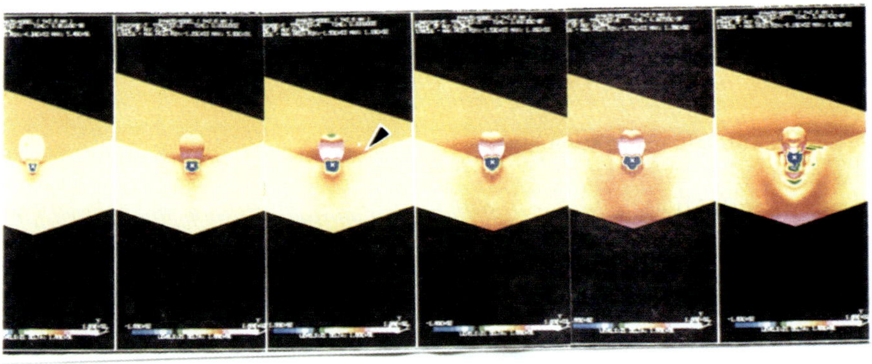

Figure 5 Stress analysis using DYNA3D(2mm thick specimen). (Arrow; surface stress related to ring crack formation)

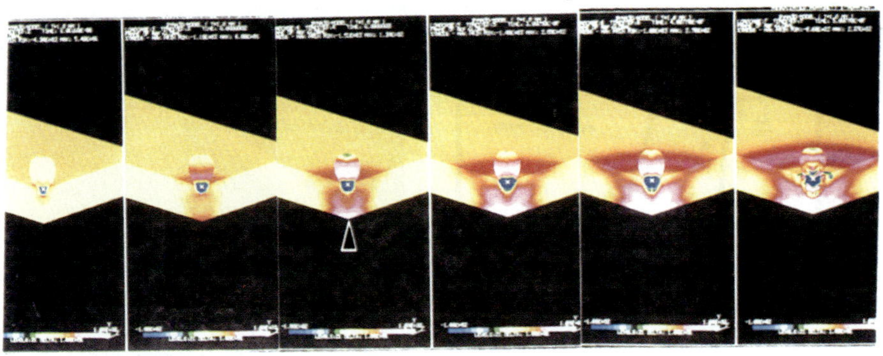

Figure 6 Stress analysis using DYNA3D(1mm thick specimen). (Arrow; stress at back face corresponds to the materials strength)

Materials Science and Engineering Serving Society
S. Sōmiya, R.P.H. Chang, M. Doyama and R. Roy (Editors)

Low-Temperature Growth of Epitaxial Barium Titanate Thin Films by Irradiation of Low-Energy Positive Oxygen Ions

Yurika Ishibashi, Takaaki Tsurumi, Naoki Ohashi and Osamu Fukunaga

Department of Inorganic Materials, Tokyo Institute of Technology
2-12-1 Ookayama, Meguro-ku, Tokyo 152, JAPAN

Abstract Barium titanate ($BaTiO_3$) thin films were deposited by a molecular beam epitaxy (MBE) system. The epitaxial temperature of $BaTiO_3$ thin films on strontium titanate ($SrTiO_3$) single crystals decreased from 600 °C to 400 °C by irradiation of positive oxygen ions with low energy generated from an electron cyclotron resonance source (ECR-gun). The energy of ionized oxygen fluxes was controlled by two voltages: the accelerating voltage (V_a) and the bias voltage (V_b). At 400 °C, the $BaTiO_3$ thin films were epitaxially grown at V_a=+50 V and V_b=100 V. The irradiation of low-energy positive oxygen ions was effective for both oxidation in a high vacuum and reducing crystallization temperature.

1. INTRODUCTION

In the silicon process, it is indispensable to decrease the substrate temperature of oxide thin films. However, they cannot be synthesized at low temperatures because of reduction in the efficiency of oxidation in a high vacuum and the surface migration of atoms at low temperatures. In order to increase the degree of oxidation in a high vacuum, the oxygen ion has been more useful than the oxygen radical [1,2]. In addition, the low-energy ion bombardment during the film growth has been utilized to modify the film properties [3,4]. The assist of charged oxygen fluxes with low energy during the film growth seems to have an effect on the enhancement of both the oxidation and the migration on the film surface at low temperatures.

$BaTiO_3$ thin films, for example, have been deposited epitaxially on $SrTiO_3$ (001) only above 600°C [5-7]. In this paper, we report the results of the effect of low-energy positive oxygen ion irradiation on $BaTiO_3$ thin films prepared by a molecular beam epitaxy (MBE) method in term of crystallinity at the temperature as low as below 600°C.

2. EXPERIMENTAL PROCEDURE

2.1. Sample preparation

$BaTiO_3$ thin films were deposited by means of an MBE apparatus equipped with an electron cyclotron resonance source (ECR-gun). The MBE system (ULVAC Co., Ltd.) has been shown in elsewhere [7]. The film thickness was kept to 80 MLs (ML: monolayer). The mirror-polished $SrTiO_3$ (001) single crystals were used as the substrates and were treated by buffered NH_4F-HF (BHF) solution (pH=4.6) for the purpose of flattening of the surface before the deposition of the films [8]. The reflection high-energy electron diffraction (RHEED) pattern was observed. The typical depositing condition is listed in Table 1.

During the deposition, the growing films were irradiated directly by low-energy positive

Table 1
Typical depositing condition of the samples

Background pressure	7.0×10^{-9} Torr
Working pressure	3.5×10^{-6} Torr
Substrate temperature	400°C
K-cell parameters	
Deposition rate	0.5～1.0 ML/min
Temperature	490～530°C
EB-gun parameters	
Deposition rate	0.5～1.0 ML/min
Emission current	12～20 mA
ECR-gun parameters	
O_2 flow rate	0.2 sccm
μ-wave power	200 W

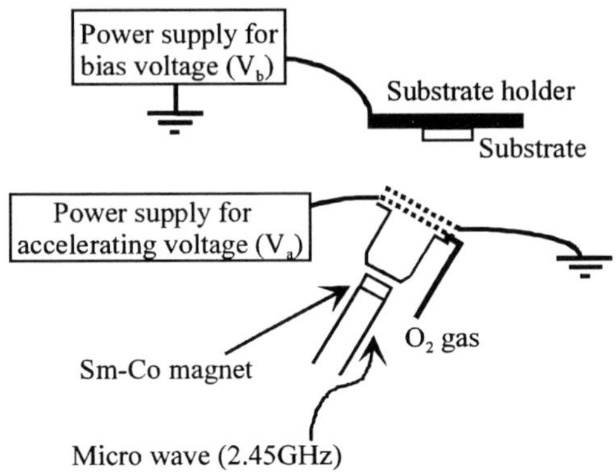

Figure 1. Schematic illustration of ECR-gun

oxygen ion (O_2^+) beam produced from ECR-gun. Two applying voltages were employed to control the energy of ionized oxygen fluxes: the one was applied to the grid of ECR-gun (accelerating voltage, V_a); the other to the substrate holder (bias voltage, V_b), shown in Fig. 1. The accelerating voltage was changed from +50 V to +150 V, while the bias voltage was from 0 V to –500V.

2.2. Characterization of the $BaTiO_3$ thin films

The chemical composition of the obtained films was determined by inductively coupled plasma atomic emission spectrometer (ICP-AES, SRS1500VR, Seiko Co., Ltd.). The crystallographic orientation of the obtained $BaTiO_3$ films was verified by means of an x-ray diffractometer (45 kV, 30 mA, CuKα; RAD II, Rigaku Co., Ltd.). The crystallinity of the films were evaluated from the full width at half maximum (FWHM) of θ-rocking curve measured around $BaTiO_3$ 002 peak by an x-ray diffractometer with a high resolution goniometer (55 kV, 40 mA, CuKα_1; X'Pert-MRD, Philips Co., Ltd.). The visible (VIS) spectra of typical samples were measured through a VIS spectrophotometer (Jasco V-570, Nihon Bunko Co., Ltd.), because the difference of transparency of the films enabled one to evaluate the degree of oxidation. The range of wavelength was from 380 nm to 680 nm.

3. RESULTS AND DISCUSSION

The mole ratio of Ba to Ti was determined for $BaTiO_3$ thin films prepared on MgO (001) single crystals. The amount of evaporated Ba and Ti were fixed at those of 600°C, where the Ba ration to Ti in the film was stoichiometric. Figure 2 shows the change of Ba/Ti ratio as a function of substrate temperature. The mole ratio of Ba to Ti decreases with decreasing

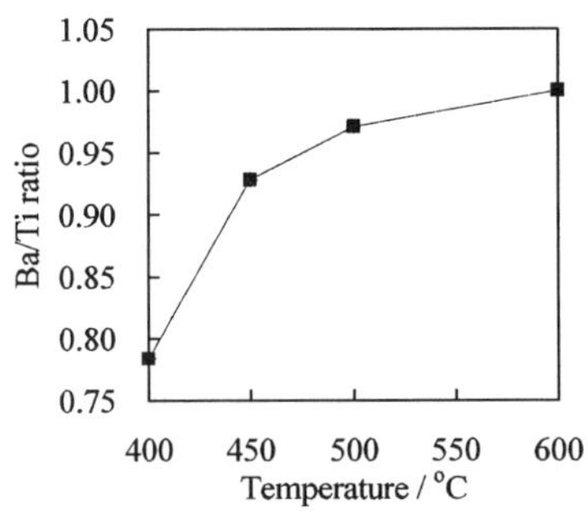

Figure 2. Mole ratio of Ba to Ti as a function of depositing temperature.

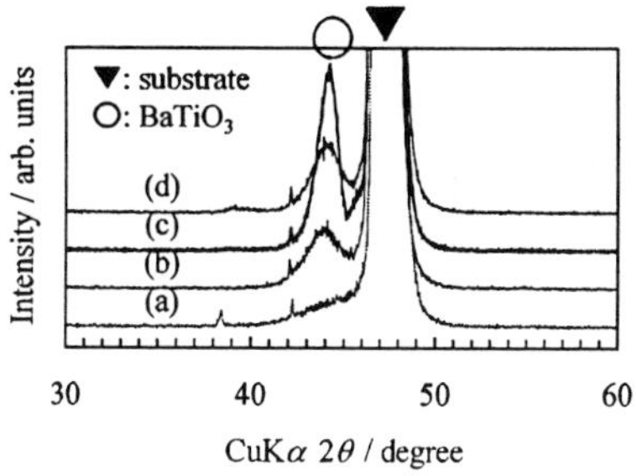

Figure 3. XRD profiles of the $BaTiO_3$ thin films measured around $BaTiO_3$ 002 peak. The films were prepared at (a) V_a=+50 V, (b) V_a=+150 V, (c) V_a=+50 V and V_b=-100 V, and (d) V_a=+50 V and V_b=-500 V.

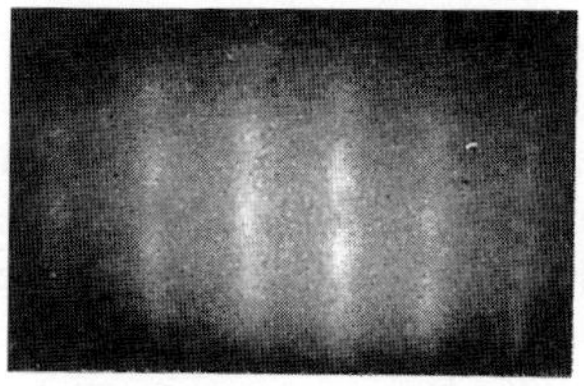

Figure 4. RHEED pattern of the $BaTiO_3$ thin film deposited at V_a=+50 V and V_b=-100 V. The pattern was observed from [100] azimuth.

substrate temperature. According to the analysis of ICP-AES, the amount of Ba- and Ti-evaporation was corrected to give a stoichiometric composition and the films were deposited on the $SrTiO_3$ (001) substrates at 400°C. Figure 3(a) shows the XRD profile of the $BaTiO_3$ thin film deposited at V_a=+50 V. This XRD profile shows that the epitaxially grown BaTiO3 thin films was not obtained at 400°C and V_a=+50 V. The XRD profile of the films prepared at V_a=+150 V, shown in Fig. 3(b), indicates that the increase in the accelerating voltage did not markedly improve the crystallinity, because the oxygen ions accelerated by the only accelerating voltage did not get enough energy to enhance the migration at low temperatures. In order to increase the energy of ionized oxygen, both the accelerating voltage and the bias voltage were applied. The total potential difference between the substrate and the ECR-gun was equal to 150 V. The XRD profile of a film deposited at V_a=+50 V and V_b=-100 V, (Fig. 3(c)) and the RHEED pattern (Fig. 4) reveal that the epitaxial growth of $BaTiO_3$ thin film was achieved at 400°C. The irradiation of positive oxygen ions using the accelerating voltage and the bias voltage leads to the reduction in the epitaxial temperature of $BaTiO_3$ to 400°C. The electric current between the substrate and the ground was measured in order to understand the difference between Fig. 3(b) and Fig. 3(c). The current observed at V_a=+150 V was 21.5 mA, while that observed at V_a=+50 V and V_b=-100 V was 95.2 mA. From the result of current measurement, it was found that the number of ionized oxygen, which reached the substrate, increased when both the accelerating voltage and the bias voltage were applied. Thus, it was found that it is necessary to supply a lot of ionized species arriving at the substrate for the epitaxial growth of $BaTiO_3$ thin films at low temperatures.

In order to investigate the influence of energy of ionized oxygen on crystallinity, the energy of oxygen fluxes was controlled by bias voltage. The accelerating voltage was fixed to +50 V. The FWHM of the film prepared at V_b=-100 V was 0.45, while that prepared at V_b=-500 V was 0.76. The result of FWHM measurements and Figs. 3(c) and 3(d) indicated that the bombardment of ionized species with high energy reduced the crystallinity because the film was damaged by irradiating high-energy oxygen fluxes.

The VIS spectra of typical samples were measured for estimating the degree of oxidation. The specimens were a film prepared at 400°C (V_a=+50 V), that prepared at 400°C (V_a=+50 V and V_b=-100 V), and that prepared at 600°C, respectively. Figure 5 shows the spectra of the films. The film obtained at 600°C was the well-oxidized sample. The spectra indicated that the film deposited at V_b=-100 V is more transparent than that deposited at V_a=+50 V. The assist of low-energy positive oxygen ions increases the efficiency of oxidation in a high vacuum and at low temperatures.

The effect of the accelerating voltage and the bias voltage is considered as follows: the accelerating voltage is necessary to choose the charged species in the ECR-gun; the bias voltage gives the direction of the species towards the substrate and accelerates them in order to supply enough energy to enhance the surface migration of atoms. The results of this study elucidate the effectiveness of positive oxygen ions rather than the radicals because the energy of ions was controlled by the application of the accelerating voltage and the bias voltage. Further improvement of quality of the films may be possible by choosing optimum number and energy of positively oxygen ion fluxes.

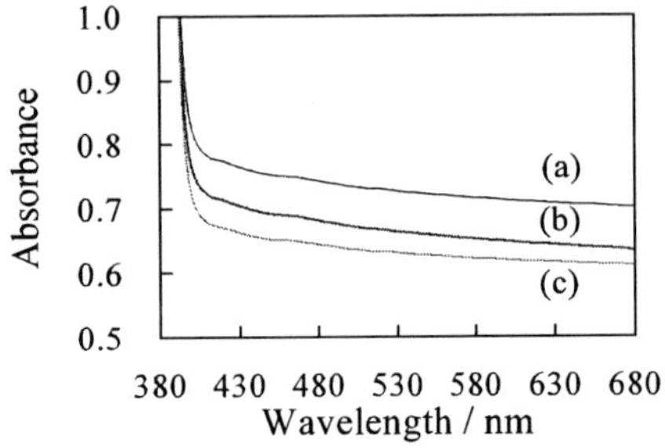

Figure 5. VIS spectra of the $BaTiO_3$ thin films (a) without irradiating oxygen ions, (b) with oxygen ion assisting (V_a=+50 V and V_b=-100 V), and (c) deposited at 600°C.

4. CONCLUSION

This paper showed the advantage of irradiating positive oxygen ions in the fabrication of epitaxial $BaTiO_3$ thin films on the $SrTiO_3$ (001) substrates at 400°C using low-energy positive oxygen ions. The $BaTiO_3$ thin films were epitaxially grown at V_a=+50 V and V_b=-100 V. The number of oxygen ions near the substrate increased when both the accelerating voltage and the bias voltage were applied. Irradiation of low-energy positive oxygen ions was effective in both reducing the crystallization temperature and oxidation in a high vacuum.

REFERENCES

1. T. Aida, A. Tsukamoto, K. Imagawa, T. Fukuzawa, S. Saito, K. Shinda, K. Takagi and K. Miyauchi, Jpn. J. Appl. Phsy. 28 (1989) L635.
2. R. Kita, T. Hase, R. Itti, M. Sasaki, T. Morisihta and S. Tanaka, Appl. Phys. Lett. 60 (1992) 2684.
3. H. Liungcrantz, S. Benhenda, G. Håkansson, I. Ivanov, L. Hultman, J. E. Greene and J.-E. Sundgren, Thin Solid Films 287 (1996) 87.
4. C.-J. Peng and S. B. Krupanidhi, J. Mat. Res. 10 (1995) 708.
5. H. Tabata, H. Tanaka and T. Kawai, Appl. Phys. Lett. 65 (1994) 1970.
6. H. Shigetani, K. Kobayashi, M. Fujimoto, W. Sugiura, Y. Matsui and J. Tanaka, J. Appl. Phys. 81 (1997) 693.
7. T. Tsurumi, T. Suzuki, M. Yamane and M. Daimon, Jpn. J. Appl. Phys. 33 (1994) 5192.
8. M. Kawasaki, K. Takahashi, T. Maeda, R. Tsuchiya, M. Shinohara, O. Ishiyama, T. Yonezawa, M. Yoshimoto and H. Koinuma, Science 266 (1994) 1540.

Materials Science and Engineering Serving Society
S. Sōmiya, R.P.H. Chang, M. Doyama and R. Roy (Editors)
1998 Elsevier Science B.V.

Exaggerated Anisotropic Growth in Barium Titanate Ceramics

D. Kolar and A. Rečnik

Jožef Stefan Institute, Jamova 39, 1000 Ljubljana, Slovenia

Abstract

Barium titanate ceramic sintered under reducing conditions above the $Ba_6Ti_{17}O_{40}$-$BaTiO_3$ eutectic temperature exhibits exaggerated anisotropic growth. The anisotropically grown grains form a hexagonal polymorph and the growth mechanism is dissolution of small matrix grains and precipitation on the hexagonal grains which grow preferentially in the direction of prism planes.

Microstructural characteristics such as size, anisotropy (aspect ratio) and grain size distribution are strongly influenced by the preparation conditions of the ceramics. The important parameters are a reducing atmosphere, the purity of the starting barium titanate powder, the presence of hexagonal nuclei, the amount of liquid phase at the sintering temperature, sintering temperature and time.

The analysis of anisotropic grain growth phenomena in $BaTiO_3$ may serve as a model system for anisotropic exaggerated grain growth in ceramics in general.

1. INTRODUCTION

Abnormal or discontinuous grain growth [1] (sometimes called secondary recrystallization or exaggerated grain growth) is a frequently observed phenomenon during sintering of ceramics. It is characterized by rapid growth of a limited number of grains to sizes much larger than those of the average grain population. Rapid grain growth results in coarse heterogeneous microstructures with captured pores within the grains, which limits the degree of densification. Therefore the prevention of abnormal grain growth is of great importance in manufacturing of dense, fine grained ceramics.

In some systems, the abnormal grains exhibit preferential growth in a particular direction, giving rise to rod-like or plate-like morphology. It has recently been pointed out that exaggerated anisotropically grown grains with a higher aspect ratio improve the fracture resistance of Si_3N_4 and SiC engineering ceramics in a similar way to the addition of whiskers, i.e. by crack deflection and crack bridging mechanisms [2-7].

In spite of its considerable importance, the general phenomenon of abnormal grain growth and anisotropic grain growth in particular is not well understood. A classic phenomenological explanation is that pore-grain boundary separation creates the condition suitable for abnormal growth [8-9]. Other possible causes include a wide initial particle size distribution [1], separation of grain boundaries from pinning segregated impurities or particles [10-12], surface energy anisotropy [13] and anisotropic grain-boundary mobilities [14-15] and the presence of a small amount of liquid phase [16-19].

Among unresolved questions remain the nucleation of anisotropic grains and the faceted growth habit of exaggerated grains in some systems.

In a recent analysis of the phenomenon of anisotropic grain growth in alumina ceramics, Seabaugh et al [20] examined several intrinsic and extrinsic motives for anisotropic grain growth. Among intrinsic motives the authors listed the crystallographic and thermodynamic properties of crystals, while extrinsic motives included the kinetic and driving forces which arise from the microstructure, composition and temperature of the sample.

$BaTiO_3$ is one of the first ceramic materials for which anomalous grain growth was recorded [21]. Sintering of $BaTiO_3$ based materials is normally carried out with a small excess of TiO_2 as a sintering aid. The TiO_2 excess reacts with $BaTiO_3$ to produce $Ba_6Ti_{17}O_{40}$ which forms with a eutectic melt $BaTiO_3$ at about 1320°C. The liquid phase triggers discontinuous grain growth which results in a coarse-grained structure.

The accepted mechanism for microstructure development is nucleation and growth from the liquid phase, similar to the Ostwald ripening process [16, 22, 23]. In an air atmosphere, the crystal structure of $BaTiO_3$ is cubic up to 1430°C [24] when it transforms to the hexagonal form. A cubic structure at the temperature of rapid grain growth (~ 1350°C) results in isometric grains.

Recently, we observed anisotropic abnormal grain growth in TiO_2 doped $BaTiO_3$ ceramics during sintering in a reducing atmosphere [25]. In the present review, we offer the experimental evidence for some intrinsic and extrinsic motives for anisotropic grain growth. We believe that the system $BaTiO_3$-TiO_2 may well serve as a model system for anisotropic exaggerated grain growth in general.

2. EXPERIMENTAL

The starting material was high purity commercial grade $BaTiO_3$ (Transelco Code 219/9). The supplier's specification lists max. impurities of Al_2O_3 as 50 ppm, CaO, SrO and SiO_2 100 ppm each, BaO+SrO/TiO_2 0.995-1.000 and the Coulter average grain size 1.3-1.5 μm. Various amounts of $Ba_6Ti_{17}O_{40}$, prepared by calcining $BaTiO_3$ and TiO_2 at 1250°C and finely ground were added to the starting material and homogenized. The powder was pressed into pellets (ϕ = 6 mm, h = 3 mm) without binder and sintered in an Ar-8 v/o H_2 gas mixture, moistened by passing through H_2O at 25°C. After 1 hour soaking at 1300°C (to ensure densification) the samples were sintered at 1340-1370°C for various times and rapidly cooled. The samples were characterized by XRD, optical and electron microscopy. Digital processing and analysis was carried out to evaluate the microstructural features quantitatively.

Hexagonal $BaTiO_3$ seed powder was prepared by heating $BaTiO_3$ powder in a reducing atmosphere at 1400°C. The sintered powder was crushed and gently milled. The finest fraction was separated by classification in a water column and used for seeding.

2. RESULTS AND DISCUSSION

2.1. Liquid phase assisted grain growth

For exaggerated grain growth, two conditions are necessary: (a) the presence of grains which are able to grow faster than others ("nuclei") and (b) favorable conditions for rapid growth. In addition for anisotropic abnormal grain growth, conditions favoring preferential growth in particular direction must be present.

Our experiments demonstrate that particularly favorable combination for anisotropic growth are the simultaneous presence of grains with a tendency for anisotropic growth and a liquid phase which accelerates grain growth. Without liquid phase the grain growth was limited, although the fine $BaTiO_3$ powder densified to 96.6 % of theoretical density (Fig. 1A). The appearance of liquid phase triggered exaggerated growth by a dissolution - precipitation mechanism. In an air atmosphere, the crystal structure of $BaTiO_3$ at the temperature of rapid growth is cubic, and grains grow isotropically. The result is a coarse, equiaxial grain microstructure with a broad grain size distribution (Fig. 1B).

In a hydrogen atmosphere, the cubic crystal structure of $BaTiO_3$ transforms to hexagonal already at the temperature where the eutectic liquid appears (~ 1330°C) [26] and grains grow anisotropically, resulting in a coarse microstructure with extremely elongated grains (Fig. 1C). Elongated grains on 2-dimensional micrographs are in reality platelets [25]. High resolution transmission electron microscopy revealed that the grains grow preferentially in the direction of hexagonal prism planes. The growth rate in the preferential direction is very rapid, initially exceeding 120 μm/min at 1350°C and 180 μm/min at 1355°C. After impingement with other rapidly growing grains, growth in the length direction was reduced due to steric hindrance. The thickening of faceted boundaries in the [0001] direction is slow, causing a high initial average aspect ratio of about 14. After grain impingement, the thickening continued and the aspect ratio decreased to ~ 8.

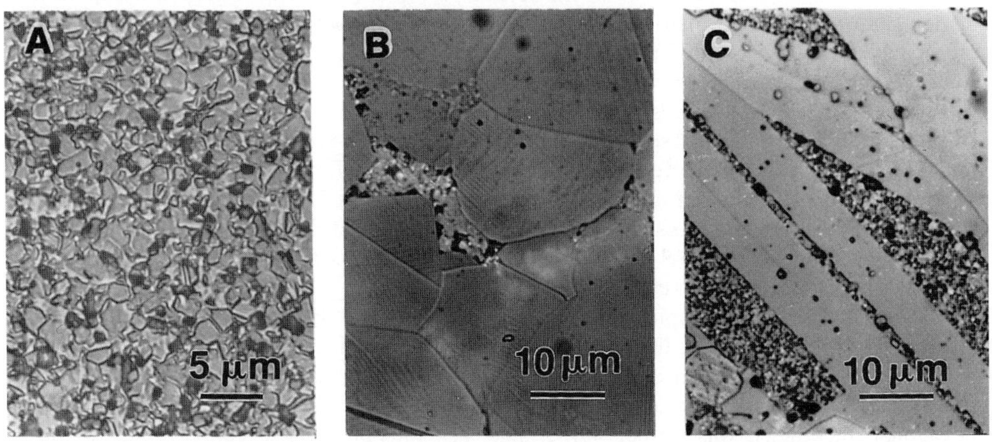

Fig. 1: Microstructure of $BaTiO_3$ - 2 m/o $Ba_6Ti_{17}O_{40}$ sintered ceramics (A) 1310°C, 2 hours, H_2/N_2, 96.6 % T.D. (B) 1350°C, 2 hours, air (C) 1350°C, 2 hours, H_2/N_2

If the conversion of cubic to hexagonal structure is inhibited by impurities such as Si, Fe or Sr [26] the $BaTiO_3$ grains grow in reducing atmosphere isotropically in a similar manner as in air. Fig. 2B shows a microstructure consisting of coarse isometric grains, developed during H_2/N_2 sintering at 1350°C of a $BaTiO_3$-2 m/o $Ba_6Ti_{17}O_{40}$ composition doped with 0.1 w/o SiO_2. Microstructure without SiO_2 additive is shown for comparison, Fig. 2A.

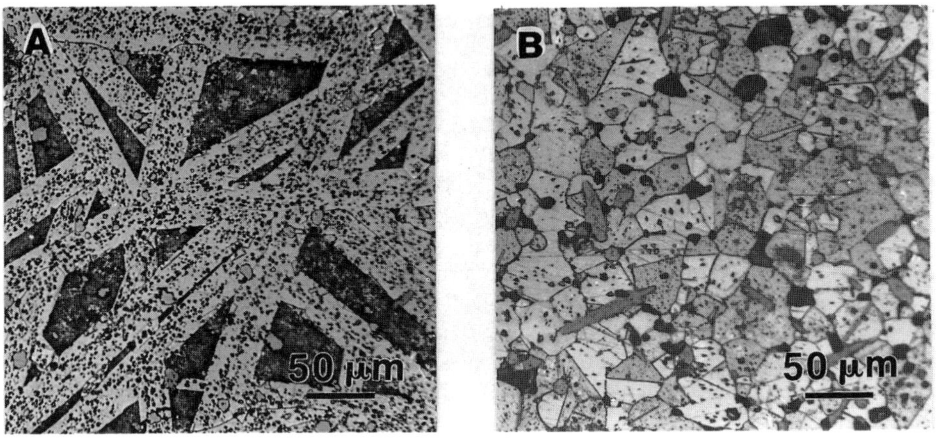

Fig. 2: Influence of impurities on microstructure development of $BaTiO_3$ - 2 m/o $Ba_6Ti_{17}O_{40}$ ceramics in H_2/N_2 atmosphere (1340°C, 30 min) (A) without additives (B) with 0.1 w/o SiO_2

2.2. Nucleation

Nuclei for abnormal growth may be present at the beginning of rapid growth, or be created during the microstructural development. In the first case one expects a narrow size distribution of exaggerated grains, in the second a broad size distribution. Our experiments demonstrate narrow size distribution (Fig. 3A) We ascribe the observed increase in nuclei density in the early stage of microstructure development to redistribution of liquid phase, which seems to occur rapidly.

Possible nucleation sources for exaggerated growth are the crystallographic defects in growing grains. Defects represent active sites for accelerated precipitation of dissolved matter in the liquid phase. Active defects like twin boundaries may form due to local grain boundary misorientation, as discussed for Al_2O_3 by Rödel and Glasser [15], or exist as special defects in the initial powder, as proposed for exaggerated growth in $BaTiO_3$. It was suggested that the exaggerated growth of parallel (111) twins in cubic $BaTiO_3$ is pronounced along the layers of face-shearing octahedra at so-called reentrant angles [27].

For exaggerated anisotropic grain growth of hexagonal $BaTiO_3$, we suggested similar mechanism (25). Comparing the structures of hexagonal $BaTiO_3$ in the [1120] projection and cubic $BaTiO_3$ with a parallel combination of (111) twins in [110] projections, it was shown that a substantial similarity exists between the atomic arrangements in the vicinity of hexagonal stacking. It was further shown that Ti^{3+} ions are very effective in stabilizing the hexagonal structure. We suggested that Ti^{3+} ions initiate the formation of face-sharing octahedra in the form of Ti_2O_9 groups which are nucleation sites for hexagonal stacking. Hexagonal nuclei, thus formed, grow rapidly in the preferential direction, assisted by a rapid supply of matter through the liquid phase.

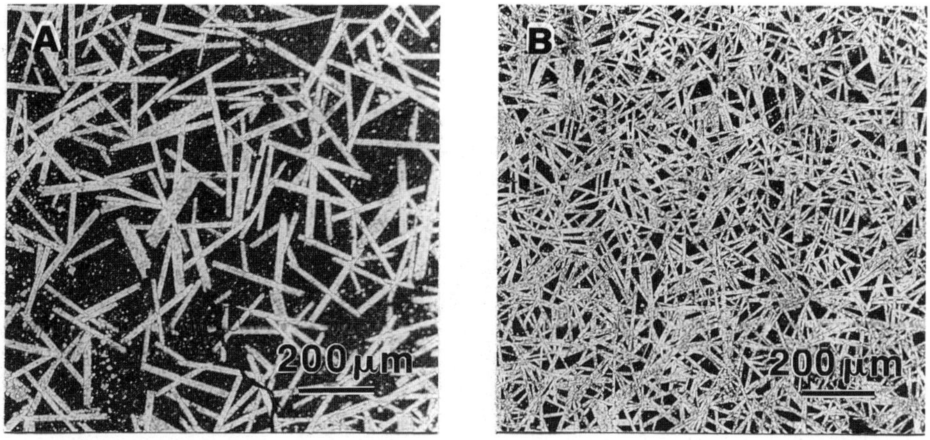

Fig. 3: Influence of seeding with hexagonal $BaTiO_3$ on microstructure development of $BaTiO_3$ - 2 m/o $Ba_6Ti_{17}O_{40}$ in a H_2/N_2 at 1340°C, 10 minutes (A) without seeding (B) with 0.01 % hexagonal $BaTiO_3$.

Seeding with hexagonal nuclei is very effective in provoking development of an anisotropic microstructure as demonstrated in Figs. 3A and 3B. The possible existence of hexagonal nuclei in the starting powder cannot be excluded. They may stem from local reduction during $BaTiO_3$ synthesis or from particular impurities. Heat treatment at high a temperature in air at temperature where the hexagonal phase is unstable should decrease the density of hexagonal nuclei and this was actually observed, as demonstrated by Figs. 4A and 4B. However, the heat treatment also changes other powder properties, such as morphology, so that the change in microstructure caused by high temperature treatment in air may have various origins.

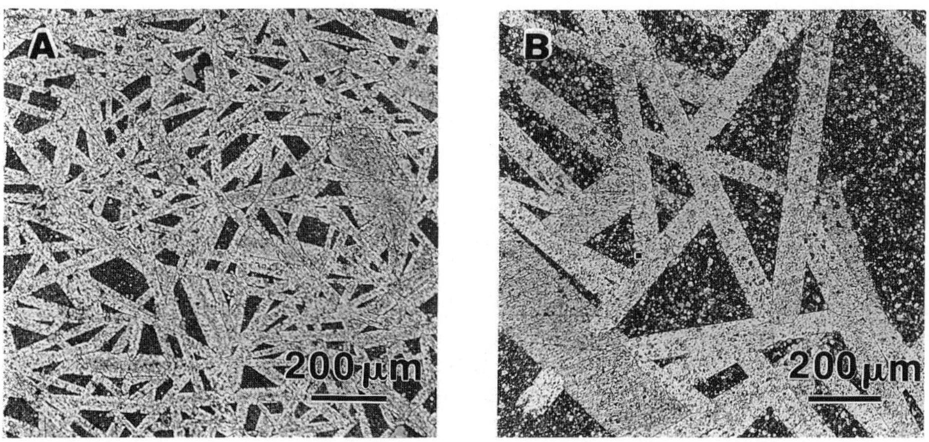

Fig. 4: Influence of calcination pretreatment in air (1200°C, 4 hours) on microstructure of $BaTiO_3$ - 2 m/o $Ba_6Ti_{17}O_{40}$ sintered in H_2/N_2 at 1340°C, 30 minutes (A) without calcination (B) with calcination.

An increasing amount of liquid phase promotes nucleation of anisotropically growing grains. High nuclei density results in smaller final grain size due to faster impingement, as demonstrated in Figs. 5A-C. With limited amount of liquid phase, the rate of recrystallization increases with increasing amount of liquid. However, above ~ 2 % of liquid, the rate of recrystallization becomes constant, indicating that the recrystallization is not controlled by the diffusion through the liquid phase.

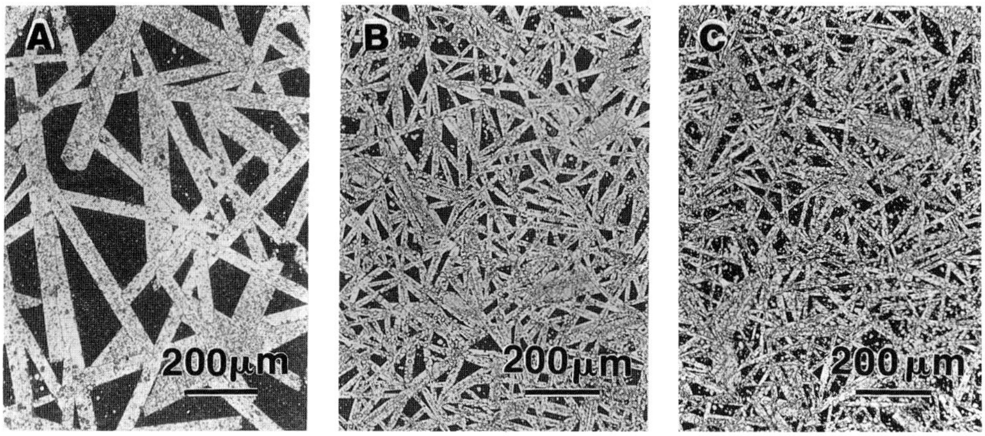

Fig. 5: Influence of the amount of liquid phase on nucleation and microstructure development of $BaTiO_3$ in H_2/N_2, 1350°C, 30 minutes (A) 0.5 m/o, (B) 4 m/o (C) 10 m/o $Ba_6Ti_{17}O_{40}$

2.3. Faceted Boundaries

Anisotropically growing $BaTiO_3$ grains are faceted. Faceted boundaries of preferentially growing grains are typical of grain growth in the presence of a limited amount of liquid phase. Actually, it was suggested [28] that faceted grain boundaries of exaggerated grains may be regarded as a strong indication of the presence of liquid phase, although the liquid phase may not be detected or expected from phase equilibrium data of a particular ceramic composition. The liquid phase may be due to impurities, or, as suggested by Kayser et al [19] may not be present initially, but appears in the later stage of grain growth at some critical grain size when the capacity of grain boundaries to absorb the impurities is exceeded. On the other hand, computer simulation of anisotropic grain growth [13] showed that the grains which grow anisotropically due to surface energy anisotropy exhibit faceting, provided the lower anisotropic surface energy is less than that of the matrix. The presence of liquid phase was not specifically simulated. Yang et al [29] calculated the surface energy of all the grains from a hypothetical Wulf plot and concluded that faceted anisotropic grains may develop from an initially equiaxial matrix, if the surface energies anisotropic.

3. CONCLUSIONS

1. The results confirm the crucial role of the crystallography of the material for anisotropic grain growth. Cubic barium titanate with an isotropic crystal structure develops a microstructure with isometric grains, whereas hexagonal barium titanate with an asymmetric unit cell develops a microstructure with platelike shaped grains with a high aspect ratio.
2. The results further confirm the crucial role of liquid phase on rapid anisotropic grain growth. The growth rate in the preferred direction is constant and extremely rapid (180 μm/mm) until the grains impinge on other fast growing grains. The growth rate in the thickening direction is much slower (almost 2 orders of magnitude). The growth mechanism is dissolution/precipitation.
3. Seeding with hexagonal nuclei (0.1 w/o) is very effective in provoking development of homogeneous anisometric microstructures with low aspect ratio.
4. An increasing amount of liquid phase strongly increases nucleation rate, i.e. the number of anisotropically growing grains. The larger number of nuclei result in a smaller grain size due to faster impingement. However, the rate of recrystallization does not decrease with increasing amount of liquid phase, indicating that the dissolution-precipitation process is boundary reaction controlled and not controlled by diffusion through the liquid phase.

References

1. M. Hillert, Acta Metall., 13 (3) (1965) 227-38.
2. P. F. Becher, J. Am. Ceram. Soc., 74 (2) (1991) 255-69.
3. E. Tani, S. Umebayashi, K. Kishi, K. Kobayashi, and M. Nishijima, Am. Ceram. Soc. Bull., 65 (9) (1986) 1311-15.
4. F. Wu, H. Zhuang, L. Ma, and X. Fu, Ceram. Eng. Sci. Proc., 14 (1-2) (1993) 321-32.
5. N. Hirosaki, Y. Akimune and M. Mitomo, J. Am. Ceram. Soc., 76 (7) (1993) 1892-94.
6. N. P. Padture, J. Am. Ceram. Soc., 77 (2) (1994) 519-23.
7. N. P. Padture and B. R. Lawn, J. Am. Ceram. Soc., 77 (10) (1994) 2518-22.
8. R. J. Brook, Controlled Grain Growth, pp. 331-64 in Treatise on Materials Science and Technology, Vol. 9, Ceramic Fabrication Processes, F. F. Wang (ed.), Academic Press, New York, 1976.
9. C. H. Hsueh, A. G. Evans, and R. L. Coble, Acta. Metall., 30 (1982) 1269-79.
10. R. L. Coble and J. E. Burke, Sintering in Ceramics, pp. 197-251 in Progress in Ceramic Science, Vol. 3, J. E. Burke (ed.), Pergamon, Oxford, 1963.
11. W. S. Patrick and I. B. Cutler, J. Am. Ceram. Soc., 48 (19) (1965) 541-42.
12. D. W. Susnitzky and C. B. Carter, J. Am. Ceram. Soc., 68 (11) (1985) 569-74.
13. U. Kunaver and D. Kolar, Acta Metall. Mater., 41 (8) (1993) 2255-2263.
14. M. F. Yan, Mater. Sci. Eng., 48 (1981) 53-71.
15. J. Rödel and A. M. Glaeser, J. Am. Ceram. Soc., 73 (11) (1990) 3292-301.
16. D. Kolar, Discontinuous Grain Growth in Multiphase Ceramics, pp. 529-45 in Sintering of Advanced Ceramics, Ceramic. Trans. 7, C. A. Handwerker, J. E. Blendell and W. A. Kayser (eds.), Am. Cer. Soc., Westerville, OH, 1990.
17. J. Wong, J. Appl. Phys., 51 (8) (1980) 4453-59.
18. C. Kooy, Sci. Ceram. 1 (1962) 21-34.

19. W. A. Kayser, M. Sprissler, C. A. Handwerker, and J. E. Blendell, J. Am. Ceram. Soc., 70 (1987) 339.
20. M. Seabaugh, D. Horn, I. Kerscht, S. H. Hong, and G. L. Messing, Anisotropic Grain Growth in Alumina Ceramics, pp. 341-348 in Sintering Technology, R. M. German, G. L. Messing and R. G. Cornwall (eds.), Dekker, New York, 1996.
21. J. E. Burke, Grain Growth in Ceramics, pp. 109-116 in Kinetics of High Temperature Processes, W. D. Kingery (ed.), Wiley, New York (1959).
22. D. G. K. Hennings, R. Jansen and P. J. L. Reynen, J. Am. Ceram. Soc., 70 (1) (1987) 23-27.
23. J. Chen, W. Jin and Y. Yao, Ferroelectrics, 142 (1993) 153-159.
24. K. W. Kirby and B. A. Wechsler, J. Am. Ceram. Soc., 74 (1991) 1841-47.
25. A. Rečnik and D. Kolar, J. Am. Ceram. Soc., 79 (4) (1996) 1015-18.
26. R. M. Gleister and H. F. Kay, Proc. Phys. Soc., 76 (1960) 763-771.
27. H. Schmelz and H. Thomann, CFI, Ceram. Forum Int., 61, (1984) 199-205.
28. F. M. A. Carpay and A. L. Stuijts, Sci. Ceramics, 8 (1975) 23-38.
29. W. Yang, G. L. Messing, and L. Q. Chen, Mater. Sci. Eng., A195 (1995) (179-187).

Materials Science and Engineering Serving Society
S. Sōmiya, R.P.H. Chang, M. Doyama and R. Roy (Editors)

Growth mechanism and structural control of epitaxially grown single crystal $PbTiO_3$ thin films

Kiyotaka Wasa

Faculty of Science, Yokohama City University,Yokohamashi,236 Japan*

This paper overviews a basic deposition process and microstructures of perovskite ceramics thin films and describes recent experiments on the growth of perovskite $PbTiO_3$ (PT) thin films with controlled microstructures. It is shown that the deposition on a miscut substrate is essential to make perfect single crystal perovskite PT thin films of a single domain having full in–plain alignment together with the formation of a continuous film throughout a large area.

1.INTRODUCTION

Wide varieties of nobel ceramics have been developed for manufacturing electronic and/or photonic devices. These ceramics were developed in a form of bulk structure. Extensive works have been done to fabricate thin films of ceramics in order to meet a recent requirement of high frequency operation and/or integration to Si ICs. Thin films of perovskite ceramics have been extensively studied, since compound perovskite such as PLT and PLZT shows a series of interesting and useful properties associated with the structural system[1]. Continuous single crystal thin films of perovskite materials are important not only for the fabrication of layered perovskite thin film devices but also for the understanding of their physical and/or chemical properties. However, most of the perovskite thin films include non–homogeneous microstructures, even if the films show three–dimensional epitaxy. Systematic studies on a deposition of basic perovskite such as PT are essential for an achievement of a fine control of the microstructure[2]. This paper will overview the basic deposition process of ferroelectric PT thin films and describe recent experiments on the sputtering deposition of PT thin films with controlled microstructures.

2.THIN FILM DEPOSITION

Several deposition processes have been used for the deposition of perovskite ceramics including thermal evaporation, sputtering, laser ablation, and chemical vapour deposition. Among these deposition processes sputtering is now widely used for the deposition of perovskite ferroelectric ceramics and high–Tc superconductors of layered perovskite[3]. Thin films of amorphous, polycrystal, and/or single crystal phase are provided by

*contact address: Kiyotaka Wasa
FAX 81–742–44–8565, e–mail kiyotkw@hi–ho.or.jp

suitable selection of substrates and substrate temperature during film growth. Single crystal thin films of perovskite structure are provided by a heteroepitaxial growth on a foreign single crystal substrate shown in Table 1. The epitaxial temperature of perovskite thin films is 500 to 600°C. The heteroepitaxial films essentially include dislocation and/or microstructures due to the thermal expansion and/or lattice mismatch between thin films and substrates. As seen in Table 1, $SrTiO_3$ (ST) substrates show a small lattice mismatch for the growth of c–domain. Large lattice mismatch, i.e. MgO and sapphire, induces a growth of dislocation and/or multi–domains above a critical film thickness in order to release the stress in the epitaxial films[4]. The critical film thickness is about 100nm[5]. The non–uniformity of initial crystal growth also induces the microstructures including multi–domain growth[6]. Uniform layer growth with a small lattice mismatch under stoichiometric composition will be essential for the deposition of continuous single crystal perovskite thin films[7]. The lattice deformation accompanied by the growth of microstructures will be enhanced during the cooling period from the epitaxial temperature to a room temperature.

Table 1
Lattice parameters of substrrates

	Crystal system	Structure	Lattice constant (A)	Thermal expansion 1/K(ppm)
PT	tetra.	perovskite	3.903(a)	16.1
			4.152(c)	-54.2
MgO	cubic	NaCl	4.203(a)	13.8
ST	cubic	perovskite	3.905(a)	10.8
$LaAlO_3$	pseudo cubic	perovskite	3.792(a)	10
Sapphire	trigonal	corundum	4.763(a)	7.5–8

3.CONTROL OF MICROSTRUCTURES

PT thin films were epitaxially grown on (001)ST substrate by rf–planar magnetron sputter. PT powder target was used for the deposition. The chemical composition of the sputtered films are controlled by the composition of the powder target. Miscut ST substrates with miscut angle of 0.2 to 10 degrees were used for the epitaxial growth. Typical sputtering conditions are shown in Table 2. Sputtered thin films were cooled down to 200°C within 30minutes.

Table 2
Sputtering conditions

Sputter system	rf–magnetron
Target	PT powder
Substrate	(001)ST*
Growth temp.	600°C
Sputter gas	Ar/O_2 (20/1, 0.5Pa)
Growth rate	4nm/min

* well–oriented and miscut ST

Below the critical thickness, thin films exhibit uniform crystal growth without any microstructure, if the thin films uniformly grow on an atomic scale. The miscut substrates will enhance the uniform crystal growth. To confirm this substrate effect, first we tried to deposit thin films of PT on the well–oriented and the miscut ST substrates under epitaxial growth conditions. The film thickness was below the critical thickness,i.e. ‹100nm. Stoichiometric composition was kept throughout the experiments so as to suppress the formation of secondary phases during film growth. Typical value of miscut angle was 1.7 degree. The direction of miscutting was parallel to ‹010› ST. Surface model of the miscut ST is shown in Figure 1. Surface crystal steps will act as a nucleation center. The surface comprises step lines and terraces. The height of the steps is around 0.4nm corresponding to a crystal unit of ST substratc[9]. The average terrace width is 14nm for the miscut angle of 1.7 degree. X–ray diffraction analysis suggests that these sputtered thin films exhibit highly c–axis oriented single crystals with the epitaxial relation, (001) PT/(001) ST. The surface SEM image of thin films on well–oriented substrates shows the presence of microstructures comprising island structure at the initial growth probably due to the three–dimensional nucleation. The islands grow with the film thickness increases under Volmer–Weber mode. When the film thickness reached a critical value, i.e. 100nm, sharp cracks were developed. Surface of sputtered PT thin films on the miscut ST substrates comprises periodic striped patterns as seen in the AFM image of Figure 2. The periodic striped pattern will result from the initial surface steps on the miscut substrates. The spacing observed was larger than the spacing of the initial surface terrace. The spacing increased with increasing the film thickness.

The cross sectional TEM images suggested that the surface of the terrace was atomically smooth as shown in Figure 3. The step height was 1 to 3nm.Typical spacing for the miscut angle of 1.7 degree was 50 to 150nm and the value showed saturation at thick film thickness. The X–ray diffraction patterns include satellite oscillation due to X–ray interference fringes probably due to the smooth film structure as shown in Figure 4.

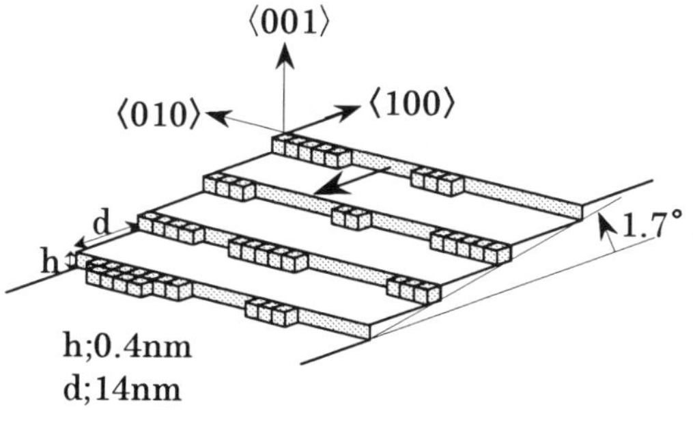

Figure 1. Surface structure of miscut ST substrate.

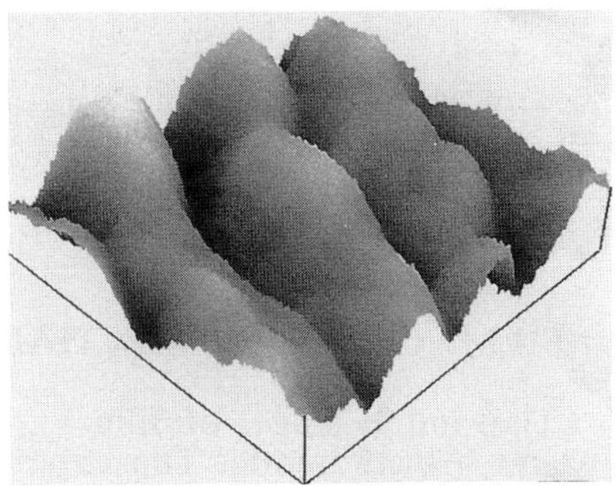

500nm x 500nm
Max 2.28nm

Figure 2. Surface AFM image of PT thin film on miscut ST, miscut 1.7^{o},100nm in thickness.

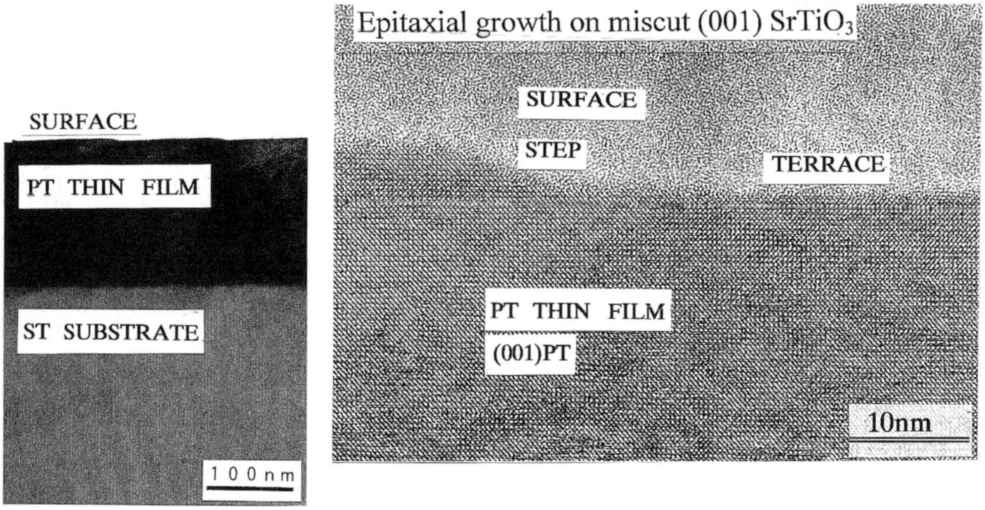

Figure 3. Cross sectional TEM images of PT thin film on miscut ST, miscut 1.7°,160nm in thickness.

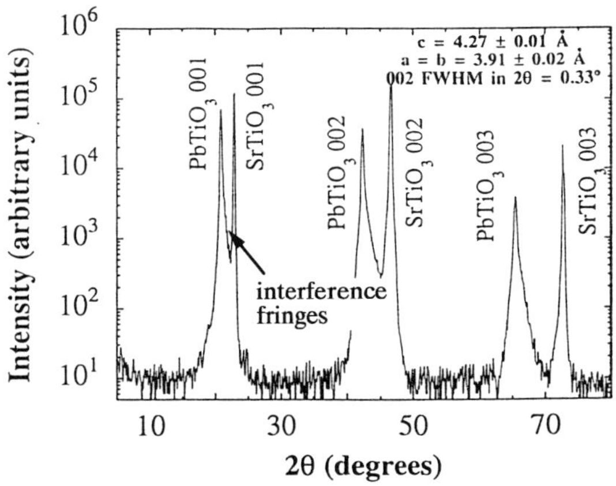

Figure 4. X−ray diffraction pattern of PT thin film on miscut ST, miscut 1.7°, 40nm in thickness.

4.FILM GROWTH MECHANISM

The initial stage of film growth on well−oriented substrates show the island growth mode. The experimental results suggested that the film growth on the miscut ST substrates was governed by a layer growth of Frank−van der Merwe mode. The layer growth was achieved by the step−flow growth from the steps on the miscut substrates[10]. At the initial stage of film growth step bunching is observed. Both terrace width and step height increased with film thickness. The step−flow growth mode is essentially governed by the surface diffusion of adatoms on the terrace.

Higher mobility of adatoms and/or narrow width of terrace will stabilize the step–flow growth mode. The commercially available well–oriented ST substrates also include crystal steps and terraces, since the error of crystal cutting angle is generally 0.3 degree which having the terrace width of › 80nm. At present deposition condition the terrace width, › 80nm, will be so long that the step–flow growth mode could not be stabilized. The critical miscut angle, i.e. minimum miscut angle, was 0.9 to 1.7 degree[11]. The stability of the layer growth of step–flow growth mode was affected by the partial oxygen pressure during film growth. The higher oxygen partial pressure during the sputtering will prevent the two dimensional nucleation accompanied by a layer growth and induces the three–dimensional crystal growth. A critical partial oxygen pressure was present for the layer growth[12].

Heteroepitaxial films show three types of lattice deformation; relaxed, strained, and dislocated lattice deformation. Thick films usually show dislocated type. Ultrathin films, thickness less than the critical thickness, show strained lattice deformation. At the growth temperature cubic PT (a=3.97A) films are epitaxially grown on ST (a=3.93A) under a compressive force due to the lattice mismatch. The unit cell of PT thin films will be prolonged. The compressive force will be maximum at growth temperature, since a–lattice mismatch will decrease due to the difference of thermal expansion coefficient. During the cooling stage c–domain stabilization will be further enhanced due to the intrinsic phase transition from cubic to tetragonal phase at Curie temperature, resulting in large tetragonal deformation as shown in Table 3.

X–ray analysis suggested that the thin films comprise a single c–domain structure, even if the film thickness increases up to 300nm. The PT thin films are continuous structure without any grain boundary. The PT thin films are extremely smooth. The PT thin films will be tightly bonded onto the substrate, so the PT thin films show a small variation of c–and/or a–lattice parameters over Curie temperature[13].

Table 3
Lattice parameters of (001)PT thin films on miscut (001)ST [RT]

	Thin Films [thickness]				Bulk PT
	10nm	40nm	180nm	250nm	
c(A)	4.26	4.27	4.26	4.26	4.14
a(A)	3.90	3.91	3.90		3.89
c/a	1.09	1.09	1.09		1.06

* a(ST)=3.905A

5.SUMMARY AND CONCLUSION

The heteroepitaxial thin films of perovskite ceramics including PT, PLT and PLZT on well–oriented substrates comprise microstructures comprising multi–domain. The growth of microstructure is governed by initial stage of film growth and stress induced during film growth. The miscut substrate modifies the initial film growth. Deposition of PT on a miscut ST substrate prohibits an island growth and induces a layer growth of step–flow type. The resultant PT thin films show a continuous single crystal/ single c–domain structure, although the films shows strained deformation. The thickness of the continuous perfect single crystal films is up to 300nm[14]. The epitaxial growth on miscut substrates is essential to fabricate perovskite thin films with controlled microstructures. A basic planar rf–magnetron sputtering is available for the deposition of perovskite thin films with controlled microstructures on an atomic scale. The sputtered thin films will be useful for the understanding of physical and or chemical properties of ceramics materials.Nobel ceramics electronic and/or photonic devices will be developed by using sputtering technology. It is also noted that the sputtering thin film technology is a promising environment benign ceramics technology, since the sputtering process reduces the growth temperatures of the ceramics and saves the energy consumption[15].

The author thanks Dr.H.Adachi and T.Sato (Matsushita Elec.) for their support for a deposition of thin films, Prof.D.G.Schlom, S.Trolier–McKinstry (Penn State) C.B.Eom (Duke), and Miss Y.Haneda (RITE) for their structural analyses, Dr.R.Kogawa (Shimadzu) for his AFM measurement Prof.E.Cross, R.Roy, R.E.Newnham, K.Uchino(Penn State) for their helpful discussion, Dr.J.Kondo, (RITE) for his continuous encouragement.

REFERENCES

1.G.H.Haertling and C.E.Land, J.Am.Ceram.Soc.,54(1971)1.
2.K.Wasa, H.Adachi,and M.Kitabatake, IMF–8,Proc. (Washington DC, 1993) in Ferroelectrics,151(1994)1.
3.K.Wasa and H.Hayakawa, Handbook of Sputter Deposition Technology, Noyes, N.J.,1992, p.162.
4.B.S.Kwak, A.Erbil, B.J.Wilkens, J.D.Budai, M.F.Chisholm, and L.A. Boatner,Phys.Rev.Lett.,68(1992)3733.
5.C.D.Theis, Thesis for Master of Science (Penn State Univ.,1996) p.104.
6.K.Wasa, T.Sato, K.Tabata, H.Adachi, Y.Yabuuchi, and K.Setsune, J. Mater. Res.9(1994)2959.
7.K.Wasa, H.Adachi, Y.Ichikawa, K.Hirochi, T.Matsushima, Enokihara, K.Mizuno,H.Higashino,and K.Setsune, Sci. and Technol. of Thin Film Supercon. 2,ed. by R.D.McConnel and R.Noufi,Prenum,N.Y.,1990,p.1.
8.T.Sato, K.Wasa, K.Tabata,H.Adachi, Y.Ichikawa,and K.Setsune, J.Vac. Sci.Technol.A13(1995)1022.
9.K.Wasa, T.Sato, H.Adachi, K.Setsune,and S.Trolier–McKinstry, ISSP–3 Proc.,1995,Tokyo,p.85.
10.K.Wasa, T.Sato, H.Adachi, K.Setsune, S.Trolier–McKinstry,and D.G. Schlom, Mater.Res.Soc.Symp.Proc.,401(1996)151.
11.K.Wasa, Y.Haneda, T.Satoh, H.Adachi,and K.Setsune, ISSP–4,Proc. 1997, Kanazawa,p.145.
12.K.Wasa, Y.Haneda, T.Sato, H.Adachi, I.Kanno, K.Setsune, D.G.Schlom S.Troiler–McKinstry,and C.B.Eom, IMF–9,Proc. (Seoul,1997)I18–FRC04.
13.K.Wasa, Y.Haneda, T.Satoh, H.Adachi,and K.Setsune, J.Vac.Sci.Technol. A15 (3)May/Jun,(1997)1185.
14.D.G.Schlom(private communication).
15.K.Wasa, Bull. Mater. Sci. India, 16(1993)643, ibid. 18(1995)937.

Materials Science and Engineering Serving Society
S. Sōmiya, R.P.H. Chang, M. Doyama and R. Roy (Editors)

Magnetic thin films with anisotropic grain-growth in the 1 or 3-dimensional direction

Masayoshi. Hiramoto, Nozomu. Matsukawa,
Hiroshi. Sakakima, Yo. Ichikawa
Kouichi. Kugimiya

Central Res. Labs., Matsushita Electric Industrial Co., Ltd.,
Seika-cho, Kyoto 619-02, Japan

Abstract Nanocrystalline magnetic materials, made of small grains under 20nm, have been investigated to realize soft magnetic materials with high saturation magnetic flux density, Bs. We investigated the magnetic, structural and corrosive properties of sputtered FeSiAlTiON. Though the general nanocrystalline materials are obtained from the amorphous state with isotropic grain growth by thermal treatment, the developed materials have fine crystalline textures as prepared state. According to the TEM observations , the microstructures of the film annealed at 500°C are consisted of needle shape fine grains in 1-dimensional growth and/or dendritic in 3-dimensional growth. The typical grain has a size over 200nm of height and 20nm of width. The developed materials exhibit high corrosion resistance with high Bs as well as good soft magnetic properties caused by their compositions and structures. The developed materials have Bs of 1.3T, permeability of 3000-8000, respectively, and have higher corrosion resistance than FeSiAl with Bs of 1T.

1. INTRODUCTION

Various type of magnetic thin films have been investigated to miniaturize the electric devices or to achieve a high density magnetic recording. The films are classified according to the structure as shown in Figure1. The polycrystalline materials made of the columnar grains having a typical grain size around 300nm of a width, represented by FeSiAl film, show good soft magnetic properties by small magnetocrystalline anisotropy. It is well known that the magnetic properties become good with increasing the grain size in those of materials. On the contrary, nanocrystalline materials made of granular structure with small grains under 20nm and amorphous materials in the system Fe(TaNbZr)N 1) and CoNbZr 2), exhibit good properties with decreasing the grain size . The grain size dependence of the soft magnetic properties has been successfully explained by so-called random anisotropy model 3) 4). However, There are few reports about a boundary structured materials which has the grain structure between a kind of granular and columnar structure. The boundary structure is expected to produce new materials which has an intensive properties controlled by magneto crystalline anisotropy and random anisotropy effect. In this work, preparation of the boundary material was investigated for the sputtered FeSiAlTIO(N) films. The prepared films were confirmed having an anisotropic grains, which were satisfied the random anisotropy condition only in the film plain.

FAX: +81-774-98-2585, e-mail: hiramoto@crl.mei.co.jp

Research supported by NEDO, under the Synergy Ceramics Project of ISTF program promoted by AIST, MITI, Japan.

2. EXPERIMENTAL

FeSiAlTiO(N) films of 3 μm thickness were prepared by r.f.-magnetron sputtering in Ar+N2 or Ar+N2+O2 mixed gas. Samples were sputtered using alloy target (3 in. in diameter or 5×15 in.) onto ceramic substrates. The substrates were located on the position which have little affection of the stray field of the target. Thermal treatment were performed in vacuum without magnetic field at 520°C. The Bs, the initial permeability, μ_i, the saturation magnetostriction, λ_s, were measured by using a vibrating sample magneto-meter, VSM , a figure 8 coil apparatus and a optical cantilever method, respectively. The coercivity and the anisotropy filed were measured with B-H loop tracer at 60Hz. The microstructure was observed by using a transmission electron microscopy, TEM and X-ray diffraction, XRD. The film composition are determined by using a inductive coupled plasma emission spectrometry ,ICP and a Rutherford back scattering, RBS. All the measurement were carried out at room temperature.

3. RESULT AND DISCUSSION

3.1. Design of a microstructure

Formation mechanism of a granular structure from an amorphous state was studied in a view point of a thermodynamic 5). The result allows us to show a qualitative guidance to form boundary structured materials. A granular, Fe(NbZrHf)N formed from amorphous state and columnar, polycrystalline, FeSiAl were discussed. Figure 2 shows Gibbs free-energy difference, $\Delta G\nu$, between Fe-N(Fe_4N) and several nitrides, M-N (M= Si, Al, Ti, Nb, Zr) per one nitrogen molecule at 500°C. Figure 3 shows diffusion coefficient, D of M (M=Si, Al, Ti, Hf) , N and O in α-Fe at 1372°C, respectively. The difference of $\Delta G\nu$, which gives a easiness of segregation between Fe-N and M-N products is small in each M element. On the other hand, D has a remarkable difference between Hf and other elements. In classical nucleation theory, the larger D gives a faster nucleation rate. Thus, we may expect that FeSiAlTiN has large grain compared with FeNb(Hf)N. In comparison with FeSiAl, additive, N(O) element interrupts the diffusion of Fe, Si, Al and Ti by chemical reaction. Therefore ,the boundary structured materials were expected by the composition of FeSiAlTiN(O). Further, considering the formation of thermally stabilized structure, we must suppress the grain growth by thermal treatment. According to the crystal growth theory, the driving force of the grain growth is the interfacial energy. In a previous work, we showed that the FeSiAl/FeSiAlO multilayers showed good thermally stabilized structure up to 700°C. The result indicated that the FeSiAl/FeSiAlO interface have the low interfacial energy6). Above all, we chose the O element as a thermally stabilized additive for FeSiAlTiN.

3.2. Microstructure and growth process

Figure 4 shows the cross sectional dark field TEM micro-graphs of 520°C annealed FeSiAlTiON film prepared by using 3 in. target. The microstructure is consisted of needle shape fine grains grown perpendicularly to the film plain. The typical grain has a size over 200nm of a height and 20nm of a width. The grains are randomly orientated crystallographically in the plain. The XRD profile of the films indicate the lattice constant is almost the same as that of Fe. It means that the grains are mixed by Al or Si. Further, The film had (110) orientation normal to the plain. The grain growth was little after 520°C thermal treatment. Figure 5 shows the TEM micro-graph of 520°C annealed FeSiAlTiON prepared by using 5×15 in. target. On the contrary to the 3in. target sample, the microstructure is consisted of dendritic grain. The branches of the dendrite are formed by the obliqued incident particles from the large target area. Thus, the FeSiAlTiON films show the anisotropic grain growth affected by the direction of incident particle (i.e. the material source direction) , while those have dense structures. Generally, a large columnar structure like as sputtered FeSiAl are obtained by the condensation of several island-like nucleus on the surface diffusion of particles. However, in the case of spattered FeSiAlTiON, a dangling bond of O, or N that are chemically absorbed on the film surface make a barrier potential to interrupt the surface diffusion. Owing to the barrier, the mean surface diffusion length is reduced and the FeSiAlTiON films have an anisotropic grain growth in the 1 or 3-dimensional direction.

3.3. Magnetic and other properties

The FeSiAlTiON film which had the structure composed of needle shape fine grains showed relatively isotropic μ_i of 3000-8000 in plain direction and λ_s of 1×10^{-6}. Though the fine grains had an anisotropic shape grown in the 1-dimensional direction , a perpendicular anisotropy caused by shape anisotropy was not observed in the hysteresis curve. The good softness of the materials indicates that the grains have a strong exchange interaction each other in the plain direction. According to the random anisotropy model, when the orientation fluctuation length of randomly distributed local easy axis (in the case of granular structure, that is a grain size.) is smaller than the exchange correlation length ,Lo, magnetocrystalline anisotropy averages out. From the analysis of the ICP and RBS, the film composition is determined as $(Fe_{76}Si_{19}Al_3Ti_2)\,O_1N_8$. On assumption that the grains are consisted of Fe-Si alloy, the Lo is estimated as over 30nm. In this film, since the average grain width is smaller than the Lo , and the grains are randomly orientated in the plain. That is, the films is satisfied the random anisotropy condition only in the plain. Therefor, we concluded that the good softness of the film was caused by 2-dimensional random anisotropy effect, while the granular films were categorized as 3-dimensinal case.

The corrosion resistivity of the film was studied. $(Fe_{76}Si_{19}Al_3Ti_2)\,O_1N_5$ film with Bs of 1.3 Tesla and $Fe_{74}Si_{17}Al_9$ film with Bs of 1Tesla were immersed in the 0.5N NaCl solution for 100 hours. The change of the masses were measured using the VSM. While the mass of the FeSiAl film reduce to 90%, the FeSiAlTiON film was not reduced at all. The high corrosion resistance was realized by the relatively small grain as well as additives with low Gibbs free-energy and relatively high diffusion coefficient, which were easy to form a passivation.

4. CONCLUSION

The thermodynamic consideration of the magnetic material lead to the preparation of the boundary structured material of the sputtered FeSiAlTiON films. The element with low Gibbs free-energy and relatively high diffusion coefficient are effective for high corrosion resistance as well as the formation of the boundary structure. The film has Bs of 1.3T, permeability of 3000-8000, respectively, and has higher corrosion resistance than FeSiAl with Bs of 1T. The good soft magnetic properties are caused by 2-dimensional random anisotropy effect.

REFERENCE

1 H. Sakakima, IEEE Trans. Magn., 9 (1983) 131
2 K. Nago, H.Sakakima and K, Ihara, IEEE Trans. J. Magn in Jap., 7 (1992) 119
3 R. Alben, J. J. Becker and M. C. Chi, J.Apple.Phys.,49 (1978) 1653
4 G. Herzer, IEEE Trans. Magn., 26 (1990) 1397
5 K, Lu, Phys. Rev. B. vol.51, No.1 (1995) 18
6 M. Hiramoto, O. Inoue, K. Kugimiya, IEEE Trans. Magn., 30 (1994) 4884

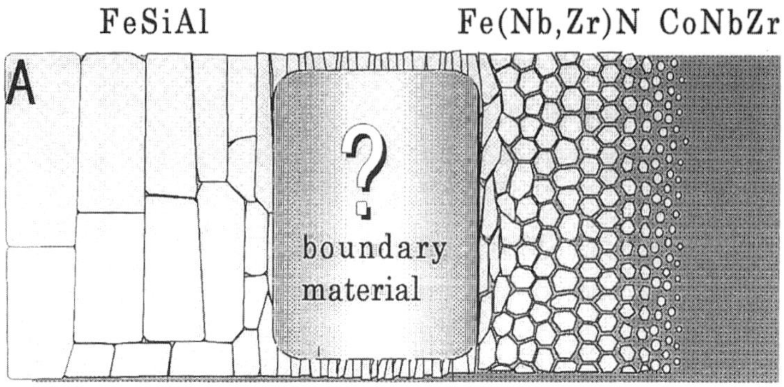

Fig.1. Various magnetic thin films have been investigated. The materials are classified according to the grain size.

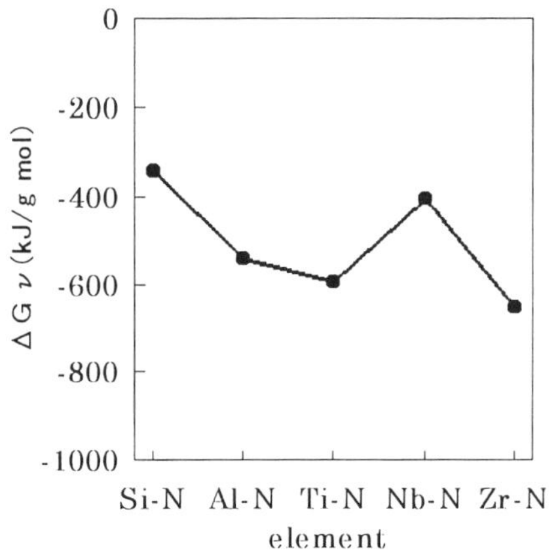

Fig.2. Gibbs free-energy difference btween 1/2(Fe4N) and other nitride.

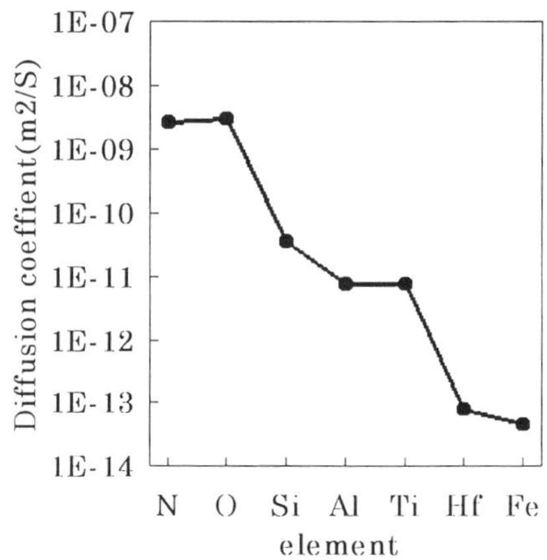

Fig.3. Diffusion coeffient,D of M (=Si,Al,Ti,Hf) in alpha-Fe.

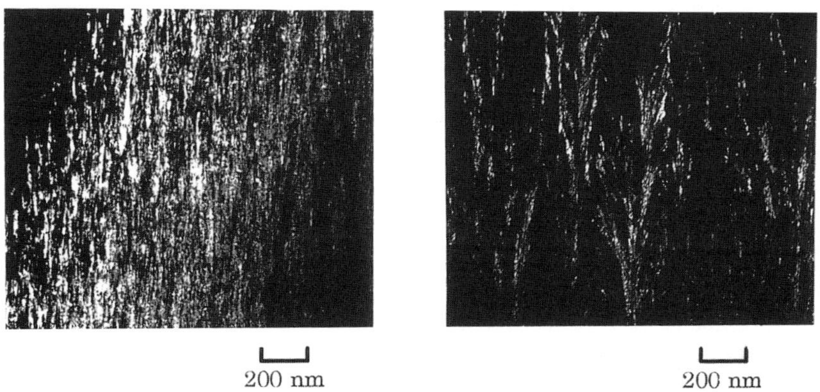

Fig.4. The cross sectional dark field TEM micro-graphs of 520 ℃annealed FeSiAlTiON films.

Materials Science and Engineering Serving Society
S. Sōmiya, R.P.H. Chang, M. Doyama and R. Roy (Editors)

Re-distribution of oxide sintering aids in liquid-phase-sintered silicon carbide due to heat treatment

H. Miyazaki[a], A. Kawaguchi[a], T. Yano[b] and T. Iseki[a]

[a]Department of Inorganic Materials, Tokyo Institute of Technology, 2-12-1, O-okayama, Meguro-ku, Tokyo 152-8552, Japan
[b]Research Laboratory for Nuclear Reactors, Tokyo Institute of Technology, 2-12-1, O-okayama, Meguro-ku, Tokyo 152-8550, Japan

Silicon carbide compacts were pressurelessly sintered with additives of alumina and yttria at 1900°C for 1h in a powder bed of which composition is the same as the compact. Both Al and Y were homogeneously distributed in the as-sintered specimens. After the specimens were heat-treated at 1900°C for 4h in a SiC powder bed, Al was homogeneously distributed and its concentration reduced by half, whereas Y concentrated at surface area. Grains near the surface of the sintered body were small and irregular, but those inside grew large and platelike.

1. INTORODUCTION

Sintering of silicon carbide to the theoretical density needs an addition of sintering aids. However, presence of impurities which are introduced into SiC as sintering aids and/or originally exist in a starting powder could deteriorate superior properties of SiC, such as thermal conductivity and mechanical strength at high temperature. Therefore, reduction of an amount of impurities in SiC is desirable to improve those properties. The purpose of this study is to clarify the effect of heat-treatment on re-distribution and removal of oxide additives in sintered SiC.

2. EXPERIMENTAL PROCEDURE

Fine β-SiC (Ibiden, Japan, Ultrafine grade), Y_2O_3 (Kojundo Chemical Laboratory, Japan) and Al_2O_3 (Sumitomo Chemical, Japan, AKP-50) powders were used as starting materials. Powder mixture containing SiC (91 mass%), Y_2O_3 (3.6 mass%) and Al_2O_3 (5.4 mass%) was prepared by wet milling using methanol as immersion medium for 24h in a polyethylene jar with alumina balls as grinding medium. The milled slurry was dried, uniaxially pressed at 50 MPa, and isostatically pressed at 200 MPa. The green compact was enclosed in a graphite crucible with loose SiC powder of the same composition as the compact, and sintered at 1900°C for 1h in flowing Ar atmosphere using a graphite furnace. Some of the as-sintered specimens were heat-treated at 1900°C for 1, 4 or 8h. During the heat-treatment, the specimens were embedded in the SiC powder containing neither Al_2O_3 nor Y_2O_3.

Density of the specimens was measured using Archimedes method. After etching with the Murakami's reagent, microstructure of the specimens was investigated by optical microscopy, SEM and EDX (energy dispersive X-ray spectroscopy).

3. RESULTS AND DISCUSSION

The bulk density and weight loss of the as-sintered specimen and those of the specimens after heat-treatment are shown in Table 1. The relative density of the specimen was calculated on the assumption that theoretical density of the specimen was 3.28 and did not change by heat-treatment. The relative density of the as sintered specimen was 96%. After the heat-treatment for 1 and 4h, decrease in the relative density was smaller than the weight loss. Slight shrinkage of 1~2% in length of these specimens was observed. It indicates that an amount of matter was transported from specimen to outside, but increment in porosity was suppressed by shrinkage of the specimen when the specimen was heated in a powder bed.

The optical microscope photographs of the etched surface of the specimens heat-treated at 1900°C for 0, 4 and 8h are shown in Fig. 1. Small equiaxed-grains of which size was less than 1μm were observed both on the surface and inside of the as-sintered specimens. After heat-treatment, it was observed that there were differences in microstructure between the surface and inside of the specimen. The surface consisted of smaller grains than inside. After heat-treatment for 4h, shape of grains at inside of the specimen became platelike and the grain size increased markedly(Fig.1d), whereas on the surface of the specimen, grains grew markedly but the shape was irregular (Fig. 1c). After the heat-treatment for 8h, the grains on the surface grew and became platelike, but their size remained still smaller than that at inside.

Fig.2 and 3 show results of quantitative analyses of the additives on the surface and inside of the specimen by EDX. The concentration of Al on the surface and inside of the as-sintered specimen were almost the same. After the heat-treatment for 4h, both the concentrations of Al reduced by half. Al was homogeneously distributed in the heat-treated specimens. From Fig.3, the concentration of Y on the surface and inside was almost the same in the as-sintered specimen. Inside the specimen heat-treated for 4h, the concentration of Y decreased from 2.7mol% to 1.8mol%, whereas Y concentration on the surface slightly increased. Then heterogeneous re-distribution of Y occurred in the heat-treated specimen.

The microstructural heterogeneity shown in Fig.1 is thought to be due to the difference in the concentration of Y between on the surface and inside of the specimen. Resemble heterogeneous structure in a liquid-phase sintered SiC and re-distribution of liquid phases near the surface were reported [1-4]. Lee et al. indicated that microstructure variation between surface and inside of a specimen might arise from heterogeneous distribution of oxide liquid phase [3]. The reason of segregation of Y on the surface aria is not clear. Misra reported that reaction of SiC and Al_2O_3 to produce volatile components from a sintered body started at 1800°C, whereas the reaction of SiC and Y_2O_3 hardly occurred even at 1820°C[4]. It seems that one of the driving forces for the migration of oxide liquid phase may be evaporation of the volatile components at the surface of the specimen. More experiments would be needed to clarify the mechanism of the migration of Al and Y.

4. CONCLUSION

Microstructure variation was observed on the surface and inside of the SiC ceramics which was sintered with additives of Al_2O_3 and Y_2O_3 and heat-treated at 1900°C for 4h in a powder bed. Grains on the surface were small and irregular, but those inside were large and platelike. The variation could be attributed to the concentration gradient of Y between surface and inside of the specimen. The amount of additives could be reduced by the heat-treatment with a little increase in porosity.

REFERENCES

1. M. Omori and H. Takei, J. Am. Ceram. Soc., 65 (1982) C92.
2. M. Omori, A. Sakuma and T. Hirai, “Ceramics Today-Tomorrow’s Ceramics”, Ed. by P. Vincenzini, Elsevier Science Publ. (1991) pp.1327-35.
3. J. K. Lee and H. Tanaka, J. Ceram. Soc. Japan, 103 (1995) 1193-96.
4. A. K. Misra, J. Am. Ceram. Soc.,74 (1991) 345-51.

Table 1. Weight loss and bulk density after heat-treatment with a powder bed at 1900°C

Heat-treatment time (h)	Weight loss (%)	Bulk density (g/cm^3)	Relative density (%)
0	-	3.15	96.0
1	5.1	3.10	94.5
4	7.5	3.08	93.9

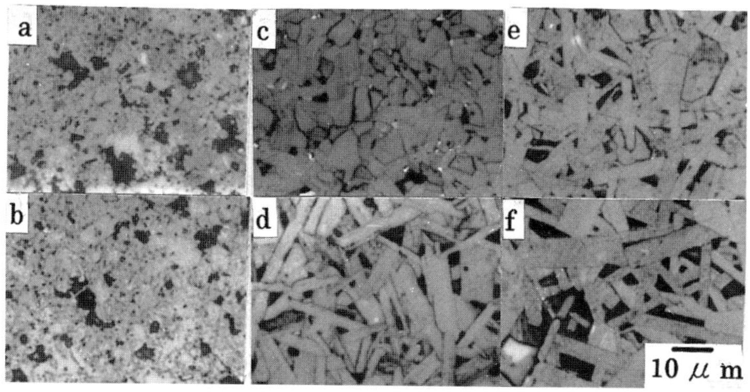

Fig.1 Optical micrographs of sintered and heat-treated specimens. Top 3 photographs show microstructures on the surface of the specimen, and bottom ones are those inside of the specimen. Heat-treatment times are 0, 4 and 8 h in (a)and (b), (c) and (d), (e)and (f), respectively.

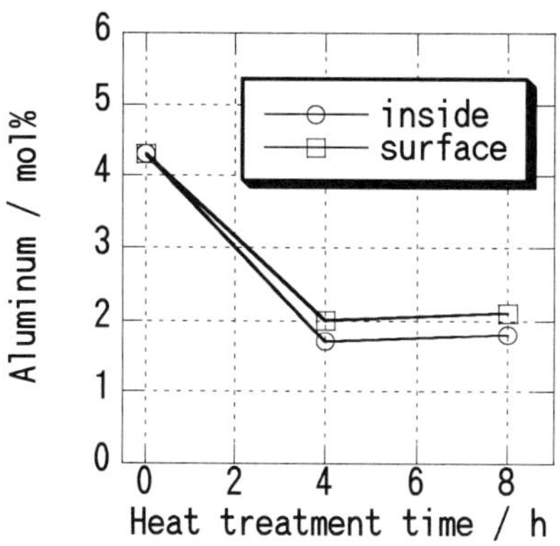

Fig. 2 Concentration of Al on the surface and inside of the specimen after the heat-treatment for 0, 4 and 8h.

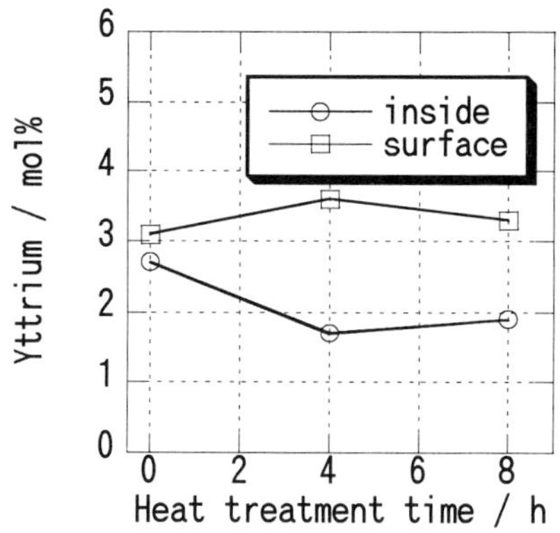

Fig. 3 Concentration of Y on the surface and inside of the specimen after the heat-treatment for 0, 4 and 8h.

Materials Science and Engineering Serving Society
S. Sōmiya, R.P.H. Chang, M. Doyama and R. Roy (Editors)

Initial Growth Process of Epitaxially Chemical-vapor-deposited Titanium Dioxide Crystallites

Hidetoshi SAITOH*, Hideki SUNAYAMA, Norio TANAKA and Shigeo OHSHIO
Nagaoka University of Technology
Kamitomioka, Nagaoka, Niigata 940-21, Japan

Abstract

Titanium dioxide films were chemically prepared on (001) single crystalline $SrTiO_3$ by a vapor deposition apparatus operated under atmospheric pressure with titanium tetra-isopropoxide. At a substrate temperature of 400°C, titanium dioxide grows epitaxially with a manner of lateral growth of the steps as well as multi-nucleation during initial growth process. Finally, multi-nucleation growth becomes dominant and promotes to grow a number of highly oriented crystallites having well developed facets. X-ray diffractmetry revealed that these crystallites consisted of <001>-oriented anatase type structure.

1. Introduction

In recent years, there has been considerable interest in rapid film formation techniques of metal oxides. This interest derives from the outstanding optical and electric properties of these films consisted of randomly oriented nano- or micro-crystallites. An air-opened reactor operated under an atmospheric pressure[1),2)] was designed for the rapid deposition of polycrystalline titanium dioxide films. Using this apparatus, not only rapid deposition rate of ~45 nm/s but also preferential orientation of the growth direction are achieved. At relatively low vaporizing temperature of titanium tetra-isopropoxide $Ti(O\text{-}i\text{-}C_3H_7)_4$ as a starting material, the films consisted of <001> preferentially grown columnar anatase crystals. In contrast, <100> and <211>-oriented crystallites grew at relatively high vaporizing temperature. These results provide a motivation to use a single crystalline substrate in obtaining epitaxial growth of anatase.

We have little understanding of the epitaxial growth processes which are involved in and which control metal oxide crystal nucleation and growth in chemical vapor deposition (CVD) processes. According to classical kinetic model[3)], precursors which enter a surface of the single crystalline substrate is immediately decomposed with heating and settled down at a certain site on terraces or lattice steps as shown in Fig. 1. Multi-nucleation growth on the terraces occurs as a first step towards polycrystalline growth. In contrast, lateral growth of the lattice steps develops single crystalline film.

In this study, titanium dioxide films were grown epitaxially on the single crystalline substrate by an air-opened reactor operated under an atmospheric pressure. The aim of the present work is to understand growth kinetics in promoting epitaxial growth of anatase. Initial growth process of titanium dioxide is described and discussed by atomic force microscopy.

*e-mail address: hts@nagaokaut.ac.jp

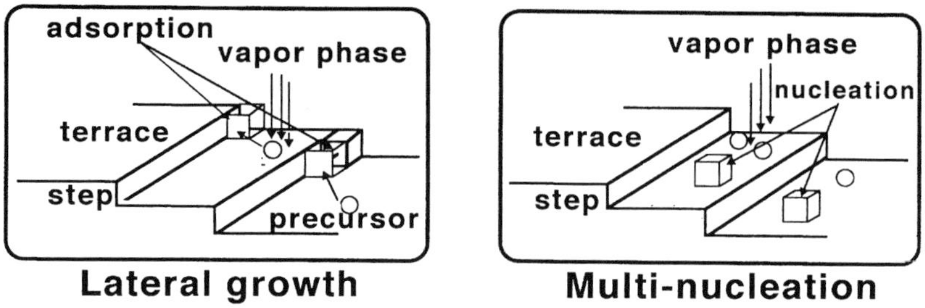

Fig. 1. Lateral growth and multi-nucleation mechanisms based on Kossel model.

2. Experimental

Titanium dioxide films were prepared by the atmospheric CVD apparatus, previously employed in obtaining polycrystalline anatase films, with titanium tetra-isopropoxide (Nakarai Tesque) as source complex. A single crystalline wafer of (001)$SrTiO_3$ were cut into a size of 5 x 5 mm^2 and used as the substrate materials. The lattice constant of $SrTiO_3$ and anatase are a_0= 0.3904 nm, and a_0= 0.378 nm and c_0= 0.951 nm. As a lattice mismatch is as low as 3.05 %, there is a extremely high possibility of that <001>-oriented anatase crystallites grow on (001) $SrTiO_3$ substrate. Prior to the deposition, lattice steps with a step hight of ~0.39 nm and a terrace width of ~150 nm were formed on the surface of the crystallites by etching in NH_4F-HF solution The reactant Ti(O-i-C_3H_7)$_4$ was loaded in a vaporizer and vaporized at 77 °C. The reactant vapor was carried by nitrogen gas with a gas flow rate of 1.5 dm^3/min and sprayed from the metalic nozzle to the substrate. The deposition duration was maintained between 5 s and 40 s for each experiment. The substrate was heated up to 400°C by an electric heater settled behind the substrate. The distance between the nozzle and the substrate was kept constant at 25 mm through the experiments. X-ray diffractmetry (using RAD No.2038, Rigaku Co.), was conducted to reveal the crystal structure and the growth orientation. The surface morphology of these films were observed by an atomic force microscope (AFM ; SPI-3700, Seiko Co.).

3. Results and discussion

3.1 Epitaxial growth of the films

Titanium dioxide films were obtained on $SrTiO_3$ substrate with a growth rate of approximately 1 nm/s. The growth rate is 100 times greater than those obtained by conventional CVD techniques operated at low pressure. Figure 2 shows the x-ray diffraction pattern of the sample. All these samples had anatase structure having the strongest relative intensity at 2θ of 37.8° corresponding to (004) diffraction peak, suggesting the preferential orientation toward the direction of <001> of anatase. As indicated in Fig. 3, x-ray diffraction using ϕ scan technique also revealed the epitaxial relation in plane as following,

$$TiO_2[100] // SrTiO_3[100] \text{ and } TiO_2[010] // SrTiO_3[010].$$

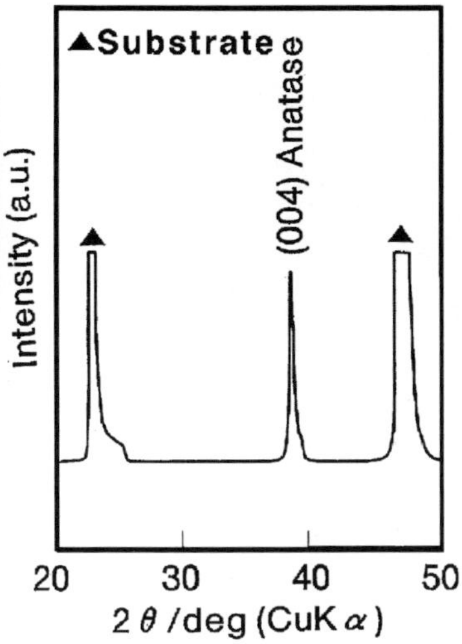

Fig.2. X-ray difraction pattern of the sample.

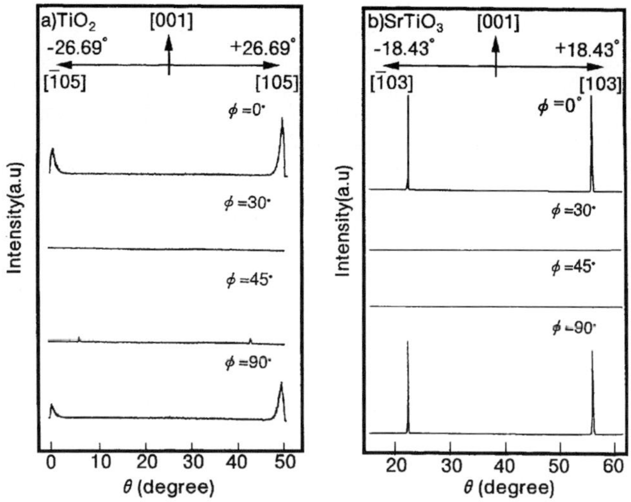

Fig. 3. Rocking curves from (a) film and (b) substrate. Sharp peaks were obtained at same ϕ, implying epitaxy relation in plane.

3.2 Initial growth process

Figure 4 shows AFM micrographs of the surface of the sample deposited for 1 s to 40 s. (a) Lattice steps with a step hight of ~0.39 nm and a terrace width of ~150 nm were seen on the surface of the $SrTiO_3$ crystallite by etching in NH_4F-HF solution. (b) At a substrate temperature of 400°C, titanium dioxide grew epitaxially with a manner of lateral growth of the steps as well as multi-nucleation growth. Note that wandering of the step front showing evidence of step instability induced by super saturation of precursors. (c) After 15 s, the step images were completely disappered. The surface morphology of <001>-oriented films indicated existence of a number of nano clusters and unclear crystalline facets. No evidence of the orientation growth of the crystal was obtained in the micrograph. (d) Finally, multi-nucleation growth becomes dominant in the growth process. The films showed that the crystalline columns having well developed facets aggregated densely. Two directions of the edges of the flat surface developed parallel to the surface of the substrate are parallel toward the directions [100] and [010] of the substrate, resulting that the edge directions are [100] and [010] of anatase. This strongly supports the results of x-ray diffractimetry showing the evidence of the epitaxial growth.

4. Conclusion

This investigation revealed the epitaxial growth and the initial growth process of anatase films deposited by the atmospheric pressure CVD apparatus. Titanium dioxide films grew epitaxially on $SrTiO_3$ substrate with a growth rate of 1 nm/s. In the initial growth process, both lateral step growth and multi-nucleation proceed at same time.

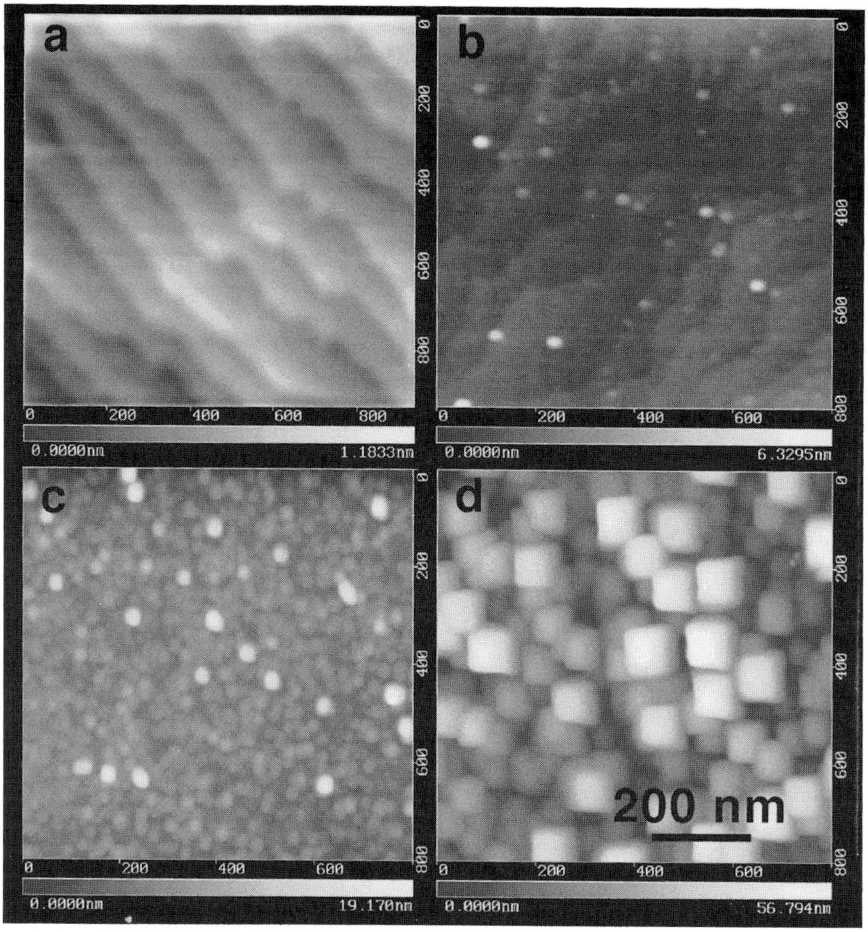

Fig. 4. Atomic force micrographs of the sample. (a) surface morphology of $SrTiO_3$ substrate. (b)-(d) titanium dioxide crystallites grown for 5 s, 15 s and 40 s.

Acknowledgment

This work was supported by a Grant-in-Aid of The Ministry of Education, Science, Sports and Culture in Japan under Contract No. 09750808.

References

1) K. Kamata, J. Nishino, S. Oshio, K. Maruyama and M. Ohtsuki, J. Am. Ceram. Soc., **75** (1992) 1167.

2) N. Tanaka, S. Ohshio and H. Saitoh, J. Ceram. Soc. Jpn., **105** (1997) 551.

3) "Hand book of crystal growth", edited by D.T.J. Hurle, Elsevier, pp.23-24.

Materials Science and Engineering Serving Society
S. Sōmiya, R.P.H. Chang, M. Doyama and R. Roy (Editors)
1998 Elsevier Science B.V.

Growth and characterization of $LiTaO_3$ single crystals by the floating zone method

Jeong Ho Ryu[a], Chang-Sung Lim[b] and Keun Ho Auh[a]

[a]Department of Ceramic Engineering, Hanyang University, Seoul 133-791, Korea

[b]Ceramic Processing Research Center, Hanyang University, Seoul 133-791, Korea

1. INTRODUCTION

$LiTaO_3$(LT) has been used in SAW device in early days[1], and the other applications are attractive in the fields of optical waveguides, electromodulators, infrared image sensors and infrared detectors by utilizing exellent piezoelectric, pyroelectric and electrooptic properties[2-4]. Because of a high nonlinear refractive index and a high optical damage resistance, LT is applied in a solid state laser and SHG(second harmonic generation) devices in these days[5,6].

To the present studies, LT single crystals with a large diameter were grown by the conventional czochralski method using Ir or Pt-Rh crucibles[7]. But the persistant problem was a fluctuation of the compositon in the growth due to a vaporization of Li_2O and diffusions of the crucible materials. In contrast to the czochralski method, the floating zone method has many advantages in crystal growing. The diffusion of the crucible materials could be avoided. The fluctuation of the Li_2O composition is smaller than that of the czochralski method. And high purity crystals could be obtained in various atmospheres.

In this work, we studied the growth characteristics of LT single crystals using the halogen floating zone method and characterizations of the grown crystals. Calcination and sintering parameters were established. Optimum crystal growth conditions were investigated by a controlling of growth rates, rotation speeds and atmospheres.

2. EXPERIMENTAL

LT polycrystalline powders of a congruent melting composition were synthesized using Li_2CO_3 and Ta_2O_5 powders with a purity of 99.99 %. The raw powders were ball-milled for 20 h and dried with a halogen lamp. The dried powders were calcined at temperatures between 750 and 950 ℃ for 10 h. LT feedrods with a circular column form were prepared in a silicon rubber mold using cold isostatic press with a 1800 Kg/cm^2, and then sintered at 1450 ℃ for 10 h.

LT single crystals were grown using a FZ SS-10W system using the feedrod with a diameter of 7mm. During the crystal growth, the optimum growth conditions were investigated by controlling the powers, growth rates, rotation speeds and atmospheres. After crystal growth, annealings were conducted at 1200 ℃ for 10 h in order to decrease a thermal stress and defects. Curie temperatures of the annealed crystals were examined in section of the part top, body and tail. Transmittance of the grown crystals was measured using a spectrophotometer.

3. RESULTS

In Fig. 1, a weight loss of CO_2 in calcined powders of Li_2CO_3 and Ta_2O_5 was indicated at temperatures between 720 ℃ and 800 ℃. It is assumed that the reaction between Li_2CO_3 and Ta_2O_5 proceeded completely at 800℃. In Fig. 2, peaks of the unreacted Li_2CO_3 and Ta_2O_5 coud be observed at 2θ = 28.32°, 28.9° and 36.7° on the X-ray diffraction patterns of the powders calcined at 750 ℃. At 850 ℃ the unreacted peaks disappered, while the peaks of $LiTaO_3$ was observed on the XRD patterns.

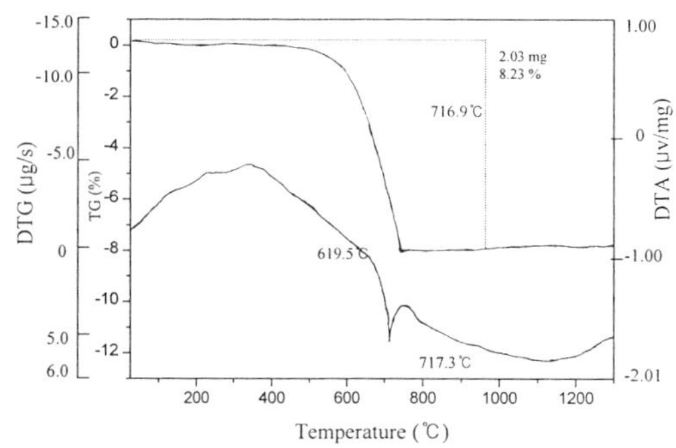

Figure 1. DT-TGA graph of the $LiTaO_3$ powders.

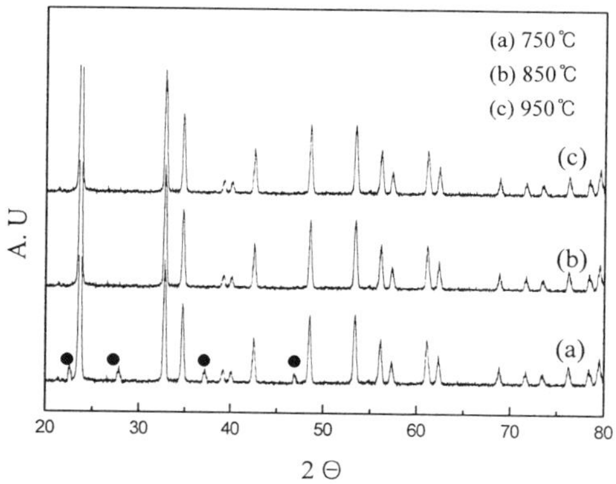

Figure 2. XRD patterns of the $LiTaO_3$ powders synthesized at temperatures of 750, 850 and 950℃.

For the growth, the feedrods were prepared as sintered at 1450 ℃ for 10 h. Fig. 3 shows a photograph of the typical LT single crystal. The grown crystals were about 20 mm in length and 8 mm in diameter. The color of the grown crystals was light brown. The crystal color was vanished after annealing at 1200 ℃ for 10 h in an air atmosphere. During crystal growth of LT, melting aspects of the LT feedrods in FZ system were very abnormal, because of a low thermal conductivity and a low infrared absorption coefficient of LT. Especially, low surface tension of the LT made easily melt flow to the direction of seed crystals. In case of $LiNbO_3$, the stability of the molten zone could be increased using Pt suspender[8]. However, in the case of LT, the stability of the molten zone is unstable because of a high melting temperature of LT. During the crystal growth, we manipulated to stabilize the molten zone by controlling the growing powers, growing rates, rotation speeds and atmospheres.

In growth of the LT crystals in air or N_2 atmosphere, stable conditions could not be found. Because the feedrod melted far above the normal melting point of LT, the melting. aspect was irregular and the shape of molten. zone was convex to the direction of feedrod. However in Ar atmosphere, the feedrod melted at a normal melting temperature of LT. The melting aspect in Ar was regular and more stable than in air or N_2 atmosphere because the shape of the molten zone was almost flat. The optimum conditions for the crystal growth were established as follows :

· Crystal growth rate : 3-5 mm/h
· Rotation speed : upper 16-20 rpm, lower 18-20 rpm
· Atmosphere gas : Ar
· Flow rate of atmophere gas : 1 L/min

To investigate the growth direction of the LT crystals, we analyzed Laue back reflection pattern. Fig. 4 shows that the growth orientation of the LT single crystal was identified to be a Y-axis. The LT crystals have a nonstoichiometric composition, and the curie temperature increases depending on the [Li]/[Ta] ratio[9]. The fluctuation of the Li_2O composition could be investigated by measuring curie temperatures in section of the part top, body and tail. Table 1 shows the Curie temperatures in the sections. The curie temperature fluctuation(ΔT_c) depending on the section of the part of the grown crystals ranges between 0.5 and 1.5 ℃. In Fig. 5 the transmittance of the grown crystal indicates between 75-80 % in visible and infrared wavelengths.

Figure 3. Laue back reflection pattern of grown the $LiTaO_3$ crystal grown in Y-axis orientation.

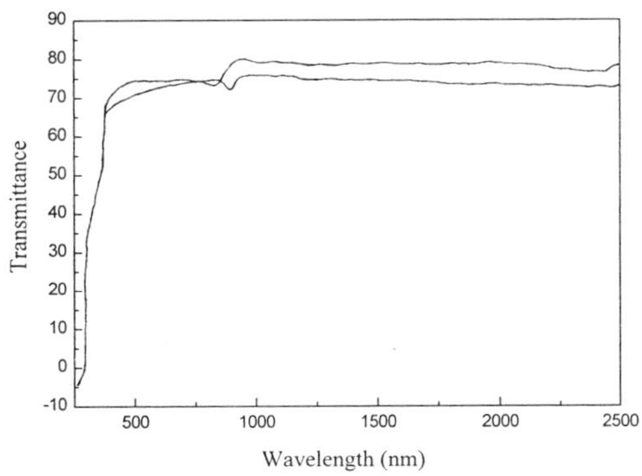

Figure 4. Transmittance of the $LiTaO_3$ wafers.

4. CONCLUSIONS

$LiTaO_3$ single crystals with a diameter of 8 mm and a length of 20 mm were grown by the floating zone method. The calcined temperature and time of the $LiTaO_3$ powders was established to be at 850 ℃ for 10 h. The optimum sintering parameter for the growth was at 1450 ℃ for 10 h. The optimum conditions for the crystal growth were 3-5 mm/h for the growth rate, 16-20 rpm for the upper rotation speed and 18-20 rpm for the lower rotation speed. The melting aspect in Ar was regular and more stable than in air or N_2 atmosphere because the shape of the molten zone was almost flat. The color of the grown crystal was light brown. The crystal color was vanished after annealing at 1200℃ for 10 h. Curie temperature fluctuation on the section of the parts of the grown crystals was within 1.5℃. The transmittance of the grown crystals indicates between 75-80 % in visible and infrared wavelengths.

Table 1
Curie temperatures in section of the part top, body, and tail of the grown crystals

Sample	Top(℃)	Body(℃)	Tail(℃)	⊿Tc(℃)
1	612	613	611	1.5
2	611	612	611	1
3	612	614	613	1.5
4	609	610	610	0.5

ACKNOWLEGEMENT:
This research was financially supported by the Ceramic Processing Research Center(CPRC) at Hanyang University, Seoul, Korea.

REFERENCE

[1] T. Fukuda, S. Matsumura, H. Hirano, and T. Ito, J of Crystal Growth, 46 (1979) 179.
[2] J. L. Jackel, Appl. Opt. 19 (1980) 1996.
[3] M. Okuyama, Y. Togami, J. Ohnishi and Y. Hamakawa, Ferroelectrics, 91(1989) 127.
[4] R. T. Smith, Appl. Phys. Lett. 11 (1967) 146.
[5] S. Thaniyavarn, T. Fundakly, D. Booher and J. Moe, ibid. 46(1985) 933.
[6] K. Mizuuchi and K. Yamamoto, ibid. 60 (1992) 1283.
[7] S. Matsumura, J. of Crystal Growth, 51 (1981) 41.
[8] K. Kitamura, S. Sawada, S. Kimura and M. Gobbels, 9th Intern, Conf, on Crystal Growth (ICCG-9), Sendai, 1989.
[9] R. L. Barns and J. R. Carruthers, J. Appl. Crystallogr. 3 (1970) 395.

Materials Science and Engineering Serving Society
S. Sōmiya, R.P.H. Chang, M. Doyama and R. Roy (Editors)

Preparation of Doped TiO_2 by Spray Reaction Method and Their Photocatalytic Behavior

N. Ichikuni, K. Ohtsuka, S. Shimazu and T. Uematsu

Department of Applied Chemistry, Faculty of Engineering, Chiba University, Inage-ku, Chiba 263, Japan

Doped TiO_2 photocatalysts were prepared by spray reaction method. The photocatalytic isomerization of *cis*-2-butene to 1-butene at 323 K has been employed to study a photocatalytic properties. The applicability of spray reaction method to the preparation of TiO_2 and doped-TiO_2 photocatalysts were proposed.

1. INTRODUCTION

Titania is well known to have remarkable high photocatalytic activity. Many efforts were devoted to enhance the activity of titania photocatalysts. There are two main ideas to improve the photocatalytic activity; the introducing dopant to titania [1-5] and the minimizing the catalyst particle size to cause quantum size effect. Fine particles prepared by spray reaction (spr) method show the following features; the easiness of preparation, the relatively strong synergetic effects between the components and the controllability of the particle size and the crystallite size [6]. Furthermore, spr proceeds very rapidly with quick heating and quenching rates in a flow system. Thus, the sprayed particles might be in metastable states. Because photocatalytic process contains the activation of the surface site, sprayed catalysts can exhibit more active photocatalysis than conventional catalysts. In the present paper, the applicability of spray reaction method for the preparation of doped TiO_2 and the characteristics in the photocatalysis were discussed.

2. EXPERIMENTAL

The 13-group element doped TiO_2 photocatalysts were prepared by spray reaction method as follows. The hydrochloric acid solution of $TiCl_4$ (0.5 $mol \cdot l^{-1}$) and the nitrate of dopant were ultrasonicated to produce the mist. The molar ratio of the dopant (M_2O_3) to TiO_2 was 5. The mist was aspirated through the quartz tube situated in the electric furnace to the glass filter under an ambient condition. The mist of the mixed solution was rapidly heated, dried, decomposed to the oxide and calcined in the furnace for less than a second, followed by quenching at glass filter. The spray reaction temperature (T_{spr}) was varied from 673 K to 1073 K. Sprayed catalysts have been characterized by X-ray diffraction, BET surface area and diffuse reflectance spectroscopy. The photocatalytic isomerization of *cis*-2-butene to 1-butene at 323 K was performed in a Pyrex glass closed circulating system. Irradiation was carried out by using 75 W-Hg lamp. The reactant and the products (1-butene and *trans*-2-butene) were separated by sebaconitrile and detected by FID.

3. RESULTS AND DISCUSSION

3.1. UNDOPED SPR TiO_2

The control of the crystallite size and the crystal phase of undoped TiO_2 have been achieved by varying the T_{spr}. XRD analysis revealed that the spr TiO_2 showed the only anatase phase in the temperature range of 673 K $\leq T_{spr} \leq$ 973 K. As T_{spr} raised to 1073 K, growth of the rutile peaks was observed in the XRD pattern. The crystallite size of TiO_2 is varied from 10 to 24 nm as T_{spr} changed from 673 K to 1073 K. The BET surface areas of these catalysts ranged from 98 m^2g^{-1} for T_{spr} = 673 K down to 14 m^2g^{-1} for T_{spr} = 1073 K. These physical properties are listed in Table 1.

Table 1 Crystallite sizes and surface areas for spr TiO_2 catalysts

T_{spr} / K	Crystallite size / nm	BET / m^2g^{-1}
673	10	98
773	10	29
873	11	21
973	18	15
1073	24	14

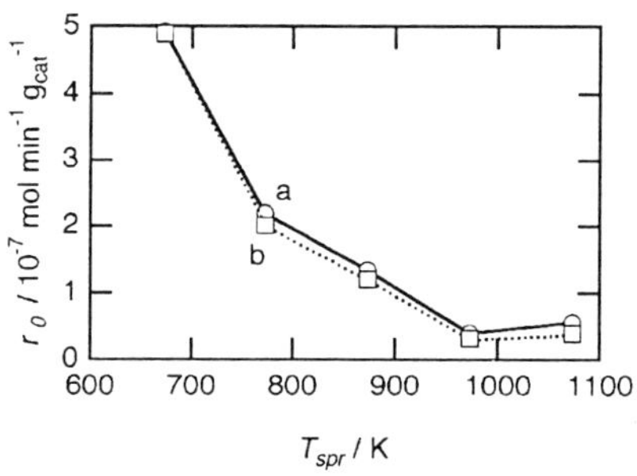

Figure 1. Rates of photoisomerization of *cis*-2-butene on spr TiO_2 at 323 K; (a) 1-butene, (b) *trans*-2-butene.

Figure 1 shows the photoisomerization results on spr TiO_2 catalysts. Since the thermal process of isomerization also proceeded at 323 K, photocatalytic activity was defined as the difference between the initial rate under irradiated and the initial rate under dark reaction. The activity was lowered as increasing T_{spr}. This phenomena can be explained by the reduction of the surface area and the growth of rutile phase, as increasing T_{spr}.

3.2. DOPED SPR TiO_2

It is said that the photoisomerization of butene on TiO_2 proceeded from the hole produced by irradiation [7]. The improvement of photocatalysis was expected by doping 3^+ element to Ti^{4+}, thus, the group 13 elements doped TiO_2 were prepared by spr. In contrast to undoped spr TiO_2, the opposite tendency was observed on Al doped TiO_2 catalysts as shown in Fig. 2. The higher the T_{spr} was raised, the higher the photocatalytic activity on spr Al_2O_3/TiO_2 catalyst became.

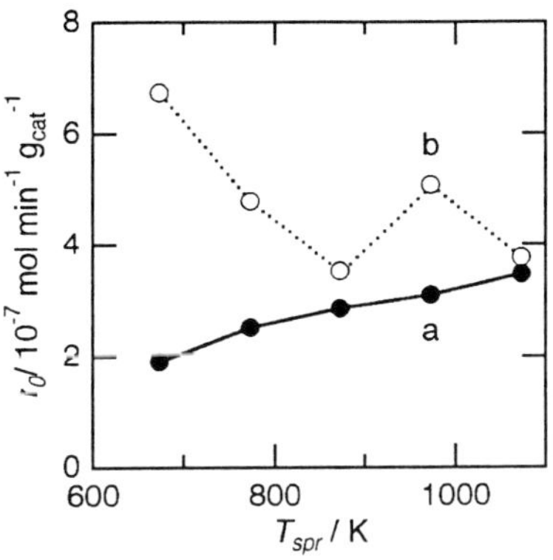

Figure 2. Rates of photoisomerization of *cis*-2-butene on spr Al_2O_3/TiO_2 at 323 K; (a) 1-butene, (b) *trans*-2-butene.

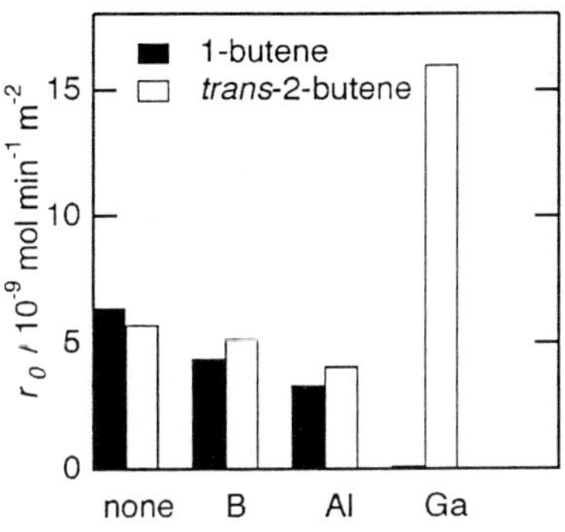

Figure 3. Rates of photoisomerization of *cis*-2-butene on spr M_2O_3/TiO_2 at 323 K, T_{spr} = 873 K.

XRD pattern for group 13 elements doped TiO_2 catalysts showed that only anatase TiO_2 peak was observed in the range from 673 K to 1073 K except for In doped catalyst. It can be said that by using spr method, the dopant is highly dispersed in the TiO_2 particle. The photocatalytic activity on spr M_2O_3/TiO_2 (T_{spr} = 873 K) is shown in Fig. 3. B_2O_3/TiO_2 and Al_2O_3/TiO_2 showed less activity than undoped TiO_2. In these cases, the BET surface areas were 2 to 4 times larger than those for undoped catalysts (67 m^2g^{-1} for B_2O_3/TiO_2, 87 m^2g^{-1} for Al_2O_3/TiO_2, 67 m^2g^{-1} for Ga_2O_3/TiO_2 and 36 m^2g^{-1} for In_2O_3/TiO_2). Increase in BET surface area was caused mainly by the dopants. Therefore, it is probable that the catalysts surface is partially occupied by B_2O_3 or Al_2O_3, and that effective TiO_2 surface area is smaller than the original one. As increasing T_{spr}, the doped catalysts showed higher activity than the undoped did. It is to be noted that the pronounced doping effect on spr M_2O_3/TiO_2 catalyst was realized by increasing T_{spr}. The Ga doped catalyst showed no photo-activity for 1-butene as shown in Fig. 3. In this case, though the rate under irradiation is faster than on undoped catalyst, the dark reaction is much faster than the irradiated reaction. These facts suggest that doping of Ga into TiO_2 causes strengthening of surface OH acidity and thus increases mainly the catalytic activity for thermal catalysis than for photocatalysis.

4. CONCLUSION

TiO_2 catalyst and group 13 elements doped TiO_2 catalysts were prepared applying spray reaction method. Undoped catalysts showed the tendency that the photo-activity was lowered as increasing T_{spr}. By doping B or Al into TiO_2 catalysts could improve the photocatalytic activity. On the other hand, doping of Ga into TiO_2 catalyst enhanced thermal activity as well as photo-activity.

REFERENCES

1. S. Kodama, H. Nakaya, M. Anpo and Y. Kubokawa, Bull. Chem. Soc. Jpn., 58 (1985) 3645.
2. M. Anpo, T. Kawamura, S. Kodama, K. Maruya and T. Onishi, J. Phys. Chem., 92 (1988) 438.
3. H. Inoue, T. Matsuyama, B. Liu, T. Sakata, H. Mori and H. Yoneyama, Chem. Lett., (1994) 653.
4. K. E. Karakitsou and X. E. Verykios, J. Phys. Chem., 97 (1993) 1184.
5. S. Sato, Shokubai, 33 (1991) 209.
6. T. Uematsu and S. Shimazu, Shokubai, 36 (1994) 252.
7. M. Anpo, Res. Chem. Intermed., 11 (1989) 67.

Materials Science and Engineering Serving Society
S. Sōmiya, R.P.H. Chang, M. Doyama and R. Roy (Editors)

Preparation of Multi-Component Fine Particles by Spray Reaction Method; Their Characterization and Catalytic Properties

N. Ichikuni, D. Murata, K. Terauchi, S. Shimazu and T. Uematsu

Department of Applied Chemistry, Faculty of Engineering, Chiba University, Inage-ku, Chiba 263, Japan

Spray reaction (spr) method was applied to prepare Co/Al_2O_3 catalysts and Ni/Al_2O_3 catalysts and clarified synergetic effects between metal and mixed-oxide, and between metal oxide and mixed-oxide playing important roles for catalysis; NO-CO reaction on spr Co/Al_2O_3 and CO hydrogenation on spr Ni/Al_2O_3. The various applicability of the catalyst design by using spr was discussed.

1. INTRODUCTION

Several techniques have been developed and applied to the preparation of the multi-component catalyst possessing high synergetic effects. Among these methods coprecipitation and hydrolysis of a mixed solution of the components were extensively applied to provide homogeneous dispersion of the active site on the surface. Though spray reaction (spr) method can be expected as one of the excellent methods to prepare multi-component fine particles, there were little examples of the application to catalysts preparation [1-3]. Spr proceeds in rapid heating rate and quenching in a flow system, and thus spr particles might be obtained in metastable states and leading to the active surface structures. By choosing suitable precursors, oxidation state of composites is easily controlled in various catalytic systems. We have applied this method to the preparation of several type of catalysts; oxide supported oxide catalysts, oxide supported metal catalysts, multi-component metal oxide catalysts, and so on, to find their characteristic physical and chemical properties.

In this paper, we introduce two types of catalysts; oxide supported oxide catalysts and oxide supported metal catalysts to find their characteristic physical and chemical properties. The former is exemplified by Co/Al_2O_3 catalyst and the latter by Ni/Al_2O_3 catalyst.

2. EXPERIMENTAL

Spr Co/Al_2O_3 catalyst and spr Ni/Al_2O_3 catalyst were prepared from aqueous solutions of the corresponding metal nitrates. Spr was carried out as follows; the aqueous solution containing each components was ultrasonicated to produce the mist containing Co or Ni and Al species. The mist was aspirated through the quartz tube situated in the electric furnace to the glass filter under ambient condition. The reacting droplets were dried, decomposed to the oxide and calcined in the furnace for less than a second. Spray reaction temperature (*Tspr*) was situated to 773 K for Co/Al_2O_3 catalyst and 973 K for Ni/Al_2O_3 catalyst unless otherwise stated. Conventional impregnation catalysts (imp) were also prepared for comparison.

The characterization of the catalysts were carried out by means of SEM, TEM, XRD, XPS, temperature programmed reduction (TPR) and adsorption measurements.

Catalytic activities were tested by using NO-CO reaction on Co/Al_2O_3 catalyst and CO hydrogenation on Ni/Al_2O_3 catalyst.

3. RESULTS AND DISCUSSION

3. 1. SUPPORTED OXIDE CATALYST

Spr Co/Al_2O_3 catalyst was prepared as a example of the oxide supported oxide catalysts. Molar ratio of Co to (Co+Al) was varied from 30 to 100. The notation of the catalyst was represented as spr 30, spr 100, respectively. Hence, cobalt oxide and cobalt aluminate were hardly able to distinguish from powder XRD peak analysis, TPR profiles were carried out to qualify the catalyst composition. Both XRD and TPR analysis revealed that as-sprayed catalysts consisted of cobalt oxide, cobalt aluminate and aluminum oxide.

Surface areas of catalysts are listed in Table 1. As increasing cobalt ratio, the BET surface areas decreased from 220 m^2g^{-1} down to 67 m^2g^{-1}. The catalysts were applied to NO-CO reaction in oxides form. Sprayed catalysts showed higher activity for NO-CO reaction than the conventional impregnation catalysts.

Table 1
BET surface areas and initial rates of NO-CO reaction

catalyst	BET / m^2g^{-1}	r_0* / 10^{-5} mol·min^{-1}·g_{Co}^{-1}		
		N_2O	N_2**	CO_2
spr 30	220	14	0.0	13
spr 50	180	40	0.3	42
spr 70	82	33	1.5	40
spr 80	76	65	55	210
spr 90	67	76	32	260
spr 100	67	57	23	190
imp 30	91	5.1	0.0	4.4

* N_2 was produced in the secondary reaction represented in following steps;

$$2NO + CO \rightarrow N_2O + CO_2,\ N_2O + CO \rightarrow N_2 + CO_2.$$

*P_{NO}=P_{CO}=3.3 kPa, T=423 K.

As increasing cobalt molar ratio above 50%, the activity dramatically increased while the surface area decreased. These high cobalt containing catalysts included cobalt aluminate phases, $CoAl_2O_4$ and/or Co_2AlO_4. The catalytic activities of these cobalt aluminates were investigated and revealed that Co_2AlO_4 possessed high activity same as spr 30 though $CoAl_2O_4$ showed less activity. The highest activity was achieved at the Co molar ratio of 80. Spr 80 catalyst does not possess $CoAl_2O_4$ but Co_2AlO_4, which was detected from TPR and XPS measurements. Assuming that the binary oxide species in spr 80 was Co_2AlO_4, the catalysts consisted of Co_3O_4 and Co_2AlO_4 in the ratio of 1:1. Thus, the combined structure from cobalt oxide and cobalt aluminate constructs the main active site for NO-CO reaction. Conventional impregnation Co/Al_2O_3 catalysts showed less activity than sprayed catalysts. This result means that the pronounced synergetic effects were induced by applying spr method to Co/Al_2O_3 catalysts.

3. 2. SUPPORTED METAL CATALYST

As for the case of oxide supported metal catalyst, we prepared Ni/Al_2O_3 catalysts by spr method. In this case, the catalysts were reduced before use for CO hydrogenation to produce nickel metal. The as-sprayed catalysts consisted of oxides such as NiO, Al_2O_3 and $NiAl_2O_4$. However, multi-component oxides such as $NiAl_2O_4$ are hardly reduced [4], and hence have less activity toward hydrogenation reaction. They seem to be able to provide a stabilizing effect for supported Ni components [5]. Catalysts are denoted as spr 15 where the Ni molar ratio to (Ni+Al) was 15. The composition of the ternary oxide mixture can be controlled by changing the Ni : Al ratio. As increasing the Ni ratio, the NiO phase being dominant in the particle as shown in Fig. 1.

From XRD analysis and TPR measurement, it was suggested that NiO existed both on alumina and the aluminate. The NiO formed on the alumina was easily reduced with hydrogen at around 620 K as shown in Fig. 2. On the other hand, the higher temperature at around 870 K was required to reduce the NiO supported on the aluminate. It can be said that the NiO contacting with only the alumina is less stabilized than the NiO contacting with the aluminate.

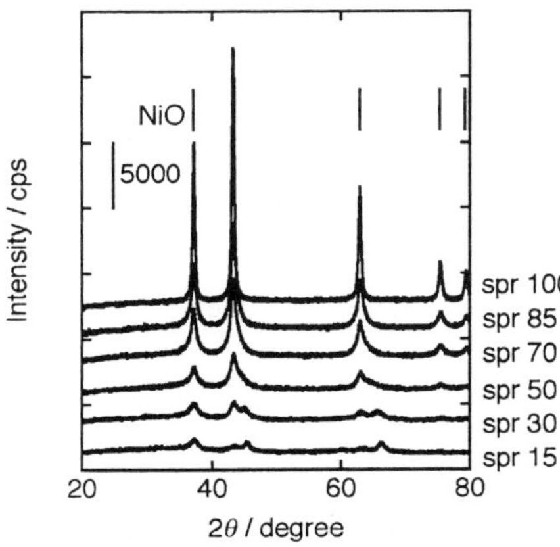

Figure 1. XRD patterns for as-sprayed Ni/Al_2O_3 catalysts.

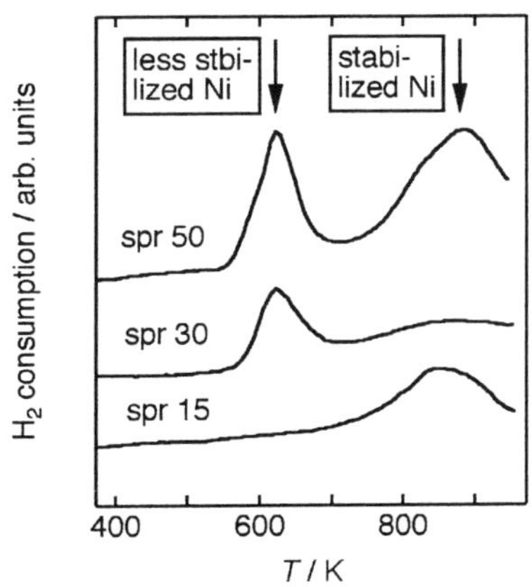

Figure 2. TPR profiles for spr Ni/Al_2O_3 catalysts.

Table 2
Activities of CO hydrogenation and crystallite sizes

catalyst	Ni crystallite size / nm	r_0 */ 10^{-5} mol·min^{-1}·g^{-1}	TOF** / min^{-1}
spr 15	8.2	0.66	1.0
spr 30	7.7	4.7	1.6
spr 50	16	5.8	1.4
spr 70	14	15	2.5
spr 85	17	170	11
spr 100	31	0.75	0.12

*P_{CO}=6.7 kPa, P_{H2}=20 kPa, T=513 K.
**initial rates divided by the amount of adsorbed hydrogen.

Catalysts listed in table 2 treated with hydrogen were composed of Ni metal, alumina and the aluminate determined from XRD analysis. The crystallite size of Ni metal and the activity are listed on Table 2. The obtained catalysts are classified into two groups in terms of the crystallite size; one group is about 8 nm and the other is about 15 nm for the catalysts with the Ni ratio below or above the aluminate stoichiometry, respectively.

Though the crystallite size increased drastically between spr 30 and spr 50, their TOFs for CO hydrogenation were in a similar level indicating that the crystallite size did not affect the activity. Considering the dramatically high activity of spr 85, it is suggested that the Ni site contacted with both alumina and the aluminate takes a significant role for the catalysis. In other words, the binary oxide support of aluminate enhances the catalysis in spr Ni/Al_2O_3 catalysis.

4. CONCLUSION

Several types of multi-component catalysts prepared by using spr method were investigated. The pronounced synergetic effect was achieved in spr catalysts. In this sense, the spr method is one of the most important technique to design the active catalysts.

REFERENCES

1. S. Shimazu, T. Uematsu, S. Matsuda and Y. Kuwasawa, Modern. Phys. Lett., B2 (1988) 501.
2. T. Uematsu, M. Fukuda, T. Iijima and S. Shimazu, Proc. 3rd Intern. Congr. Spillover, (1993) 397.
3. T. Uematsu and S. Shimazu, Shokubai, 36 (1994) 252.
4. J. M. Rynkowski, T. Paryjczak, M. Lenik, Appl. Catal. A: General, 126 (1995) 257.
5. J. A. Peña, J. Herguido, C. Guimon, A. Monzón and J. Santamaría, J. Catal., 159 (1996) 313.

Materials Science and Engineering Serving Society
S. Sōmiya, R.P.H. Chang, M. Doyama and R. Roy (Editors)

Electronic properties in $La_{n-nx}Sr_{1+nx}Mn_nO_{3n+1}$ (x=0.4: n=1, 2, ∞) with low dimensional crystal structure

Minoru Takemoto, Takeshi Ogawa and Hiroyuki Ikawa

Department of Applied Chemistry, Faculty of Engineering, Kanagawa Institute of Technology, Shimo-Ogino 1030, Atsugi-shi, Kanagawa 243-02, Japan

The Ruddlesden-Popper type manganates of $La_{n-nx}Sr_{1+nx}Mn_nO_{3n+1}$($x$=0.4: n=1, 2, ∞) have been prepared. Their electronic properties depends on their dimension of crystal structure. The n=1 manganate with two dimensional structure shows semiconducting and paramagnetic behavior below room temperature. The n=2 manganate with pseudo-two dimensional structure shows the electric transition from semiconductor to metal at 127 K when reducing temperature. The n=∞ manganate with three dimensional structure shows metallic behavior below room temperature. Ferromagnetism and giant magnetoresistance are found when the metallic conduction occurred, *i.e.* in the n=2 manganate below the electric transition temperature and in the n=∞ manganate below temperature.

1. INTRODUCTION

Many structures of cuprate superconductor have been found, since the first cuprate superconductor $La_{2-x}Ba_xCuO_4$ was discovered[1]. Two dimensional structure is considered to be required for high temperature superconductivity in the cuprates, since there are two dimensional CuO_2 layers in most cuprate superconductors. Some of them were based on Ruddlesden-Popper (RP) type structure[2], which has the general composition of $(AO)(ABO_3)_n$ and consists of stacking of rock salt layers AO and n layers of perovskite blocks $(BO_6)_n$. The above-mentioned cuprate superconductor, $La_{2-x}Ba_xCuO_4$ has the n=1 RP structure A_2BO_4. The $La_2SrCu_2O_6$ type cuprate superconductor has the structure based on the n=2 RP structure $A_3B_2O_7$[3-5].

The 3d transition metal oxides except cuprate, on the other hand, are semiconductive, when they have two dimensional structure such as the n=1 RP structure; $Sr_{1+x}La_{1-x}CrO_4$[6], $La_{1-x}Sr_{1+x}FeO_4$[7] and $La_{2-x}Sr_xCoO_4$[8]. They often have metallic conduction when they have three dimensional structure such as perovskite structure ; $SrCrO_3$[9], $SrFeO_3$[10], $La_{1-x}Sr_xCoO_3$[11]. The RP structure has higher dimension with increasing n and the perovskite structure corresponds to the RP structure with n=∞. A series of strontium vanadates $Sr_{n+1}V_nO_{3n+1}$ (n=1, 2, 3, ∞) has the RP type structure[12]. The n=1 vanadate Sr_2VO_4 which has two dimensional structure is an insulator. The n=2 vanadate $Sr_3V_2O_7$ and the n=3 vanadate $Sr_4V_3O_{10}$, with pseudo-two dimensional structure and the n=∞ vanadate $SrVO_3$ with three dimensional structure have metallic conduction. The dimension of crystal structure, therefore, seems to be a significant factor to control the electronic properties in the 3d transition metal oxides.

In the present study, manganates were prepared with the RP structure $(SrO)[La_{1-x}Sr_xMnO_3]_n$, that is, $La_{n-nx}Sr_{1+nx}Mn_nO_{3n+1}$ (n=1, 2, ∞) with x=0.4. The electric and magnetic properties were measured to study the effect of dimension of their crystal structure on the electronic properties.

2. EXPERIMENTAL

The samples of $La_{n-nx}Sr_{1+nx}Mn_nO_{3n+1}$ (x=0.4: n=1, 2, ∞) were prepared from the starting materials, La_2O_3, $SrCO_3$ and Mn_3O_4 by a solid state reaction method. The materials were mixed thoroughly and calcined for 12 h at 1350 - 1400°C in air atmosphere for the n=1 and 2 samples or in O_2 gas for the n=∞ sample. The calcined samples were pulverized well and were isostatically pressed into pellets under pressure of 1 GPa. The pellets were sintered for 24 h at 1350 - 1500°C in N_2 gas for the n=1 and 2 samples or in O_2 gas for the n=∞ sample.

Crystalline phases were identified and lattice constants were measured using X-ray powder diffractmeter (RINT2500VHF, Rigaku Co., Ltd., Japan). Oxygen content of the samples was determined by a iodometric titration method. Electric conductivity was measured below room temperature under magnetic field up to 15 kOe by a conventional dc four probe method. Magnetic properties were measured using a vibrating sample magnetometer (BHV-55LHCS, Riken Denshi Co., Ltd., Japan).

3.RESULTS AND DISCUSSION

The oxygen contents measured are listed in Table 1. All samples have smaller oxygen contents than the stoichiometric ones. Average valence of manganese ion calculated from the oxygen contents are also listed in Table 1. All samples have the same average valence of manganese ion, +3.37.

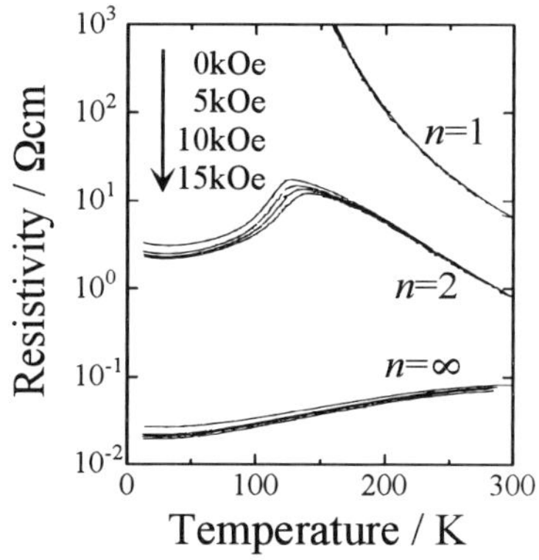

Figure 1 Temperature and magnetic field dependence of electric resistivity.

Figure 1 shows temperature dependence of electric resistivity for the n=1, 2 and ∞ samples under applied magnetic field up to 15 kOe. The n=1 sample has semiconductive behavior and the n=∞ sample has metallic one below room temperature. The n=2 sample is semiconductive near room temperature and shows a transition from semiconductor to metal at 127 K. The transition temperature agreed well with the temperature of 126 K derived by Moritomo *et al*[13]. The electric resistivity decreases with increasing magnetic field in the n=∞ sample below room temperature. In the n=2 sample, the electric resistivity decreases remarkably below and decreases slightly above the transition temperature. On the other hand, negligible decrease in the electric resistivity is observed in the n=1 sample below room temperature under the magnetic field up to 15 kOe. The magnetoresistance MR is defined here as MR=$(\rho_H - \rho_0)/\rho_0$, where ρ_H and ρ_0 are the electric resistivity under magnetic field of H and under no magnetic field, respectively. The values of MR are listed in table 2. The MR of the n=2 sample is two and half times as much as that of the

Table 1
Oxygen content, y and average valence of manganese ion, p in the samples of $La_{n-nx}Sr_{1+nx}Mn_nO_y$ (x=0.4).

n	composition	y (stoichiometric)	y (experimantal)	p in Mn^{p+}
1	$La_{0.6}Sr_{1.4}MnO_y$	4	3.99	3.37
2	$La_{1.2}Sr_{1.8}Mn_2O_y$	7	6.97	3.37
∞	$La_{0.6}Sr_{0.4}MnO_y$	3	2.98	3.36

n=∞ sample at 120 K which is just below the transition temperature of the former sample.

Figure 2 indicates logarithmic electric resistivity as functions of inverse temperature for the semiconductive samples (n=1 and 2). A linear plot is observed in the n=1 sample and a curve is observed in the n=2 sample. The linear plot indicates that the electric conduction process is explained as the thermally activated process in the n=1 sample The activation energy of the n=1 sample is estimated to be 0.15 eV from the slope of the linear plot in the figure. The electric conduction has been comfirmed to be semiconductive both parallel and perpendicular to single perovskite layer using single crystals of $La_{1-x}Sr_{1+x}MnO_4$ [14]. However, anisotropic electric resistivity has been measured parallel and perpendicular to the double perovskite layer using the n=2 single crystals with composition of x=0.3 and 0.4. In the x=0.4 crystal, both electric resistivity parallel and perpendicular to the conducting perovskite layer have been reported to be semiconductive[13]. In the x=0.3 crystal, on the other hand, the electric resistivity parallel and perpendicular to the conducting layer has been reported to be metallic and semiconductive, respectively[15]. From these reports, the temperature dependence of electric resistivity should be simple semiconducting behavior for ceramic sample with composition of x=0.4 and should be complicate for ceramic sample with composition of x=0.3. The curved plot in the fig. 2 for the present n=2 ceramic sample is due to the slight reduction of valence of manganese ions from the stoichiometric valence +3.4, as shown in Table 1. The ceramic sample has metallic conduction parallel to the conducting layer so that show the slight MR when it is semiconductive above the transition temperature.

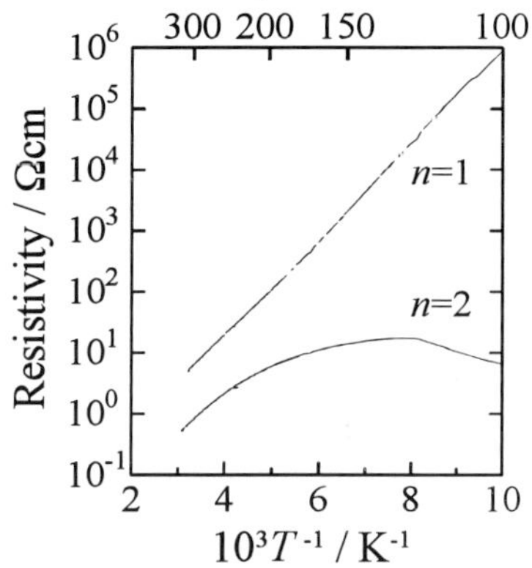

Figure 2 Logarithmic electric resistivity as functions of inverse temperature for the n=1 and 2 samples.

Figure 3 indicates the magnetization curves measured at 10 K and 200 K. Hysterisis loops are observed at 10 K and 200 K for the n=∞ sample so that the sample is confirmed to be a ferromagnet. The semiconducting n=1 sample showed no apparent hysterisis at 10 K and 200 K. The n=2 sample only has the hysterisis loop at 10 K. Therefore, the sample is also a ferromagnet at low

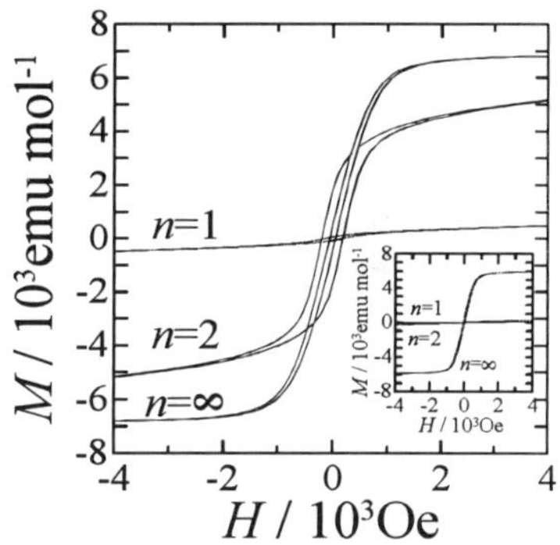

Figure 3 Magnetization at 10 K and 200 K (the inset).

Table 2
Magnetoresistance MR under magnetic field of 15 kOe for the samples of $La_{n-nx}Sr_{1+nx}Mn_nO_{3n+1}$ (x=0.4) at various temperatures.

sample (n)	MR / %			
	60 K	120 K	180 K	240 K
1	-*)	-2	-1	-1
2	-28	-51	-14	-4
∞	-24	-20	-16	-14

*) Not measured.

temperature. The n=2 sample shows a magnetic transition which was detected as an abrupt increase in magnetization at 120 K when reducing temperature. The magnetic transition therefore corresponds to the transition from paramagnetism to ferromagnetism. The magnetic transition temperature of 120 K is close to the electric transition temperature of 127 K. It is plausible that the magnetic transition temperature is coincide well with the electric transition temperature in the n=2 sample, since the coexistence of ferromagnetic and metallic properties is explained by the double exchange interaction[16].

4. SUMMARY

The manganates $La_{n-nx}Sr_{1+nx}Mn_nO_{3n+1}$($x$=0.4, n=1, 2, ∞) with perovskite related structure, have various electronic properties with changing dimension of crystal structure. The electric conduction was semiconductive when n=1 and was metallic when n=∞ below room temperature. The magnetic properties were paramagnetic when n=1 and were ferromagnetic when n=∞ below room temperature. The intermediate sample with n=2 had the intermediate properties between the n=1 and ∞ sample. It had semiconductive conduction and paramagnetism near room temperature. It showed the electric and magnetic transition simultaneously and it was metallic and ferromagnetic below the transition temperature. The samples with metallic conduction, *i.e.* the n=2 sample under the transition temperature and the n=∞ sample showed giant magnetoresistance (GMR) and the former sample had much larger GMR than the latter sample had in the vicinity of the transition temperature of the former sample.

Acknowledgment: This work was supported by the Science Research Promotion Fund by the Japan Private School Foundation and the Kanagawa Academy of Science and Technology Research Grants (No. 9971129).

REFERENCES

[1]J.G.Bednorz and K.A.Müller, Z.Phys. B64, (1986) 189.
[2]S.N.Ruddlesden and P.Popper, Acta Crystallogr. 10, (1957) 538.
[3]N.Nguyen, L.Er-Rakho, C.Michel, J.Choisnet and B.Raveau, Mater. Res. Bull. 15, (1980) 891.
[4]R.J.Cava, B.Batlogg, R.B.van Dover, J.J.Krajewski, J.V.Waszczak, R.M.Fleming, W.F.Peck Jr, L.W.Rupp Jr, P.Marsh, A.C.W.P.James and L.F.Schneemeyer, Nature 345, (1990) 602.
[5]M.Takemoto, N.Ohashi, T.Tsurumi, O.Fukunaga and J.Tanaka, Physica C 268, (1996) 279.
[6]T.Omata, S.Hoshino, H.Ikawa and T.Sasamoto, J. Ceram. Soc. Jpn. 103, (1995) 1297.
[7]T.Omata, K.Ueda, H.Hosono, M.Katada, N.Ueda and H.Kawazoe, Phys. Rev. B49, (1994) 10194.
[8]T.Miyajima, T.Omata and H.Ikawa, Ceramics transactions 71, 429, edited by K.Koumoto, L.M.Sheppard and H.Matsubara, USA, Westervill 1996.
[9]B.L.Chamberland, Solid State Commun. 5, (1967) 663.
[10]J.H.van Santen, Physica 16, (1950) 599.
[11]J.B.MacChesney, P.K.Gallagher and D.N.E.Buchanan, J. Solid State Chem. 43, (1965) 1907.
[12]A.Nozaki, H.Yoshikawa, T.Wada, H.Yamuchi and S.Tanaka, Phys. Rev. B43, (1991) 181.
[13]Y.Moritomo, A.Asamitsu, H.Kuwahara and Y.Tokura, Nature 380, (1996) 141.
[14]Y.Moritomo, Y.Tomioka, A.Asamitsu and Y.Tokura, Phys. Rev. B51, (1995) 3297.
[15]T.Kimura, Y.Tomioka, H.Kuwahara, A.Asamitsu, M.Tamura and Y.Tokura, Science 274, (1996) 1698.
[16]C.Zener, Phys. Rev. 82, (1951) 403.

Materials Science and Engineering Serving Society
S. Sōmiya, R.P.H. Chang, M. Doyama and R. Roy (Editors)

Migration behaviour of Co(II) and Fe(III) ions in an alkali-borosilicate glass during thermal phase sepration of the system.

Radhaballabh Debnath
Central Glass and Ceramic Research Institute
Calcutta- 700 032 , India

Migration behaviour of incorporated Co(II) and ubiquitous impurity Fe(III) ions in an alkali-borosilicate glass of vycor type composition has been investigated both before and after the phase separation of the glass as well as in its vycorized silica glass by absorption and EPR studies.The results show that although almost all cobalt ions migrate to the separated alkali borate phase at the time of phase separation, only the interstitial ions in the case of iron can take part in similar migration . The ferric ions in the substitutional silicon sites however, remain unaffected. The absorption and the EPR results of Co(II)) suggest that the migration of the ions from the alkali-borosilicate matrix to the alkali-borate phase does not bring about any significant change to its co-ordination structure which is typically a distorted tetrahedron both in the parent and the phase separated glass.

1. Introduction:

The method of preparation of high silica porous glass using vycorization technique is very popular[1-2]. The process involves (1) preparation of a low melting alkakli-boroslicate glass as a precursor material (2) phase separating the glass into an alkali-borate and a silicate phase and finally, (3) leaching out of the more soluble alkali-borate phase by acid solution. The silica structure left after leaching constitutes an almost pure silica (wt% SiO_2= 94-97, B_2O_3 =0.5-3, Na_2O =.05-0.1). To make the process effficient sometimes, transition metal ions like Co(II) , Ni(III) and Fe(III) are also added [2] to the composition of the precursor glass.Transition metal ions because of their high polarizing ability promotes the process of phase separation. These ions are known to migrate to the soluble borate phase at the time of phase separation and are removed from the glass on leaching. Here we report migration behaviour of promoter Co(II) and Fe(III) ions in such a glass.

2.Experimental:

Two glasses of composition in the range SiO_2= 57-60 , B_2O_3=26-32, Na_2O = 6-11, Al_2O_3 = 2-4 (wt %) , one with 0.025% Co(II) and the other without cobalt oxide were prepared by melting at1400-1450°C. The Ubiquitous iron contaminated with natural quartz used as a source of silica was taken as the iron impurity of the glass. To achieve effective phase separation , glasses were cut into pieces and then heat treated in a furnace at 500-550 °C for 30-35 hours. The vycor silica glass was obtained from the phase separated glass by leaching the latter with a solution of mineral acid. The samples were subsequently sintered at 1100-1150°C to make monolithic glass.

Absorption spectra were recorded in a Varian 2300 UV-VIS-IR absorption spectrometer. EPR spectra were recorded at 77K by using a JEOL X-band spectrometer.

3. Results and Discussions:

Fig. 1 shows the base glass corrected UV-VIS-NIR absorption spectra of the Co(II) ion in the alkali - borosilicate glass both before and after its phase separation. Both the spectra show two sets of absorption bands one in the visible region and the other in the near infrared region of light due to Co(II) ions in this glass. Iron content in the glass being very

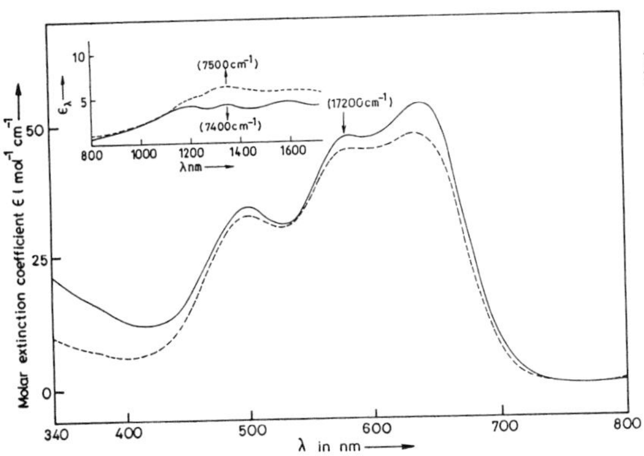

Fig.1 VIS-NIR absorption spectra of Co(II) in the alkali-borosilicate glass at 300K (__), as melted and (------), after phase separation.

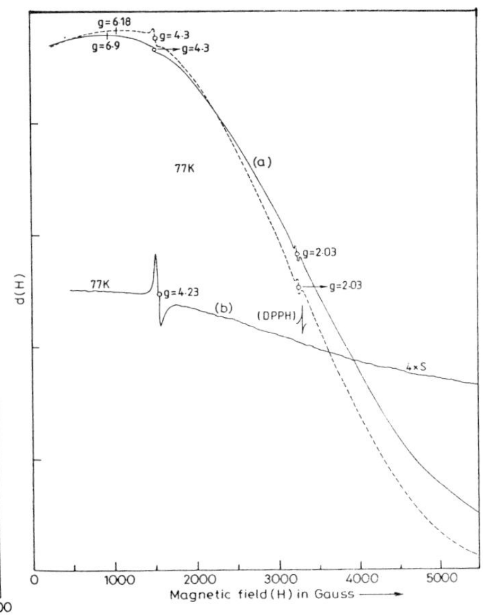

Fig. 2(a) EPR spectra of Co(II) and Fe(III) in the glass at 77K (__), before phase separation and (---) after phase separation. Experimental condition is the same for both the spectra, ν(MW) =9.256 GHz. (b). EPR spectrum of Fe(III) in the substitutional silicon sites of the structure of the silica glass

low it does not make any detectable absorption in the wavelenth range studied. Each of the two observed bands of Co(II) are found to be further split in to three components due to the spin orbit coupling (S.O.). In the case of the parent glass, these components occur around 500nm , 580nm, and 640nm within the the visible band and around 1200nm, 1500nm and 1600nm within the NIR band. In the case of phase separated glass the split components of both the visible and the NIR bands also occur in similar wavelenth regions excepting the lowest energy component of the NIR band which is not distinct in this case. It is thus clear from the spectra that the phase separation does not bring about any significant change in the absorption energies of the of the Co(II) ions of the glass. A comparison of the energies (17200cm^{-1} and 7400 cm^{-1}) and the molar extinction coefficient values (50 and 4.5 lit mol^{-1} cm^{-1}) of the visible and the NIR bands of Co(II) ion in the case of the present glass with those of Co(II) ion in known coordination suggests [4-5] that the ions in this glass are in a tetrahedral four fold coordination. We therefore, assign the NIR and the visible set of absorption bands of Co(II) ions both in the case of parent alkali- boro-silicate glass and the phase separated glass respectively to the $^4A_2 \rightarrow {}^4T_1$ (F) and $^4T_1 \rightarrow {}^4T_1$ (p) transitions.

Theoretically [4], in an ideal tetrahedral coordination maximum separation expected between the two extreme S.O. split levels of the $^4T_1$ (P) state is 800 cm^{-1} (4 λ) and that between the similar two levels of the $^4T_1$ (F) state is 1200cm^{-1} (6 λ) , where λ is the spin orbit coupling constant of free Co(II) ion. The experimental values of separation between the S.O. split levels e.g. 3400 cm^{-1} and 2000 cm^{-1} respectively for $^4T_1$ (P) and $^4T_1$ (F) states seem to be too large to be accounted for by the S.O. interaction effect alone. The result to the contrary, indicates the possibility of a trigonal distortion of the proposed tetrahedral field of the Co(II) ion in the glass. It is therefore, evident that the field structure of Co(II) ion in the case of the present glasss is a distorted tetrahedron rather than a regular one.

The electron paramagnetic resonance (EPR) spectra of Co(II) and Fe(III) in the glass recorded at 77K both before and after the phase separation are shown in Fig.2. along with that of Fe(III) in the silica glass obtained from the phase separated glass following vycorization. Each of the spectra of Co(II) consists of a broad signal showing only one differential peak .The peak ‘g’ value exhibited by the parent glass however, differ from that of the phase separated glass e.g., it is g = 6.90 in case of the parent glass while it is g = 6.2 in case of the phase separated glass. Another interesting feature of each of these spectra is the appearance of two well separated low intensity sharp resonances of respective ‘g’ values 4.3 and 2.03 overlapping on the broad signal of Co(II). The peak to peak band width of the g= 4.3 and g = 2.03 resonances are found to be of the order of ΔHpp =50 G and ΔHpp=15 G respectively. Considering the values of ‘g’ tensors and the related peak to peak widths of these two weak signals one can easily associate [9-12] them with the impurity Fe(III) ions present in the glass.

The EPR spectrum of the derived silica glass which is also given in Fig.2, does neither show any signal due to cobalt ions nor it exhibits the g = 2.03 resonance of iron. The only signal which the spectrum exhibits is the g = 4.3 resonance of iron. The results suggest that although the process of leaching of the phase separated glass removes almost all cobalt from the body of the glass it can only remove a fraction (which is responsible for g=2.03 resonance) of the total Fe(III) ions present in the glass.

It is also evident from the EPR results that the spin lattice relaxation time of Co(II) both in the parent glass and in the phase separated glass is so short that it is not only impossible to detect its EPR at room temperature but also only partially detectable even at 77K. The differential peak 'g' values of 6.9 and 6.2 of Co(II) observed respectively in the case of the parent glass and the phase separated glass support [6,7] the view that the field structure of Co(II) ions in these glasses is tetrahedral. The observed very short spin relaxation time of the ion in these hosts on the other hand, indicates [8] that the tetrahedron is not a regular one but somewhat distorted. The spin Hamiltonian that can give the EPR parameters of Co(II) under such a crystal-field can be written as[8],

$$H= \beta\, [g_{11}\, S_Z\, H_Z + g_{\perp}\ (S_Y\, H_Y + S_X\, H_X\,)\,] + D[S^2_Z - S(S+1)\,/3\,] + E\,(\,S^2_X - S^2_Y\,)$$

where D and E are two distortion parameters.

Regarding the two weak and sharp resonances of Fe(III) of 'g' values 4.3 and 2.03 obseved along with the broad signal of cobalt it is noted that the nature of the iron center with g=4.3 resonance is different from that of the iron center with g=2.03. Ferric ions being a network former as well as a network modifier , can exist in an oxide glass both in the substitutional sites as well as in the interstitial positions [10,11]. It is also known from literature that in the case of a silicate glass the tetrahedral Fe(III) ion (FeO_4), in the substitutional silicon site suffers a rhombic distortion [10] due to the presence of a compensating cation like Na^+ in its neighbour. The spin Hamiltonian of this type iron is given by the Equation [12],

$$H= g_0\, \beta HS + D\, [S^2_Z\, - S(S+1)/3\ ] + E[\, S^2_X - S^2_Y\,]$$

where g_0= 2, S= 5/2 and D and E are axial and rhombic terms respectively . When (E/D) = 1/3, an apparent isotropic value of g = 4.3 is obtained at the middle level [10]. The interstitial Fe(III) ions in such glasses on the other hand, are known to exist [12] mostly as ion pairs $[Fe(III)\,]_2$, which gives [10,11] a resonance with a 'g' tensor around 2.0.

Now considering the other result that the EPR spectrum of the leached silica glass does not show the g =2.03 resonance of the interstitial iron but very much exhibits the g=4.3 resonance due to substitutional Fe (III) ions , it is realised that the Fe (III) ions existing in the substitutional silicon sites of the network structure of the parent glass cannot migrate to the borate phase at the time of thermal phase separation of the glass and hence are not removable from the body of the glass by leaching.

Conclusions:

Use of transition metal ions like Fe(III) as a promoter of phase separation when preparing a pure silicate glass through vycorization process is not advisable. Such ions being both a network former as well as a network modifier enter in to the sustitutional sites of silica net work at the time of melting as well as at the time of phase separation of the glass. These substitutional ions in the network can not migrate to the borate phase at the time of thermal phase separation of the glass and hence are not removable by leaching.

References:

1. H.P. Hood and M. E. Nordberg, US Patent No. 2, 106 (1934) 744
2. M.B Volf, “Technical glasses” [SNTL, Publishers of technical literature ,Prague and Sir Issac Pitman and Sons. Ltd. London,(1961)]
3. C. J. Ballhousen, “ Introduction to ligand field theory” (McGraw-Hill book Company Inc, 1962)
4. C. Nelson and W. B. White, J. Mater. Res, 1(1986)130
5. R. Debnath, J. Mater. Res, (1998), (In Press)
6. Gan Fuxi, Deng He and Liu Huiming, J Non Cryst. Solids , 52 (1982)142
7. K. Kojima, H. Yano and J. Matsuda, J. Am. Cer. Soc, 73 (1990) 63
8. K Dyrek and Z. SojkaJ, Chem. Soc, Faraday Trans, 78 (1982) 3177
9. R. H Sand, Phys. Rev, 99 (1955) 222
10. T. Castner (jr), G.S. Newell, W. C. Holton. and C. P. Slichter, J. Chem. Phys. 32(1960)668
11. C. R. Kurkjian and E. A. Sigety, Phys. Chem. Glasses, 9(1968) 73
12. H. Wickman, M. P. Klein and D. A. Shirley, J. Chem. Phys. 42 (1965) 2113

Materials Science and Engineering Serving Society
S. Sōmiya, R.P.H. Chang, M. Doyama and R. Roy (Editors)

Photostimulable luminescence glasses as novel materials for 2-dimensional X-ray sensor and optical memory

Jianrong Qiu [a] and K. Hirao [a,b]

[a] Hirao Active Glass Project, ERATO, Keihanna Plaza, Seika-cho, Kyoto 619-02, Japan
[b] Faculty of Engineering, Kyoto University, Sakyo-ku, Kyoto 606-01, Japan

We report on photostimulated luminescence in sodium borate glasses doped with Ce^{3+} and Sm^{3+}. When X-ray irradiated glasses were excited by a He-Ne laser (633 nm), photostimulated luminescence at 360 nm for Ce^{3+}-doped glasses due to the 5d-4f transition of Ce^{3+} was observed. X-ray absorption spectra of glasses X-ray irradiated for various durations and photostimulated excitation spectra were measured. By comparing with the mechanism of photostimulated luminescence in BaFBr:Eu^{2+}, the luminescence is proposed to result from the photostimulated recombination of holes and electrons at traps which leave electrons in a long-lived excited state. These glasses may be promising materials for 2-dimensional X-ray sensor and optical memory.

1. INTRODUCTION

Photostimulated luminescence, a phenomenon due to the photostimulated recombination of holes and electrons at traps which leave electrons in a metastable state, has attracted considerable attentions in the recent years [1]. In photostimulated luminescence, a material excited by a light 2 after the irradiation by a light 1 e.g. X-ray, emits a light 3 whose wavelength is usually shorter than that of the excitation light 2, such that the intensity of light 3 is dependent on the exposure of the light 1. Photostimulable luminescence dosimeters using photostimulable luminescence phosphor are considered to have advantages over thermoluminescence dosimeters since they can be stimulated by photons without thermal stress and are able to be read out faster. In addition, photostimulated luminescence can be considered as a phenomenon of optical memory. The data recorded by the light 1 is read out by the light 2 in the luminescence form of the light 3. Therefore, photostimulable phosphor can be used as materials for optical memory [2].

In the present study, we report the phenomenon of photostimulated luminescence in Ce^{3+} and Ce^{3+}-Sm^{3+} doped sodium borate glasses. The mechanism of the photostimulated luminescence of these glasses is also discussed.

2. EXPERIMENTAL

The compositions of the glasses prepared were $25Na_2O \cdot 75B_2O_3 \cdot 0.50CeO_2$ (NBC) and $25Na_2O \cdot 75B_2O_3 \cdot 0.50CeO_2 \cdot 0.50SmO_{3/2}$ (NBCS) (mol%). High purity Na_2CO_3, B_2O_3, CeO_2 and Sm_2O_3 were used as starting materials. Approximately 10g batches were mixed and then melted in Pt crucibles at 1200°C for 30 min under an ambient atmosphere. The melts were then poured on a stainless plate and cooled to room temperature. The obtained glasses were then put into glassy carbon crucibles and treated under an Ar_2+H_2 (5 vol%) atmosphere at 1250°C for 60 min in a horizontal carbon furnace. The melts were then quenched to room temperature.

Photoluminescence spectra were measured by a Hitachi 850 fluorescence spectrophotometer. For the measurement of photostimulated luminescence decay curves, glass specimens were first irradiated with 40 KV/30 mA X-rays from a W target for 600 s, and a He-Ne laser (633 nm) with a power of 170 μ W/cm^2 was used as the source of excitation light. A photomultiplier tube was used as the detector. The X-ray absorption spectra of the glasses X-ray irradiated for various durations and photostimulated excitation spectra were measured. All measurements were carried out at room temperature.

3. RESULTS

Figure 1 shows the X-ray absorption spectra of the NBCS glass samples fabricated under the ambient and the reducing atmospheres. It is found that only an absorption band at 5.725 KeV, which is assigned to the absorption of Ce^{3+}, is observed in the spectrum of the sample fabricated under the reducing atmosphere. An absorption band at 5.737 KeV attributed to the absorption of Ce^{4+}, together with the band at 5.725 KeV are observed in the sample fabricated under the ambient atmosphere. Therefore, Ce ions exist in the trivalent state in the glass fabricated under the reducing atmosphere, while a part of Ce ions remains Ce^{4+} state in the sample fabricated under the ambient atmosphere. We only discuss the results of samples fabricated under the reducing atmosphere below.

Figure 2 shows the photoluminescence spectra of the NBC and NBCS glasses fabricated under the reducing atmosphere when excited by the 300 nm UV light. A broad band from 300 to 450 nm with a maximum at 360nm due to the 5d-4f transition of Ce^{3+} is observed. The emission at 560, 600 and 650 nm are caused by the transitions $^4G_{5/2}$-$^6H_{n/2}$ (n=5, 7 and 9) of Sm^{3+}.

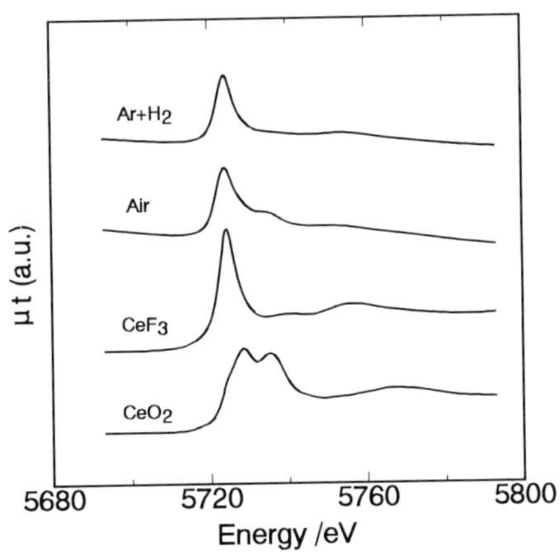

Figure 1. X-ray absorption spectra of Ce ions in the NBCS fabricated under various atmospheres.

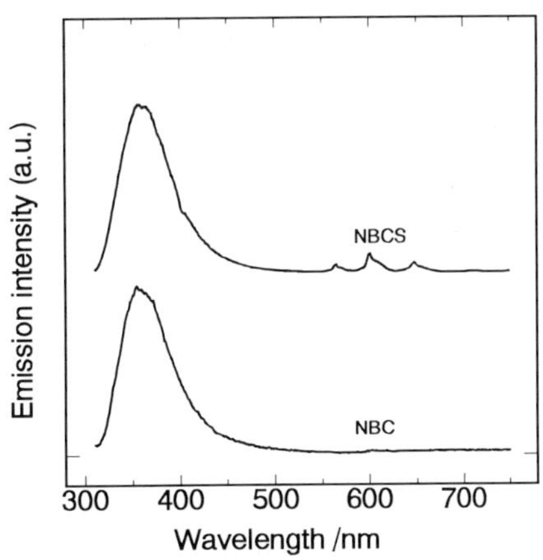

Figure 2. Photoluminescence spectra of the NBC and NBCS glasses fabricated under the reducing atmosphere. The wavelength of the excitation light is 300 nm.

When X-ray irradiated samples were excited by a He-Ne laser, samples emitted a light 360 nm for the NBC and NBCS glasses, and the intensity of the luminescences decreased quickly.

Figure 3 shows the decay curves of the luminescence at 360 nm of the X-ray irradiated samples NBC and NBCS when they were continuously excited by the He-Ne laser.

Figure 4 shows the photostimulated excitation spectra of the NBC and NBCS glass samples while the photostimulated luminescence at 360 nm was monitored. The spectra has a broad band with a maximum at about 650 nm.

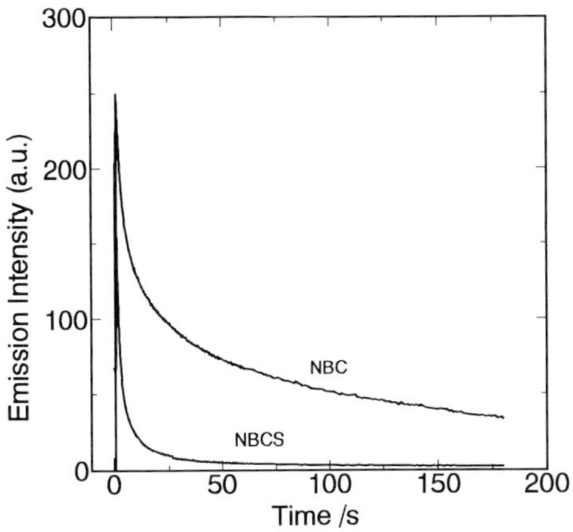

Figure 3. Decay curves of photostimulated luminescence at 360 nm for X-ray irradiated NBC and NBCS glass samples.
The excitation light is He-Ne laser (633 nm).

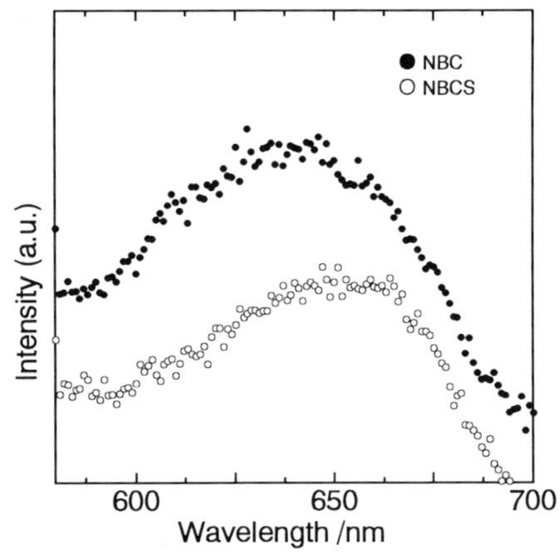

Figure 4. Photostimulated excitation spectra of the NBC and NBCS glass samples. The photostimulated luminescence at 360 nm was monitored.

Figure 5 shows the X-ray absorption spectra due to the absorption of Sm ions in the sample NBCS irradiated by 6.22 KeV X-ray for various durations together with a spectrum of a Sm^{2+}-Sm^{3+} co-doped glass sample. Little difference can be found between the absorption spectra of glasses before and after X-ray irradiation.

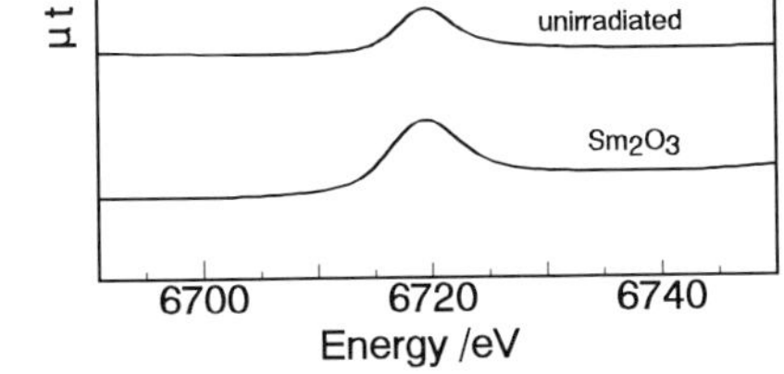

Figure 5. X-ray absorption spectra due to the absorption of Sm ions of the NBES glass irradiated by 6.22 KeV X-ray for various the durations.

4. DISCUSSION

The mechanism of the photostimulated luminescence in typical photostimulable phosphor, BaFX:Eu^{2+} (X=Cl, Br), has been investigated intensively. It is assumed that F centers act as the electron traps and photostimulated luminescence is caused by electronic transfer from $4f^65d$ level to the $4f^7$ ground state of $Eu^{2+}$when electrons are released by photostimulation and recombined with trapped holes.

ESR and absorption spectra of undoped and Ce^{3+}-doped alkali borate glasses before and after the irradiation of X-ray have been investigated extensively [7]. Ce^{3+} is considered to act as strong hole traps in the glasses. Additionally, the band observed in the photostimulated excitation spectra of NBC and NBCS agrees well with the absorption of electron traps, i.e. the quasi F-centers, formed in the glasses after the irradiation of γ rays [7].

Therefore, comparing with the mechanism of the photostimulated luminescence of $BaFBr:Eu^{2+}$, the mechanism of photostimulated luminescence in the present glasses may be proposed as below. After the irradiation of X-ray, the electron and hole pairs are formed in the glasses, some of electrons were trapped by electron trap centers, and quasi F-centers were formed in the glasses. On the other hand, some of holes were caught by hole traps e.g. Ce^{3+} ions. The wavelength of the He Ne laser (633 nm) is just in the region of the absorption due to the quasi F-center. When excited by the He-Ne laser, electrons trapped by the quasi F^{+}-centers are liberated, and combined with the holes, thus lead to the characteristic Ce^{3+} emissions.

Sm^{3+} ions are considered to act as electron traps in Ce^{3+}-Sm^{3+} co-doped sulfide crystals [8]. However, no variance in the valence state and the local structure of the Sm ions before and after X-ray irradiation has been observed from the X-ray absorption spectra. Therefore, as opposed to their behavior in sulfide crystals, Sm^{3+} ions probably do not act as electron traps in the present glasses.

5. CONCLUSION

The Photostimulated luminescence phenomena are reported in the sodium borate glasses doped with Ce^{3+} and Sm^{3+}. When X-ray irradiated glasses were excited by a He-Ne laser (633 nm), photostimulated luminescence at 360 nm for NBC and NBCS due to the 5d-4f transition of Ce^{3+} was observed. X-ray absorption spectra of glasses X-ray irradiated for various durations and photostimulated excitation spectrum show that Ce^{3+} and quasi F-centers in the glasses act as hole and electron traps, respectively. These glasses may be promising materials for 2-dimensional X-ray sensor and optical memory.

ACKNOWLEDGEMENT

J. Qiu is grateful to Dr. Y. Iwabuchi of Fuji Photo Film Co. Ltd. and Dr. Y. Shimizugawa of Hirao Active Glass Project for their corporation in the experiments and valuable suggestions on the present paper.

REFERENCES

1. Y. Amemiya and J. Miyahara, Nature, 336 (1988) 89.
2. H. Nanto, Kotai Butsuri (Solids Physics), 28 (1993) 119.
3. T. Kanou, Phosphors Handbook, Ohhm Press, Tokyo, 1987, p. 120
4. A. S. Mahbub'ul Alam and B. D. Bartolo, J. Chem. Phys., 47 (1967) 3790.
5. Y. Iwabuchi, N. Mori, K. Takahashi, T. Matsuda and S. Shionoya, Jpn. J. Appl. Phy., 33 (1994) 178.
6. A. R. Lakshamanan , Phys. Stat. Sol., (a) 3 (1996) 153.
7. R. Yokota, Glass Handbook, Ed. by S. Sakka, T. Sakaino and K. Takahashi, Asakura Press, Tokyo, 1975, p. 826.
8. K. Chakrabarti, V. K. Mathur, L. F. Thomas and R. J. Abbundi, J. Appl. Phys., 65 (1989) 2021.

Materials Science and Engineering Serving Society
S. Sōmiya, R.P.H. Chang, M. Doyama and R. Roy (Editors)

XPS Studies of Graded-Composition Films of Aluminum and Nitrogen

Shingo Uchiyama, Yoshihisa Watanabe, Yoshiki Amamoto and Yoshikazu Nakamura
Department of Materials Science and Engineering, National Defense Academy, 1-10-20 Hashirimizu, Yokosuka, Kanagawa 239, JAPAN

Graded-composition films of aluminum and nitrogen, 0.5 μm Al film piled on 0.5 μm Al-N/AlN film, were prepared by an ion-beam assisted deposition method, and the chemical states of aluminum and nitrogen were analyzed by X-ray photoelectron spectroscopy (XPS). The analyses of aluminum $2p$ ($Al2p$) photoelectrons spectra reveal that (1) the concentration of the metal state gradually decreases and the nitride state supersedes the metal state in the process of sputtering of the film by Ar ions and (2) the intensity of $Al2p$ electrons with the nitride state increases together with growth of the nitrogen peak and reaches to the constant values simultaneously. From the present results, it is proposed that Al and AlN coexist in the intermediate region of the graded-composition films of aluminum and nitride.

1. INTRODUCTION

Aluminum nitride (AlN) films have attracted much attention in microelectronics fields due to their high thermal conductivity, high electrical resistivity and high mechanical strength. In addition, because the value of thermal expansion coefficient of this material is close to that of Si or GaAs, AlN films have potential applications for metal-insulator-semiconductor (MIS) materials [1] and cap layers for GaAs substrate [2].

The present authors' group has prepared AlN films by an ion beam assisted deposition (IBAD) method at near room temperature and studied the effects of deposition conditions on film microstructures and properties [3]. Recently, taking advantage of controllability of ion beam current during deposition, the same group has succeeded in preparing graded-composition films of aluminum and nitrogen, AlN/Al-N/Al, on Si substrate by changing the arrival ratio of aluminum to nitrogen at the substrate using the IBAD [4,5]. These reports have opened the door for new preparation techniques of MIS materials and functionally graded materials.

In the previous papers [4,5], the changes in microstructure and composition were mainly studied by transmission electron microscopy (TEM), energy dispersive X-ray spectroscopy (EDX) and X-ray photoelectron spectroscopy (XPS). From the photoelectron spectra, it was found that the peak energy of aluminum $2p$ ($Al2p$) electrons changes with sputtering of the film from the surface in the vicinity of the substrate. However, it is still unknown how the chemical state of aluminum changes in the film, especially in the graded region of composition.

The purpose of the present paper is to measure the photoelectron spectra of Al2*p* and nitrogen 1*s* (N1*s*) by XPS with sputtering of the films and study the change in the chemical state of aluminum inside the graded-composition films.

2. EXPERIMENTAL

The procedure of the graded-composition films is as follows: First, AlN film of 100 nm in thickness is deposited on a Si(100) substrate at near room temperature by the IBAD. Then, four Al-N films, of 100 nm each in thickness, were deposited on the AlN film by decreasing the arrival ratio of nitrogen to aluminum atoms at the substrate, yielding four piled up Al-N films with different compositions. Finally, 500 nm aluminum film was deposited on the graded-composition films by stopping irradiation with the nitrogen ion beam. Thus, approximately 1 μm in thickness film of AlN/Al-N/Al was obtained.

The chemical states of aluminum and nitrogen in the film were analyzed with an XPS (ESCA 5600 ci, Physical Electronics). The photoelectron spectra of each element were obtained by sputtering of the film with 3 keV Ar ions. The period of sputtering was 5 minutes per cycle and the reduced speed of etching for SiO_2 films was approximately 1.8 nm/min.

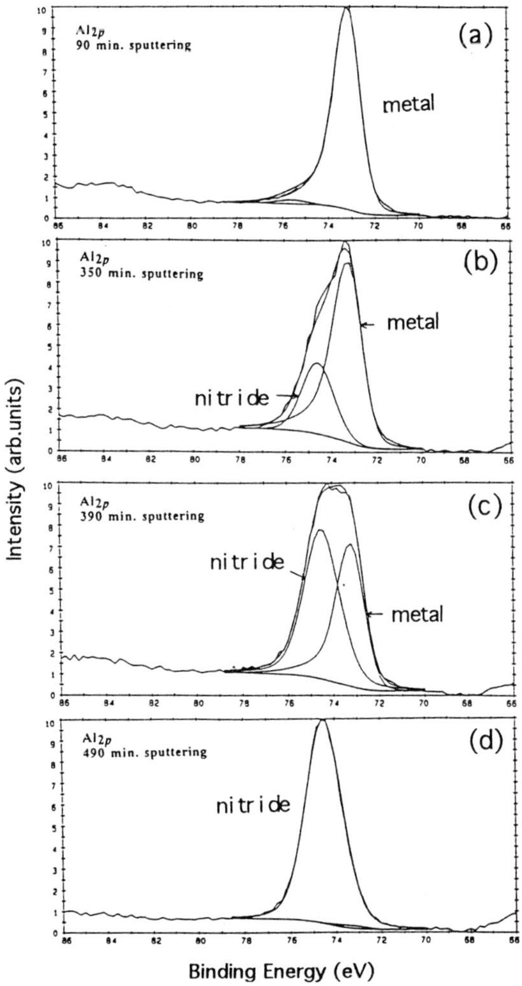

Fig.1 Aluminum 2p electron spectra observed by XPS with sputtering of (a) 90 minutes, (b) 350 minutes, (c) 390 minutes and (d) 490 minutes.

3. RESULT AND DISCUSSION

From a previous work, it has been known that (1) the peak position of Al2*p* electrons shifts from about 73 eV to 74.5 eV with sputtering of the film from the surface to the substrate and (2) the Al2*p* peaks at near half depth of the film seem to be pressed down, viz. the peak heights decrease and the peak widths broaden [5]. These results suggest that the chemical state of aluminum changes with depth and mixing states exist inside the film. The Al2*p* peaks positioned at about 73 eV and 74.5 eV are attributed to the metal and nitride states of aluminum [6]. On the other hand, the peak of N1*s* electrons appears after sputtering and it grows up with depth.

In order to study the change in the chemical state of aluminum with depth, the peaks of Al2*p* electrons were separated with non-linear least-squares fitting method. Typical results are shown from Figs.1(a) to (d). The Al2*p* spectra near the surface were not analyzed because they were affected by oxygen and contaminants. Fig.1(a) shows an Al2*p* spectrum after 90 minutes sputtering, corresponding to etching of about 160 nm SiO_2 films, and the peak centered at about 73 eV is due to the metal aluminum mentioned above. As the sputtering proceeds, the aluminum spectrum is distorted as shown in Fig.1(b), which was obtained after 350 minutes sputtering. This spectrum can be fitted to the metal peak and the nitride peak centered at about 74.5 eV. The nitride peak dominates the metal peak with proceeding of sputtering, as shown in Fig.1(c) for the sample after 390 minutes sputtering. As approaching to the substrate, the aluminum spectrum is fitted to the single nitride peak as shown in Fig.1(d). A series of peak separation studies shows that the chemical state of aluminum changes from the metal to the nitride via the coexisting region.

The concentration of aluminum with metal and nitride states was calculated from each peak area obtained by peak separation and shown in Fig.2 as a function of sputtering time. This figure shows clearly that (1) the concentration

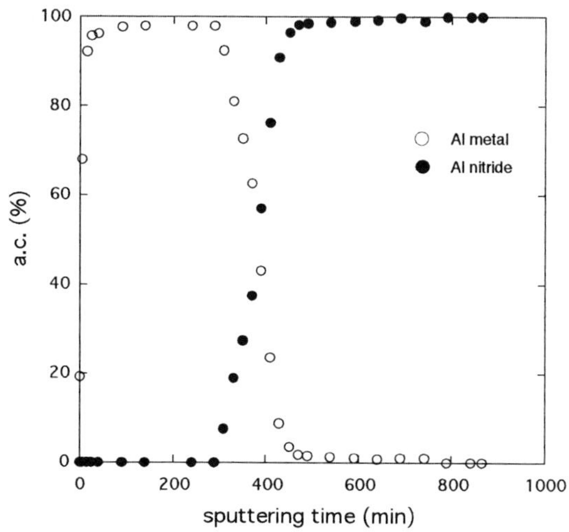

Fig.2 Relation between the concentration of aluminum 2*p* electrons with metal and nitride states, and the sputtering time.

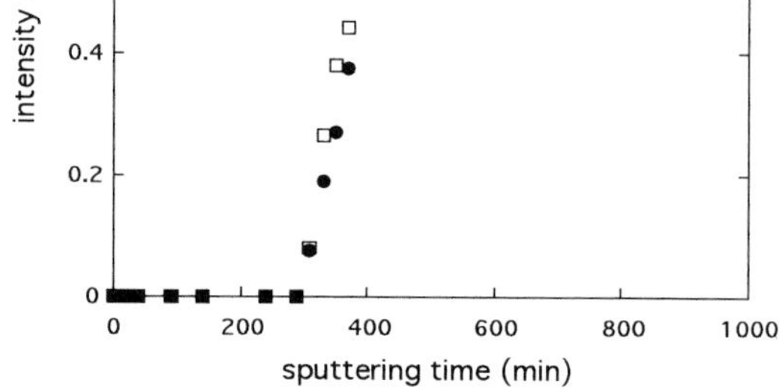

Fig.3 Relation between the relative intensities of aluminum 2*p* electrons with nitride state and nitrogen 1*s* electron, and the sputtering time.

of the metal state gradually decreases and the nitride state supersedes the metal state with proceeding of sputtering and (2) both states of the metal and nitride of aluminum coexist in the intermediate region. Recent theoretical report on the phase stability of the Al-N system describes that Al(s) + AlN(s) can exist at room temperature in the region of nitrogen concentration below 50 at. % [7]. This theoretical work supports the present results of XPS analyses .

The relative intensity of Al_{2p} electrons with the nitride state and N_{1s} electrons normalized to their maximum intensities is plotted as a function of sputtering time and shown in Fig.3. This figure shows that the intensity of aluminum with the nitride state increases together with the growth of nitrogen and reaches to the saturated values simultaneously. These results indicate that aluminum and nitrogen is combined into AlN in the coexisting region. In other words, Al and AlN coexist in the intermediate region of the graded-composition films.

4. CONCLUSIONS

Graded-composition films of aluminum and nitrogen were prepared by the IBAD method. The peak separation analyses of Al_{2p} spectra obtained by XPS reveal that the concentration of the metal state gradually decreases, and the nitride state supersedes the metal state in the process of sputtering of the film, and the intensity of Al_{2p} electrons with the nitride state increases together with growth of nitrogen peak. From the present work, it is concluded that AlN and Al coexist in the intermediate region of the graded-composition film.

ACKNOWLEDGMENTS

This work was performed using the ion vapor deposition apparatus in the Advanced Materials Laboratory in the National Defense Academy. The authors would like to thank Dr. H. Tomizuka and Ms. T. Nakamura, Ulvac-Phi, Inc., for their assistance of XPS measurements.

REFERENCES

1. S. Bengtsson, M. Bergh, M. Choumas, C. Olesen and K. O. Jeppson, J. Appl. Phys., 35 (8) (1996) 175.
2. F. C. Stedile, I. J. R. Baumnol, W. H. Schreiner and F. L. Freire, Jr., J. Vac. Sci. Tech., A 10 (5) (1992) 3272.
3. Y. Watanabe and Y. Nakamura, Ceramic International, 23 (1997) in press. References to AlN thin film works by the present authors' group cited therein.
4. Y. Amamoto, S. Uchiyama, Y. Watanabe and Y. Nakamura, Mater. Res. Soc. Symposium Proceedings, 438 (1997) 645.
5. Y. Amamoto, S. Uchiyama, Y. Watanabe and Y. Nakamura, Proceedings of the 1st joint Symposium of IEMT and IMC, Omiya (Japan), (1997) 133.
6. G. E. McGuire, G. K. Schweitzer and T. A. Carlson, Inorg. Chem., 12 (1973) 2450.
7. Siang-Chung L., J. Mater. Sci. Lett., 16 (1997) 759.

Materials Science and Engineering Serving Society
S. Sōmiya, R.P.H. Chang, M. Doyama and R. Roy (Editors)

Structural study of perovskite-type $PbZn_{1/3}Nb_{2/3}O_3$ by single crystal X-ray diffraction and EXAFS

Y. Matsushima[a], K. Suda[a], N. Ishizawa[a], N. Wakiya[b] and N. Mizutani[b]

[a]Materials and Structures Laboratory, Tokyo Institute of Technology, Nagatsuta 4259, Midori, Yokohama 226, Japan

[b]Faculty of Engineering, Tokyo Institute of Technology, O-okayama 2-12-1, Meguro, Tokyo 152, Japan

A structural study of lead zinc niobate $PbZn_{1/3}Nb_{2/3}O_3$ (PZN) was carried out by EXAFS and single crystal X-ray diffraction. The EXAFS results suggested rather large structural distortion of PZN in contrast to the past understandings. The single crystal X-ray diffraction study confirmed a presence of positional disorder of Pb atom near the ideal position in the rhombohedral perovskite-type structure.

1. INTRODUCTION

Lead zinc niobate $PbZn_{1/3}Nb_{2/3}O_3$ (PZN) is classified into perovskite-type mixed compounds with formula $AB'B''O_3$. Though PZN possesses excellent properties required for relaxor ferroelectrics, much are left unknown about its crystallography, growth mechanism, thermal stability, etc. The compound was reported to show rhombohedral symmetry at room temperature[1]. Crystals have never been synthesized through the solid state reaction under the atmospheric pressure. Some past studies assumed that the perovskite-type phase of PZN is unstable at room temperature [2,3]. However, it was recently reported that single crystals can be grown from the solution with PbO flux [4], which also enabled a detailed crystallographic study. To understand the nature of relaxor materials, knowledge of the local distortion of the structure is as important as that of the average structure. In this paper, therefore, the crystal structure of PZN was investigated by both the single crystal X-ray diffraction and EXAFS techniques.

2. EXPERIMENTAL

2-1. Sample preparation

Commercial reagents PbO, ZnO and Nb_2O_5 were mixed with PbO flux at the ratio of 2:1 to PZN compound, using a plastic mill with ZrO_2 balls for 18 hours. After calcination, the mixture was pressed into a pellet with subsequent heating in Pt crucible at 1150 °C for 5 hours, and cooling down to 900 °C at the rate of 1 °C/h. Transparent and yellow-colored single crystals were obtained and used for EXAFS and diffraction experiments.

2-2. EXAFS

EXAFS spectra were measured near the Pb L_{III}, Zn K and Nb K absorption edges at the beam line 14A Photon Factory KEK, using a synchrotron radiation. Measurements started at 12730 eV for Pb L_{III}, at 9363 eV for Zn K and at 18689 eV for Nb K absorption edges, in the range up to +1500 eV with 4 eV intervals. The dwelling time for each step was 3.0 seconds. Powders of PbO, ZnO and Nb_2O_5 were used as references in order to determine ΔE_0, the difference between the observed and theoretical absorption edges, and λ, the mean free path of electrons.

2-3. Single crystal X-ray diffraction

Single crystal X-ray diffraction data were collected using a four-circle diffractometer with AgKα radiation in the range between $1 < 2\theta < 80°$, employing 2θ-ω scan mode. A spherical crystal with 67.5 μm in radius was used. The cell dimensions were determined by 25 reflections in the range $0.522 < sin\theta/\lambda < 0.551$. The hexagonal setting of the rhombohedral cell was used in the calculation.

Table 1
Crystal data of PZN and experimental conditions

Space group	$R3m$
Cell dimensions	
rhombohedral setting	a=4.0588(3)Å, α=89.93(1)°
hexagonal setting	a=5.7364(6)Å, c=7.0389(7)Å
Number of reflection	3393
Number of independent reflections	223 (with $I > 3.0\sigma(I)$)

3. RESULT AND DISCUSSION

3-1. Pb L_{III} edge EXAFS

Fourier transforms corresponding to radial distribution functions (RDFs) were obtained from $k\chi(k)$-weighted EXAFS oscillations of PbO and PZN as shown in Fig. 1. The RDF of PbO shows contributions from neighboring O atoms around the central Pb atom at R=1.8 Å. The RDF of PZN exhibits two peaks at R=1.2 and 1.9 Å , which can be ascribed to contributions from two oxygen shells, O1 and O2. The reverse Fourier transform of the RDF in the range between R=0.69 and 2.38 Å was used for the least-squares fitting of interatomic distances, R_j, coordination numbers, N_j, and debye-waller factors, σ_j. The refinement under constraints, $Nj\ (O1) + N_j(O2) = 12$ and $\sigma_j(O1)=\sigma_j(O2)$, converged with the reliability factor of 0.28. Results are given in Table 2. It is supposed that the Pb atom is surrounded by approximately 6 oxygen atoms (O1) at a distance of 2.49 Å and 6 oxygen atoms (O2) at a distance of 2.81 Å.

Table 2
EXAFS results of PZN, showing the bond distances(Å), R_j, coordination numbers, N_j, and displacement parameters(Å^2), σ_j, of the 1st and 2nd oxygen shells, O1 and O2.

	O1			O2		
Central atom	R_1	N_1	σ_1	R_2	N_2	σ_2
Pb	2.49(1)	6.1	0.09	2.81(1)	5.9	0.09
Zn	2.09(1)	4.2	0.05	2.86(1)	4.6	0.05
Nb	1.95(1)	5.1	0.09	2.29(1)	0.9	0.09

3-2. Zn K edge EXAFS

Similar procedure to that given in **3-1** was employed for the analyses of Zn K profiles of PZN and ZnO, except that $k^3\chi(k)$-weighting scheme was used instead of $k\chi(k)$. The RDFs of ZnO and PZN are shown in Fig. 2. The peak at R=1.6 Å in the RDF of ZnO can be ascribed to the contributions of neighboring O atoms. The RDF of PZN have two peaks between 1.15 and 3.07 Å, which are considered as the contributions from two oxygen shells, O1 and O2. The fitting procedure of the reverse Fourier transform in the range between R=1.15 and 3.07 Å gave results as given in Table 2. In this procedure, $\sigma_j(O1)$ and $\sigma_j(O2)$ were assumed to be equal to that obtained for the ZnO reference material. The final reliability factor was 0.08. It is suggested that the O1 shell consists of about 4 oxygen atoms at 2.09 (1) Å and the O2 shell consists of about 5 oxygen atoms at 2.86 (1) Å.

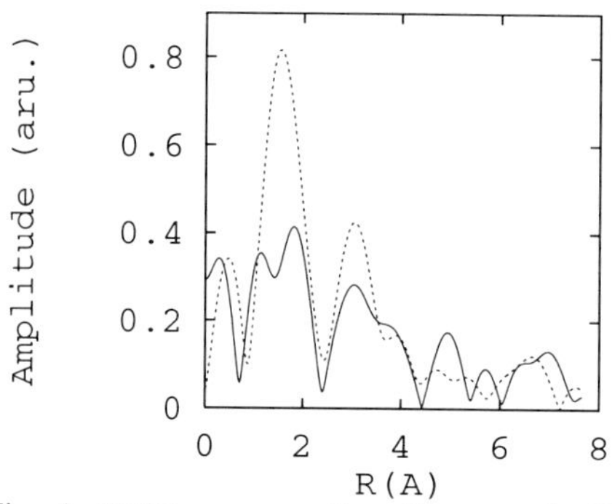

Fig. 1 RDFs surrounding Pb atom in PbO (dashed line) and PZN (solid line)

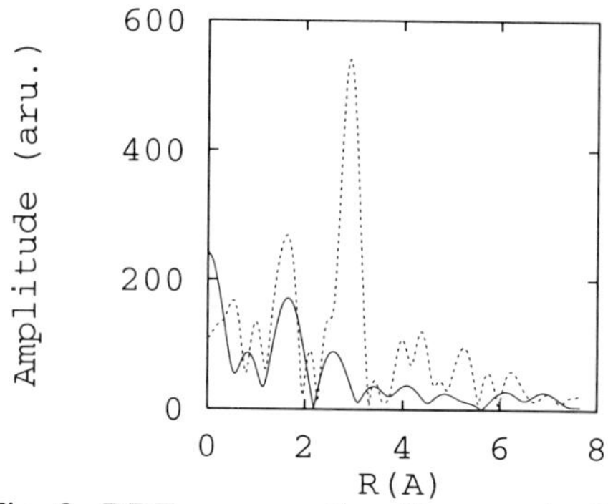

Fig. 2 RDFs surrounding Zn atom in ZnO (dashed line) and PZN (solid line)

3-3. Nb K edge EXAFS

The RDFs of Nb_2O_5 and PZN obtained from the Fourier transform of the $k^3\chi(k)$ are shown in Fig. 3. Peaks in the PZN RDF at R=1.8 and 2.2 Å represent two oxygen shells, O1 and O2. The fitting results in the range between R=1.07 and 2.45 Å with constraints, $N_j(O1)+N_j(O2)=6$ and $\sigma_j(O1)=\sigma_j(O2)$ are given in Table 2. The final reliability factor was 0.21. The Nb atom is surrounded by approximately 5 oxygen atoms at 1.94 Å and one oxygen atom at 2.29 Å.

3-4. Single crystal X-ray diffraction

In the ideal rhombohedral perovskite-type structure with $R3m$ symmetry, the metal atoms reside at 3a positions and O atoms at 9b positions. The R/wR factors became 0.094/0.048, assuming this ideal structure. The difference Fourier maps, however, revealed compilations of electrons with a height of more than 10 e/Å^{-3}, around Pb atoms. They were considered as possible statistic displacements of Pb atoms. It looked most reasonable to locate Pb atoms statistically at 9b positions such as x,-x,z with x and z having small values, to explain the observed residual electrons. The R/wR factors reduced to 0.050/0.030 under this assumption. The positional and thermal parameters are given in Table 3. The projection along three-fold axis is illustrated in Fig. 4, showing the atomic shift directions.

Since the determination of the structure of PZN is still in progress, it is difficult to establish a final structure model taking into consideration the local structure found by EXAFS. However, it is clear that the PZN has a rather distorted structure as was not expected so far. It is also interesting that the local structure distortion around Pb atom is slightly different from that found in PMN[5,6]

Table 3
Results of Atomic coordination parameters and anisotropic temperature factors

Atom	x	y	z	U11	U22	U33	U12	U13	U23
Pb	-0.0227(4)	-x	0.018(6)	0.02(1)	U11	0.05(1)	0.01(1)	0.006(1)	-U13
Zn/Nb	0.0	0.0	0.5	0.01(1)	U11	0.005(1)	0.006(1)	0	0
O	0.499(2)	-x	-0.09(3)	0.04(1)	U11	0.03(1)	0.03(1)	0.002(3)	-U13

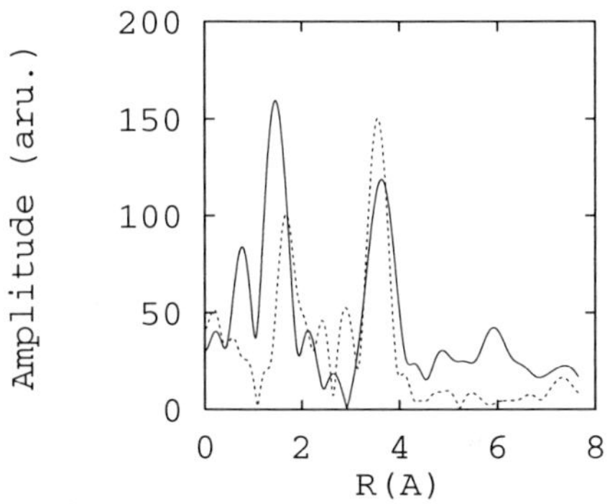

Fig. 3 RDFs surrounding Nb in Nb_2O_5 (dashed line) and PZN (solid line)

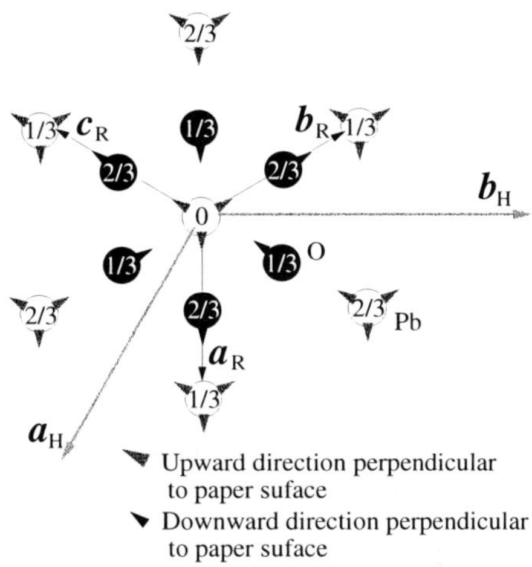

Fig. 4 Atomic shift directions determined by X-ray diffraction study. Arrows and fractional numbers indicate atomic shift directions and ideal fractional-atomic -coordinates z_H, respectively.

REFERENCE

1. Y. Yokomizo, T. Takahashi and S. Nomura, J. Phys. Soc. Jpn., **28**, 1278-1284 (1970)
2. Yoshihiro Matsuo, Yogyo-Kyokai-Shi, **78**[2] 14-25 (1970)
3. N. Wakiya, K. Shinozaki and N. Mizutani, submitted to Am. Ceram. Soc. (1997)
4. V. A. Bokov and I. E. Myl'nikova, Soviet Phys. Solid State, **2**[11] 2428-32 (1961)
5. P. Bonneau, P. Garnier, G. Calvarin, E. Husson J. R. Gavarri A. W. Hewat and A. Morell, J. Solid State Chem. **91**, 350-361(1991)
6. N. Wakiya, N. Ishizawa, A. Saiki, K. Shinozaki and N. Mizutani, J. Ceram. Soc. Jp., **102**[1] 8-12 (1994)

Materials Science and Engineering Serving Society
S. Sōmiya, R.P.H. Chang, M. Doyama and R. Roy (Editors)

Relationship between glass basicity and O1s core energy for oxide glasses

H. Segawa, T. Yano, S. Shibata and M. Yamane

Faculty of Engineering, Tokyo Institute of Technology, 2-12-1 Ookayama,
Meguro-ku, Tokyo 152 Japan

Optical basicity (using Pb^{2+} probe ion) and O1s core spectra by using XPS spectrometer were measured for R_2O-SiO_2 (R=Li, Na, K), Na_2O-P_2O_5 and Na_2O-B_2O_3 glasses. Relation between the optical basicities and the O1s spectra was discussed.

1. Introduction

Several oxygen atoms in different chemical sites such as bridging, non-bridging and double-bonded oxygens are known to exist in oxide glasses of various systems and compositions. Since these different oxygen atoms affect important glass properties, much effort has been devoted to make a scale for representing their influence on glasses. One example of glass properties which are greatly affected by these different oxygen atoms is redox equilibrium of transition metal ions in glasses.

Duffy & Ingram developed a parameter called "optical basicity" as a mean of characterizing glass properties [1], which provide a way to predict the change in the redox equilibrium. Concept of optical basicity gave us two advantages for its application: first, it can be measured experimentally from a frequency shift of absorption peaks due to s-p transition of probe ions such as Tl^{+}, Pb^{2+} and Bi^{3+} in glasses; second, it can be calculated from the glass compositions using parameters of cations. Unfortunately, however, the absorptions caused by s-p transition are located in ultraviolet wavelength region, it cannot be applicable to the glasses having an absorption edge at a longer wavelength. Moreover, differences were reported between the calculated and the measured optical basicities in sodium borate [2] and soda alminosilicate glasses [3].

On the other hand, X-ray photoelectron spectroscopy (XPS) can directly provide information on the chemical environment and the oxidation state of each element in glasses. For oxide glasses, using O1s core spectra, bridging oxygen atoms (BO) can be distinguished from non-bridging oxygen atoms (NBO) (or double-bonded oxygen atoms (DBO)) [4,5] with their binding energy differences.

In this study, both of optical basicities (using Pb^{2+} probe ion) and the O1s core spectra were measured in several binary glass systems: R_2O-SiO_2 (R = Li, Na, K), Na_2O-P_2O_5 and Na_2O-B_2O_3. Relation between the optical basicity and the O1s core spectra from XPS measurements was discussed.

2. Experimental

Silicate, phosphate (Na_2O>50mol%) and borate glasses were prepared from reagent grade of R_2CO_3 (R = Li, Na, K), $(NaPO_3)_n$, SiO_2, H_3BO_3 and PbO. Glasses of 0.024mol% Pb^{2+}-doped and Pb^{2+}-free of the same composition were made in all cases. The glasses were melted in a Pt crucible for 2hours at a suitable temperature for the glass composition in an electric

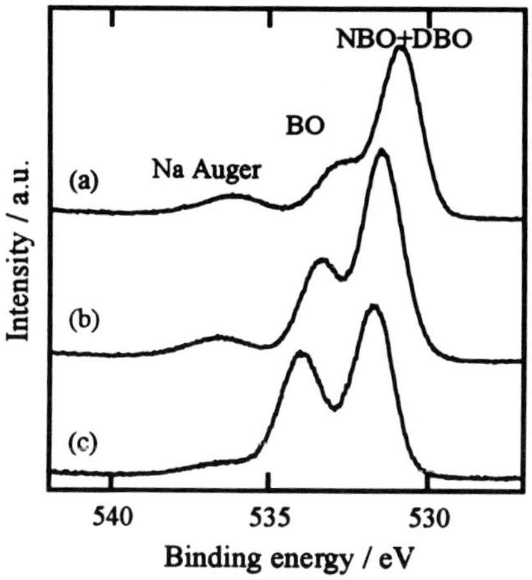

Figure 2. O1s core spectra for Na_2O-P_2O_5 glasses. (a):59.7Na_2O-40.3P_2O_5, (b):49.6Na_2O-50.4P_2O_5, (c):36.6Na_2O-63.4P_2O_5 (mol%).

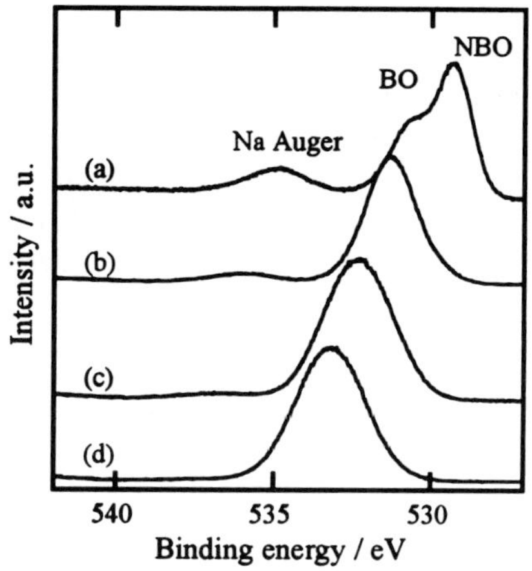

Figure 3. O1s core spectra for Na_2O-B_2O_3 glasses: (a):70Na_2O-30B_2O_3, (b):35Na_2O-65B_2O_3, (c):20Na_2O-80B_2O_3, (d):5Na_2O-95B_2O_3 (mol%).

Na_2O content, peaks shifted lower energy side and the peak ratio of (NBO+DBO)/BO increased.

In Na_2O-B_2O_3 glasses (Fig. 3), only one peak assigned to BO was observed in the region of $Na_2O \leqq 35$mol%. In 70Na_2O-30B_2O_3 glass, however, showed two peaks due to BO (higher energy) and NBO (lower energy). Peak shift to lower energy was also observed with increasing Na_2O content in borate glasses.

4. Discussion

As shown in Figs. 1~3, O1s core spectra reflect the change in the glass composition. Oxygen atoms in different chemical sites such as BO, NBO and DBO were assigned to the peaks of different binding energies, and their peak areas corresponded to molar ratio of each oxygen atoms. Since the concept of optical basicity is an average scale of glass chemistry, in order to compare the optical basicity, Λ_m, with O1s core spectra, the spectra which are consist of BO, NBO and DBO must be modified to the scale reflecting the average property. Considering that the energy value at the center of the gravity in O1s core spectra, E_G, can be used as a possible mean of characterizing average glass environment, E_G are compared with Λ_m. In Fig. 4, E_G of all glasses are plotted against their Λ_m. The E_G decreased with increasing Λ_m, and they showed close relationship. Optical basicity of oxide glasses is known to be a measure of the degree of which oxygen atoms donate negative charge to the dopant cation, Pb^{2+}. Chemical shift of O1s core spectra is also considered to be a scale for measuring the electric donation property of oxygen atoms. Fairly good relationship observed in Fig. 4 supports this explanation.

furnace. After quenching, glasses were crushed and remelted to improve homogeneity for 1hour at the same temperature as the first melting.

In case of Na_2O-P_2O_5 ($Na_2O \leqq 50$mol%), liquid H_3PO_4 was used as a source of P_2O_5 component. After drying for 1~3days at 250°C, the batch was melted in a Pt crucible at 900°C for about 7hours with N_2 gas bubbling to remove water. Remelting the glasses for 1 hour was carried out at 1000°C. Since the bubbling procedure sometimes causes changes in glass compositions, those of Na_2O-P_2O_5 glasses were analyzed with inductively coupled plasma spectroscopy (SII SRS 1500VR).

Optical absorption spectra for Pb^{2+}-doped glasses were measured with a conventional UV-VIS spectrometer (Japan Spectroscopic Co. Model V570). In all cases, samples of the same without Pb^{2+} were used for the references. Optical basicity, Λ_m, was calculated from the following equation,

$$\Lambda_m = \frac{60700 - \nu_{glass}}{31000}$$

where ν_{glass} is the measured peak frequency of Pb^{2+} due to the $^1S_0 \rightarrow {}^3P_1$ transition.

Glass samples (Pb^{2+}-doped glasses) were fractured under high vacuum of 10^{-10} Torr to obtain a fresh surface prior to the XPS measurement. Photoelectron spectra of O1s were measured by using XPS spectrometer (PHI Model 5500MT) with monochromated AlKα X-ray source. Pass energy of XPS was 5.85eV and photoelectron counts were collected at an interval of 0.025eV. A neutralization gun was used to compensate positive charges on a glass surface. Spectra were calibrated using C1s peaks at 284.6eV.

3. Results

Photoelectron spectra of O1s for Na_2O-SiO_2, Na_2O-P_2O_5 and Na_2O-B_2O_3 glasses were shown in Figs. 1, 2 and 3, respectively. As shown in Fig. 1, two different peaks were clearly observed in Na_2O-SiO_2 glasses, in which a peak at higher energy side is assigned to BO and that at lower energy side to NBO. With the increase of Na_2O content, ratio of NBO peak area to BO increased and both peaks shifted to lower energy side. Two peaks were observed in lithium and potassium silicate glasses, and change in the peak ratio and peak shift showed qualitatively similar tendency.

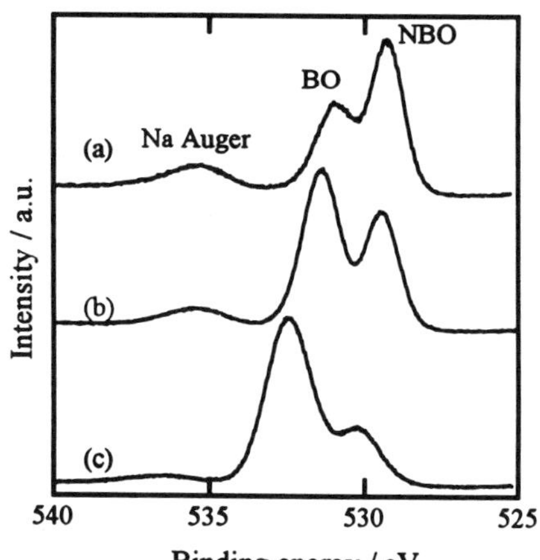

Figure 1. O1s core spectra for Na_2O-SiO_2 glasses.
(a): $50Na_2O$-$50SiO_2$, (b): $35Na_2O$-$65SiO_2$, (c): $18Na_2O$-$72SiO_2$ (mol%).

In Fig. 2, two peaks were also observed in Na_2O-P_2O_5 glasses. The peak at higher energy side is assigned to BO as well, but that at lower energy side is due to NBO and DBO. The NBO and DBO are known to be the same binding energy because of the resonance effect in phosphate glasses [5]. With increasing

It should be noticed, however, in Fig. 4 that the E_G vs. Λ_m relations of different glass systems cannot be expressed by a single line. Further examination must be carried out on the relation depending on each glass systems. On the other hand, taking account of the fact that oxygen atoms of different chemical sites can be distinguished by XPS measurement, O1s spectra have a potential to determine the optical basicity related with respective oxygens of different sites, BO, NBO and DBO in glass structures. This implies that O1s spectra can give the experimental definition of a "microscopic optical basicity" [1], which has been derived from the data of a single component material.

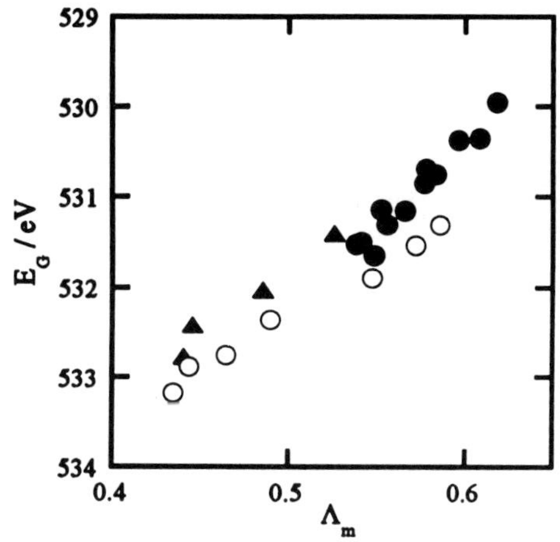

Figure 4. The energy values at the center of the gravity in O1s core spectra, E_G are plotted against their optical basicities, Λ_m. ●:R_2O-SiO_2 (R=Li, Na, K), ▲:Na_2O-P_2O_5, O:Na_2O-B_2O_3.

5. Conclusions

Optical basicities and O1s core spectra of alkali silicate, sodium phosphate and sodium borate glasses were measured by optical absorption of Pb^{2+} probe ions and XPS measurements, respectively. Different chemical sites of oxygen atoms in glasses BO, NBO and DBO could be distinguished in O1s spectra, and they changed their shapes and shifted to lower energy side with increasing alkali components.

Energy values at the center of the gravity in O1s spectra of glasses (E_G) decreased with increasing their optical basicites (Λ_m). As in the case of the probe ions, their relation suggests that O1s spectra give us a scale for measuring the degree of electric donation oxygen; the optical basicity. The relation of E_G vs. Λ_m, however, still have a problem to show some differences depending on the various glass systems.

References

1. J.A. Duffy and M.D. Ingram, *J. Non-Cryst Solids*, **21** (1976) 373.
2. J.A. Duffy and M.D. Ingram, *Physics and Chem. Glasses*, **16** (1975) 119.
3. R.M. Klein and P.I.K. Onorato, *Physics and Chem. Glasses*, **21** (1980) 199.
4. R. Brückner, H.-U. Chun and H. Goretzki, *Glastechn Ber.*, **51** (1978) 1.
5. R. Gresch , W. Muller-Warmuth and H. Dutz, *J. Non-Cryst Solids*, **34** (1979) 127.

Materials Science and Engineering Serving Society
S. Sōmiya, R.P.H. Chang, M. Doyama and R. Roy (Editors)

Grain-boundary Sliding and Interlocking of Two-Phase Ceramics.

Hiroyuki MUTO
Department of Materials Science. Tyohashi University of Technology
Tempaku-cho, Tyohashi 441, Japan

The grain-boundary sliding and interlocking at high temperatures of two-phase ceramics are examined. The Dryden-Kucerovsky-Wilkinson-Watt theory (DKWW theory) is extended to the viscoelastic deformation associated with grain-boundary sliding and interlocking of elastic grains embedded in a contiguous viscous matrix. A two-phase model-ceramic in creep deformation acts as a non-Newtonian fluid which has the viscosity with creep-strain-hardening. The constitutive relation between the strain rate and the creep strain is utilized to determine the viscosity, volume fraction of intergranular melt, and the equilibrium creep strain for grain interlocking.

1. Introduction

Upon application of a load to polycrystalline ceramics at high temperatures, the grain boundaries start to slide.[1)] However, grains that form the microstructure do not lie in ordered planes, developing serrated grain boundaries which do not allow to continue to creep indefinitely merely by sliding. The grains, thus, form a self-locking system. As sliding proceeds with relaxation in shear stresses at grain boundaries (the so-called grain-boundary Zener relaxation[2)]), elastic normal stresses build up at grain edges and corners. The stresses ultimately grow up to balance themselves with the applied load when sliding ceases by grain interlocking.

In ceramic materials which have an extremely thin layer of a viscous or a diffusive substance between the mating facets of grains, the layer may transmit all normal stresses, while allow shear stresses to relax with a characteristic relaxation time. In such a grain-boundary sliding, the accommodation of internal stresses and strains may be purely elastic[2)-4)]. In contrast, in ceramic materials represented by two-phase ceramics, the grain-boundary layer is thick enough for the grains not only to slide but also to rearrange themselves by rotation.[5),6)] After rather short time shear relaxation at grain boundaries, this anelastic accommodation process proceeds with the long-range migration (squeezing in/out) of the grain-boundary melt (molten vitreous intergranular phase) through channels of the interconnected grain-boundary network. The creep deformation in the primary stage of various structural ceramics and refractory materials will be mainly accommodated by this anelastic process.[5)] Most high-temperature structural ceramics comprise highly refractory grains bonded by vitreous intergranular phase, forming two-phase materials. The vitreous phase is important for the structural integrity of such high-temperature ceramics. For the grains to continue to slide over one another, hydrostatic stresses will be developed at grain-boundary pockets, as well as local normal stresses will also build up at points and surfaces between interlocking gains.[7)] Accordingly, the creep deformation would become more and more difficult, and then eventually stop, if the lubrication of the melt were the only one possible process.

The intent of the present study is to argue the grain-boundary sliding and interlocking at high temperatures of two-phase ceramics which are composed of elastic grains embedded in a contiguous viscous matrix. Theoretical considerations will be given on the basis of Dryden-Kucerovsky-Wilkinson-Watt theory (DKWW theory).[8)]

2. Theoretical Considerations

2.1. Viscoelastic Constitutive Equation

In theoretical considerations on the "viscoelastic'' deformation of two phase ceramics at high temperatures, rigid hexagonal grains in the DKWW-model is replaced with ''elastic'' grains which are capable of storing elastic strain energy. The flow of the grain-boundary melt and the elastic deformation of the grains are, respectively, expressed by the shear modulus G and the shear viscosity η through the following constitutive equations of a Hookean solid and a Newtonian fluid;[9)]

$$\sigma_i = -p'\delta_i + 2Ge_i \tag{1}$$

$$\sigma_i = -p\delta_i + 2\eta\dot{e}_i \tag{2}$$

where σ_{ij}, e_{ij}, $\dot{e}_{ij}$ are the ij-components of stress, strain, and the rate of strain tensors, respectively. δ_{ij} is the Kronecker delta symbol. p', p are the hydrostatic pressure in the respective materials. To make the problem tractable, we consider the two-dimensional deformation and flow of the model aggregate with hexagonal grains, as shown in Fig. 1.

The normal force F which is required to squeeze together two parallel grain facets with the squeezing rate of $\dot{H}$ is given by[10)]

$$2\dot{H} = \frac{1}{\eta}\left(\frac{H}{L}\right)^3 (F + 2Lp) \tag{3}$$

The rate of elastic deformation $\dot{L}$ of the two-dimensional grains under the normal force F is,

$$\dot{L} = \frac{1}{4G}(\dot{F} + 2L\dot{p}') \tag{4}$$

where the derivative of a variable x with respect to time is indicated by $\dot{x}$.

As shown in Fig.1, upon using the center-to-center translation vectors ($\vec{A}$, $\vec{B}$, $\vec{C}$), and the widths of the grain-boundary fluid (H_A, H_B, and H_C at the $A-$, $B-$, and $C-$facet, respectively) as DKWW have defined in their paper, the following relations are derived between the microscopic variables and the microscopic strain rates $\dot{e}_{ij}$ and the stress σ_{ij};

$$\begin{aligned} 2\dot{H}_A + 2\sqrt{3}\dot{L} &= 2\sqrt{3}L\dot{e}_{11} \\ 2\dot{H}_B + 2\sqrt{3}\dot{L} &= \sqrt{3}L(\sqrt{3}\dot{e}_{12} - \dot{e}_{11}) \\ 2\dot{H}_C + 2\sqrt{3}\dot{L} &= -\sqrt{3}L(\sqrt{3}\dot{e}_{12} + \dot{e}_{11}) \end{aligned} \tag{5}$$

and

$$\begin{aligned} F_A &= 2L\left(\frac{3}{2}\sigma_{11} - \frac{1}{2}\sigma_{22}\right) \\ F_B &= 2L\left(\sigma_{22} + \sqrt{3}\sigma_{12}\right) \\ F_C &= 2L\left(\sigma_{22} - \sqrt{3}\sigma_{12}\right) \end{aligned} \tag{6}$$

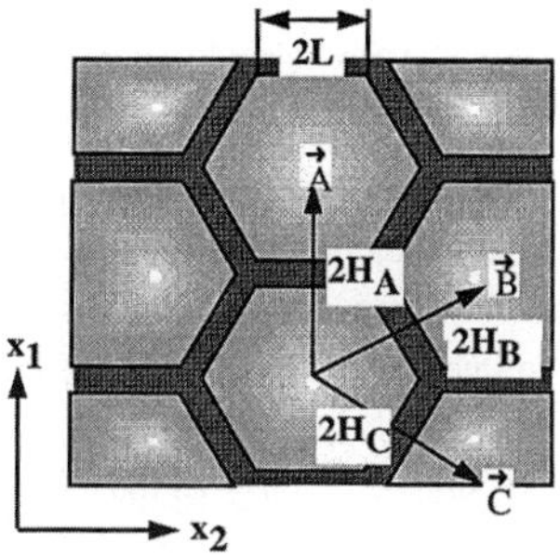

Fig. 1 A part of the model aggregate, showing a two dimensional array of hexagonal grains embedded in an incompressible contiguous fluid.

where H_I ($I = A, B, C$) and F_I ($I = A, B, C$) indicate the width and the force of the i-th type facet. The coordinate axes, x_1 and x_2, are also indicated in Fig. 1. In the derivation of Eq. 5, no change in volume for the overall deformation is assumed ($\dot{e}_{11} + \dot{e}_{22} = 0$).

Substituting Eqs. 3 and 4 into Eq. 5 and using the relations of Eq. 6, we have the following viscoelastic constitutive equations for the model hexagonal array;

$$\begin{aligned} \dot{e}_{11} &= \frac{1}{\sqrt{3}\eta}\left(\frac{H_A}{L}\right)^3\left(\frac{3}{2}\sigma_{11} - \frac{1}{2}\sigma_{22} + p\right) + \frac{1}{2G}\left(\frac{3}{2}\dot{\sigma}_{11} - \frac{1}{2}\dot{\sigma}_{22} + \dot{p}'\right) \\ \sqrt{3}\dot{e}_{12} - \dot{e}_{11} &= \frac{2}{\sqrt{3}\eta}\left(\frac{H_B}{L}\right)^3\left(\sigma_{22} + \sqrt{3}\sigma_{12} + p\right) + \frac{1}{G}\left(\dot{\sigma}_{22} + \sqrt{3}\dot{\sigma}_{12} + \dot{p}'\right) \\ \sqrt{3}\dot{e}_{12} + \dot{e}_{11} &= \frac{2}{\sqrt{3}\eta}\left(\frac{H_C}{L}\right)^3\left(-\sigma_{22} + \sqrt{3}\sigma_{12} - p\right) + \frac{1}{G}\left(-\dot{\sigma}_{22} + \sqrt{3}\dot{\sigma}_{12} + \dot{p}'\right) \end{aligned} \tag{7}$$

In the special case of infinitesimal strain where the grain-boundary channels all have the same equilibrium thickness $2H_o$, the two-dimensional hexagonal array exhibits six-fold symmetry. It is straightforward to manipulate Eq. 7 not only to solve for p and p', but also to demonstrate the equivalence between the hydrostatic pressures p on the viscous component and p' on the elastic component in equilibrium, and then result in

$$\dot{e}_{12} = \frac{1}{2\overline{\eta}_o}\sigma_{12} + \frac{1}{2\overline{G}}\dot{\sigma}_{12} \tag{8}$$

for simple shear loading ($\sigma_{12} \neq 0$, $\sigma_{11} = 0 = \sigma_{22}$). In Eq. 8, the zero-shear viscosity $\overline{\eta}_o$ and the shear modulus $\overline{G}$ of the hexagonal aggregate are, respectively, related to the shear viscosity η of the grain-boundary melt and to the shear modulus G of the hexagonal grains, as follows; $\overline{\eta}_o = 2\eta / 3f^3$ and $\overline{G} = 1/2 \cdot G$ where the volume fraction f of the contiguous melt is given by $2H_o / \sqrt{3}L$.

Under a finite strain, the asymmetry in grain-boundary channels is caused by the fact that the grain-boundary melt is squeezed out of the compressive channels, and squeezed into tensile channels, resulting in different thicknesses of H_A, H_B, and H_C from the initial thickness H_o in equilibrium. In consideration on simple shear deformation ($\sigma_{12} \neq 0$), we have the relationships of $H_A = H_o$, $2H_o = H_B + H_C$, $(e_{12})_\eta = -2(H_C - H_o)/3L$, and then $H_c = H_o[1 - \sqrt{3}(e_{12})_\eta / f]$ in terms of the internal shear strain $(e_{12})_\eta$ and the volume fraction f of the melt. Applying this consideration and derived relations to Eq. 7 to eliminate the hydrostatic pressures, p and p', the following nonlinear Maxwell constitutive equation is obtained for simple shear deformation:

$$\dot{\gamma} = \frac{1}{\overline{\eta}_s}\sigma_{12} + \frac{1}{\overline{G}}\dot{\sigma}_{12} \qquad (9)$$

with the internal strain dependent viscosity $\overline{\eta}_s$ in simple shear flow;

$$\overline{\eta}_s = \overline{\eta}_o \big/ \left(1 - \sqrt{3}/2 \cdot \hat{\gamma}_\eta\right)^3 \qquad (10)$$

where uses have been made of the engineering shear strain $\gamma \equiv 2e_{12}$ and the strain rate of shear $\dot{\gamma}(\equiv 2\dot{e}_{12})$ for convenience. The internal viscous strain $(e_{12})_\eta$ define the following "normalized" internal strain $\hat{\gamma}_\eta = \gamma_\eta / f$ with respect to the volume fraction f of the melt. The creep and the stress relaxation of the present model aggregate in simple shear deformation are discussed in the following sections.

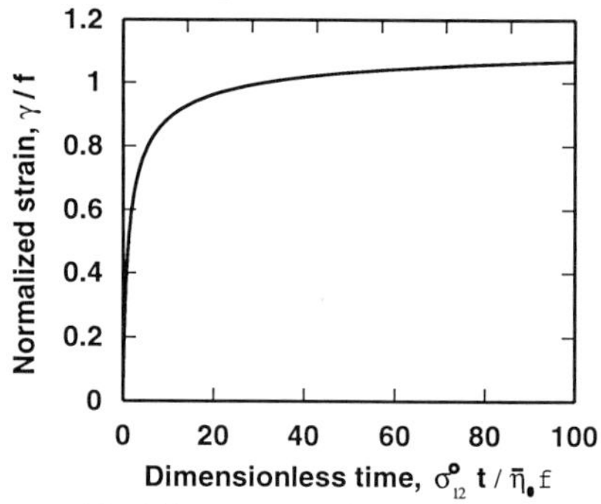

Fig. 2 Creep curve of the model hexagonal aggregate under simple shear.

2.2. Creep deformation under simple shear

Upon the application of a constant shear stress σ^o_{12} at time $t = 0$ to the model aggregate, Eqs. 9 and 10 give the shear rate of creep strain $\dot{\gamma}(t)$, as follows;

$$\dot{\gamma}(t) = \frac{\sigma^o_{12}}{\overline{\eta}_o}\left[1 - \frac{\sqrt{3}}{2}\frac{\gamma(t)}{f}\right]^3 \qquad (11)$$

Accordingly, the creep strain $\gamma(t)$ as a function of time t can, then, be derived by integrating Eq. 11 with respect to t;

$$\frac{\sigma^o_{12}}{\overline{\eta}_o f} \cdot t = \frac{1}{\sqrt{3}}\left[\frac{1}{(1 - (\sqrt{3}/2)\{\gamma(t)/f\})^2} - 1\right] \qquad (12)$$

The creep strain $\gamma(t)$ as a function of dimensionless time $(\sigma^o_{12} t)/(\overline{\eta}_o f)$ is illustrated in Fig. 2. The creep strain $\gamma(t)$ evolves monotonically with time, approaching to the equilibrium plateau of $\gamma_e (\equiv 2f/\sqrt{3})$ for grain interlocking.

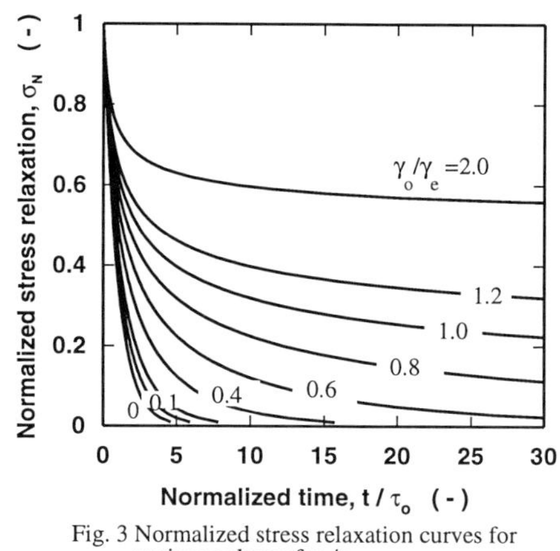

Fig. 3 Normalized stress relaxation curves for various values of γ_o/γ_e

2.3. Stress relaxation under simple shear

The model hexagonal aggregate is now subjected at time $t = 0$ to a suddenly applied shear strain γ_o which is maintained constant thereafter. Thus, the applied strain is expresses by a step function. After some straightforward calculations for the stepwise strain, the differential equation, Eq. 9, is cast into the following specific form of differential equation,

$$\frac{d\{\sigma_N(t)\}}{\sigma_N(t)[1-(\gamma_o/\gamma_e)\cdot\{1-\sigma_N(t)\}]^3} = -\frac{dt}{\bar{\tau}_o} \tag{13}$$

where the normalized stress relaxation $\sigma_N(t)$ is defined by $\sigma_{12}(t)/\sigma_{12}(0)$ using the induced shear stress $\sigma_{12}(0)(\equiv \bar{G}\gamma_o)$ at time $t = 0$ and the equilibrium shear strain γ_e at grain interlocking. Notice in the derivation of Eq. 13 that the internal viscous shear strain γ_η is given by $\gamma_o[1-\sigma_N(t)]$ in any Maxwell materials. The zero-shear relaxation time τ_o in the equation is defined by $\bar{\eta}_o/\bar{G}$. The integrations can be carried out with respect to the normalized stress $\sigma_N(t)$ in the left-hand side and with respect to time t in the right-hand side of Eq. 13, resulting in

$$\frac{t}{\bar{\tau}_o} = \frac{1}{2(1-\gamma_o/\gamma_e)} \times \frac{(\gamma_o/\gamma_e)\{1-\sigma_N(t)\}[2-(\gamma_o/\gamma_e)\{1-\sigma_N(t)\}]}{[1-(\gamma_o/\gamma_e)\{1-\sigma_N(t)\}]^2}$$
$$+\frac{1}{(1-\gamma_o/\gamma_e)^2}\frac{(\gamma_o/\gamma_e)\{1-\sigma_N(t)\}}{[1-(\gamma_o/\gamma_e)\{1-\sigma_N(t)\}]} - \frac{1}{(1-\gamma_o/\gamma_e)^3}\ln\left|\frac{1-(\gamma_o/\gamma_e)\{1-\sigma_N(t)\}}{\sigma_N(t)}\right| \tag{14}$$

The stress relaxation curves, $\sigma_N(t)$ vs. t/τ_o, give in Eq. 14 are illustrated in Fig. 3 for the normalized applied strains γ_o/γ_e ranging from 0 (corresponding to the linear Maxwell relaxation) to 2.0 (the applied strain γ_o two times larger than the equilibrium strain). A complete relaxation in stresses is observed for $\gamma_o/\gamma_e \leq 1.0$, though the time to a complete relaxation increases with increasing applied strain.

3. Application to the experimental results

The creep and stress relaxation tests under simple shear of a β-spodumene glass ceramic are conducted to scrutinize the constitutive equation, resulting in a satisfactory agreement with the theoretical predictions, and giving important rheological information. It is emphasized that stress relaxation test is one of the important test techniques which enable to study quantitatively the rheological behavior of polycrystalline ceramics with grain-boundary sliding and interlocking without any difficulties and ambiguities accompanied by stress-induced grain-boundary cavities which so often appear in conventional creep tests. The details of the geometry of tested specimen and the apparatus, and some rheorogical analysises have been reported in the literature.[11)12)13)]

4. Conclusion

The DKWW model for the viscous creep deformation of a two-dimensional array of rigid hexagonal grains embedded in an incompressible contiguous fluid has been extended in the present study to incorporate the elastic deformation of grains. The resultant viscoelastic constitutive equation has been expressed by a nonlinear Maxwell type, and successfully applied to the creep and stress relaxation processes in the deformation and flow of a β-spodumene ceramic at elevated temperatures.

References

1) A.S. Nowick and B.S. Berry, Anelastic Relaxation in Crystalline Solids; Ch. 15. Academic Press, New York, 1972.
2) C. Zener, *Phys. Rev.*, **60**, 906-08 (1941).
3) T.-S. Kê, *Phys. Rev.*, **72** [1] 41-46 (1947).
4) F. Ghahremani, *Int. J. Solids Structures*, 16, 825-45 (1980).
5) S.M. Wiederhorn, B.J. Hockey and R.F. Krause, Jr., *J. Mater. Sci.*, **21**, 810-24 (1986).
6) C.J. Gasdaska, *J. Am. Ceram. Soc.*, **77** [9] 2408-18 (1994).
7) R. Raj and M.F. Ashby, *Metal. Trans.*, **2** 1113-27 (1971).
8) J.R. Dryden, D. Kucerovsky, D.S. Wilkinson and D.F. Watt, *Acta. Metall.*, **37**, 2007-15 (1989).
9) I.H. Shames and F.A. Cozzarelli, Elastic and Inelastic Stress Analysis; Ch. 3. Prentice Hall, Englewood Cliffs, NJ, (1992).
10) J.C. Jaeger, Elasticity, Fracture and Flow, 3rd Ed., pp. 140-43, Methuen & Co. Ltd., London (1992).
11) H. Muto and M. Sakai, *J. Am. Ceram. Soc.*, in press.
12) M. Sakai and H. Muto, *J. Am. Ceram. Soc.*, in press.
13) M. Sakai , H. Muto and M. Haga, *J. Am. Ceram. Soc.*, **79** [2] 449-54 (1996).

Materials Science and Engineering Serving Society
S. Sōmiya, R.P.H. Chang, M. Doyama and R. Roy (Editors)

Aluminum deposition by using MOCVD on pitch based carbon fibers

Takakazu Suzuki

National Institute of Materials and Chemical Research
1-1 Higashi, Tsukuba 305, Japan

Mechanical and physical properties of aluminum films deposited by MOCVD are studied as a function of deposition temperature, deposition pressure, carrier gas flow rate or species. The ultimate tensile strength of the aluminum foil deposited under the condition of 573 K, 1330 Pa and 7 ml/s of Ar reached 70 MPa. This value is higher than that of pure aluminum; 47 MPa, and it is equivalent to that of annealed aluminum (1060). The maximum elongation arrived at 3% and it is equivalent to that of cold-worked one of 1100 aluminum. The lattice constant of the same conditioned foil was 4.0497Å. This value is very close to that of pure aluminum, 4.0496 Å, which means the physical properties, e.g. coefficient of thermal expansion, etc. are also very close to pure aluminum.

1. INTRODUCTION

High performance aluminum films possessing high mechanical and physical properties close to pure aluminum are expected for wide variety of the applications, e.g. electronics field; very large integrated circuit (VLSI) surface engineering and composite materials technology. CVD under the condition of reduced pressures of 100 Pa -100 KPa is interesting, because of the expectation for the stronger and purer films' production at higher deposition rate. Moreover, the excellent throwing power of CVD makes it effective to aluminum coating onto the complex shapes such as fiber bundles. The optimal conditions have been investigated to produce the strongest and most pure aluminum, and it is employed for the research of evaluating the tensile strength after annealing at 773-873K and the interface between aluminum and fibers that have wide variety of micro-texture and discussing interaction with aluminum [1], which suggested that well-controlled micro-structural fibers are useful to improve the properties of fiber reinforced aluminum composites.

2. EXPERIMENTAL

2.1 Apparatus for Metal-organic chemical vapour deposition (MOCVD)

The apparatus for preparation of aluminum consists of gas supplying system, three-zone electric furnace, thermal decomposition reaction chamber, atmospheric pressure control system, and exhausting system. MOCVD by pyrolizing tri-isobutyl aluminium under the condition of vaporized temperature (Tv) of 333-373K, deposition temperature (Td) of 553-680K, and deposition pressure (Pd) of 100Pa-80kPa in Ar at flow rate (F) of 3-30ml/s was employed.

3. RESULTS AND DISCUSSION

3.1 Optimum Condition for Aluminum Deposition

The optimum condition to deposit aluminum at higher rate in the whole chamber has already been reported in previous paper [2][3] so that these are omitted in the limited text.

3.2 Chemical composition of Deposited Aluminum

From the results of AES, In an aluminum foil deposited on a quartz under the conditions of Td= 553 K, Pd= 1330 Pa, F=7 ml/s, the carbon and oxygen contents were less than 0.3 at.% (0.1 mass%). On the other hand, the carbon and oxygen were detected much more in the greyish aluminum foil deposited under the condition of Td = 623 K, Pd = 1330 Pa, F = 7 ml/s. The difference in carbon content agrees with the results by Bent and others [4]. For oxygen, it was probably derived from water adsorbed inside stainless pipe, or from oxygen or water containing in carrier gas. The results of the aluminum foil formed under the conditions of Td = 553 K, Pd = 80 kPa, F= 3 ml/s shows that the foil is containing much more carbon and oxygen, compared with the foil under the condition of Td= 553 K, Pd= 1330 Pa, F=7 ml/s. This is because the diffusion coefficient of gas decreased with increasing Pd [5], and consequently H2O or remaining alkyl group is taken in the films. Aluminum foil formed under the condition of Td = 553 K, Pd = 1330 Pa, F= 17 ml/s indicates that the carbon and oxygen in the film tend to increase with increasing flow rate. The oxygen was derived from a trace of H2O contained by carrier gas.

3.3 Quantitative Analysis of Carbon, Hydrogen and Oxygen in Aluminum

The results of gas analysis for the carbon, hydrogen and oxygen content in deposited Aluminum have been shown in Table 1. The values determined by using AES already mentioned and a gas chromatography are also shown in Table 1 as well. The correlation between the results by using gas analysis and AES is estimated to be high. In the temperature range, Td= 600 K or more carbon, hydrogen and oxygen were remarkably increased in the foil. Hydrogen content was the highest under the pressure of Pd= 80 kPa. This is because gas products or remaining raw materials accumulated with decreasing exhaust times.

Table1. Results of quantitative analysis of carbon, oxygen and hydrogen in Aluminum

	C	H	O	Al
(T_d/P_d/F)	(mass%)	(mass%)	(mass%)	
553K, 1330Pa, 7ml/s	0.4, (0.1), 0.005*	0.004	0.1	
603K, 3990Pa, 7ml/s	(8)	0.01	9,(8)	
553K, 80kPa, 3ml/s	(4)	0.2	5	
553K, 399Pa, 17ml/s	(1)	0.01	0.6,(1)	

Values in ( ) are by using AES and value with * is by using gas chromatography.

Figure 1. SEM image of fractured surface

3.4 Hardness

SEM image of fractured surface of the deposited aluminum is shown in Figure1. From this, evidently the foil is made up of aluminum grains of about a few μ m and its fractured surface takes on an aspect of grain boundary fracture.

Hardness is a physical property that shows a resistance for a plastic deformation in materials and is most conventional testing method for estimating the material strength without destruction. The ultimate micro Vickers hardness (Hv) (applied load: 2g) of aluminum foil obtained under the condition of Td= 553 K, Pd= 1330 Pa, F= 7 ml/s was Hv= 40, the foil under the condition of Td= 673 K, Pd=1330 Pa, F=17 ml/s was Hv= 34 and it is decreasing with increasing Td or F.

3.5 Tensile Strength and Young's Modulus

The ultimate tensile strength of the aluminum foil (Td= 573 K, Pd= 1330 Pa, F= 7 ml/s) was about 70 MPa. This value is larger than that of high purity aluminum (99.996 mass%): 47 MPa, and is comparable to the tensile strength of annealed 1060 aluminium. The maximum elongation for fracture of this aluminum was about 3%. Though the value is close to that of 1100-H18 or some cast aluminium alloy, it is considerably small compared to pure aluminum. The Young's modulus was obtained as 17 GPa . The value is extremely small compared to 70 GPa for pure aluminum.

3.6 Lattice Constant

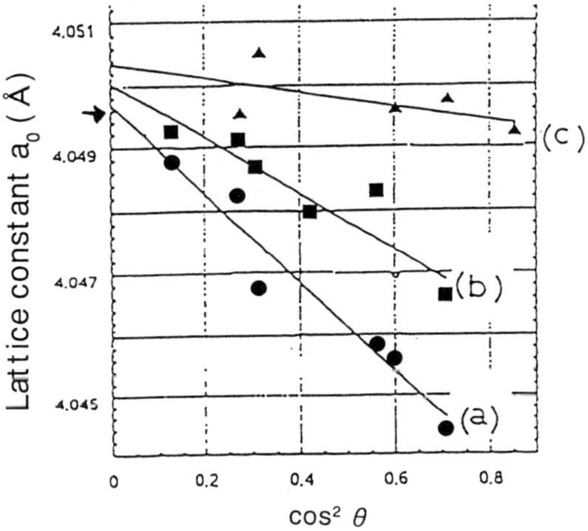

Figure 2. Variation of lattice constants of aluminum(a_0)

(a) : T_d=573 K, P_d=1330Pa, F=7 ml/s (●)
(b): T_d=603 K, P_d=3990Pa , F=7ml/s (■)
(c): T_d=553K,P_d=80kPa, F=3 ml/s (▲)

Lattice constant of each aluminum lattice plane measured by x-ray and cos2 θ are plotted in Figure 2. Lattice constant of the aluminum foil (a) deposited under the condition of Td= 573 K, Pd= 1330 Pa, F =7 ml/s was 4.0497 Å. This value is very close to a genuine lattice constant of pure aluminum, 4.0496 Å, which means that the physical and chemical properties are also very close to pure aluminum. Lattice constant of the aluminum foil (b) formed under the conditions, Td= 603 K, Pd= 3990 Pa, F=7 ml/s was 4.0500 Å, and that of the aluminum foil (c) formed under the condition of Td= 553 K, Pd= 80 kPa, F=3 ml/s was 4.0503 Å. The genuine lattice constants of foil (b) and (c), were slightly larger than that of the foil (a).

The expansion coefficient of aluminum foil prepared by the condition of Td= 573 K, Pd= 1330 Pa, F= 7 ml/s was 24.1 at 293-372K. This value is very close to that of pure aluminum value (23.9)

4. SUMMARY

In this paper, clarified was the relationships between conditions for depositing aluminum and impurities in the aluminum and the mechanical and physical properties. The data provided by this study on the mechanical properties and lattice constant of the CVD-Al films are seen to be of great use for industrial applications.

REFERENCES

1. T.Suzuki, Compos. Sci. and Technol., 56 (1996) 147.
2. T.Suzuki, H.Umehara, J. Jpn. Inst. Metals, 51 (1987) 577.
3. T.Suzuki: J. Mater. Res., 7 (1997) 1866.
4. B.E. Bent, R.G. Nuzzo and L.H. Dubois: J. Am. Chem. Soc., 111 (1989) 1634.
5. G.B. Stringfellow: Organometallic Vapor-phase Epitaxy, Ed. by G.B.Stingfellow Academic Press, San Diego, (1989) 216.

Materials Science and Engineering Serving Society
S. Sōmiya, R.P.H. Chang, M. Doyama and R. Roy (Editors)

Combustion synthesis of fine ceramics cascade

B.-W. Chen and C.-C. Chen
Department of Chemical Engineering, National Chung Cheng University, Chia-Yi 621, Taiwan

Due to the convective and radiant heat losses, there exists a maximal converted length of a dense pellet synthesized by the SHS method. In this study, we numerically investigated the possibility to increase that maximal converted length by cascading two reactant pellets in series, where an interface is naturally and artificially introduced. The impacts of both the bulk and interfacial parameters on the SHS process were estimated. The maximal converted length for a single pellet was computed. By varying the interfacial parameters, it was found that more than 10% of extra converted length was obtained by the proposed cascade arrangement. Effects of the bulk parameters on the same purpose were also addressed.

1. INTRODUCTION

The self-propagating high-temperature Synthesis (SHS), or the combustion synthesis is emerging as an attractive way to produce advanced materials, because the SHS method has the following advantage [1]: lower energy requirement, higher product purity, simpler and cheaper equipment, higher sinterability and possible nonequilibrium phases in the final products. Also, combining with the pressurized techniques such as explosives [2], HIP [3], and HPCS [4,5], the SHS method is capable of producing very dense products.

Besides the successes of producing advanced materials experimentally, theoretical studies on the SHS method are abundant in the literature. However, we are not aware of any studies on the synthesis of finite-length reactant pellets with interface present. In this study, we are interested in introducing interface by cascading two dense green pellets to increase the conversion length. We ask a question: under the same process condition, is it possible to further increase the reacted length so that the amount of final product could be raised, especially for the high density pellet? In this study, we investigate this possibility by cascading two reactant pellets to see if the overall converted length can exceed the maximal one of a single pellet.

2. MODEL EQUATIONS AND THE SOLUTION METHOD

Following the model equations of a single pellet proposed by Varma et al. [6], model equations the system of two pellets cascaded together could be described by a 3-zone model. The dimensionless equations of the 3 zones can be written as
zone 1:

$$\frac{\partial\theta_1}{\partial\tau} = \frac{\partial^2\theta_1}{\partial x^2} + \beta\,(1 - \eta_1)\exp(\gamma\,(1-\frac{1}{\theta_1})) - \alpha\,(\theta_1 - \theta_0) - \psi\,(\,\theta_1{}^4 - \theta_0{}^4\,) \tag{1}$$

$$\frac{\partial\eta_1}{\partial\tau} = (1 - \eta_1)\exp(\gamma\,(1-\frac{1}{\theta_1})) \tag{2}$$

zone 2:

$$\frac{\partial\theta_2}{\partial\tau} = \frac{\varepsilon_3}{\varepsilon_1\varepsilon_2}\frac{\partial^2\theta_2}{\partial x^2} + \frac{v}{\varepsilon_1\varepsilon_2}\beta(1 - \eta_2)\exp(\gamma\,(1-\frac{1}{\theta_2})) - \frac{\varepsilon_4}{\varepsilon_1\varepsilon_2}\alpha(\theta_2 - \theta_0) - \frac{1}{\varepsilon_1\varepsilon_2}\psi(\theta_2{}^4 - \theta_0{}^4) \tag{3}$$

$$\frac{\partial\eta_2}{\partial\tau} = (1 - \eta_2)\exp(\gamma\,(1-\frac{1}{\theta_2})) \tag{4}$$

zone 3:

$$\frac{\partial\theta_3}{\partial\tau} = \frac{\partial^2\theta_3}{\partial x^2} + \beta\,(1 - \eta_3)\exp(\gamma\,(1-\frac{1}{\theta_3})) - \alpha\,(\theta_3 - \theta_0) - \psi\,(\,\theta_3{}^4 - \theta_0{}^4\,) \tag{5}$$

$$\frac{\partial\eta_3}{\partial\tau} = (1 - \eta_3)\exp(\gamma\,(1-\frac{1}{\theta_3})) \tag{6}$$

with the initial and boundary conditions

$$\theta_1 = \theta_2 = \theta_3 = \theta_0,\ \eta_1 = \eta_2 = \eta_3 = 0;\ \tau = 0 \tag{7}$$

$$-\frac{\partial\theta_1}{\partial x} = \phi; \qquad 0 \le \tau \le \tau_\phi,\ x = 0 \tag{8}$$

$$\frac{\partial\theta_1}{\partial x} = \mathrm{Bi}\,(\theta_1\text{-}\theta_0) + \Omega\,(\theta_1{}^4\text{-}\theta_0{}^4),\ \tau>\tau_\phi\ ,x = 0 \tag{9}$$

$$\frac{\partial\theta_1}{\partial x} = \varepsilon_3\frac{\partial\theta_2}{\partial x},\ \theta_1 = \theta_2;\ x = L_1 \tag{10}$$

$$\varepsilon_3\frac{\partial\theta_2}{\partial x} = \frac{\partial\theta_3}{\partial x},\ \theta_2 = \theta_3;\ x = L_1 + \varepsilon_5 \tag{11}$$

$$-\frac{\partial\theta_3}{\partial x} = \mathrm{Bi}\,(\theta_3 - \theta_0) + \Omega\,(\,\theta_3{}^4 - \theta_0{}^4);\ \tau \ge \tau_\phi\ ,\ x = L_1 + \varepsilon_5 + L_2 \tag{12}$$

where θ, η, β, α, ψ, Ω, ϕ, γ, τ, and τ_ϕ denote the dimensionless temperature, conversion, heat of formation, Newtonian heat transfer coefficient, radiant heat transfer coefficient from the pellet surfaces, radiant heat transfer coefficient from the pellet ends, strength of external heat flux, activation energy, time, and duration of external heat source, respectively. L_1, ε_5 and L_2 represents the dimensionless length of zone 1, 2 and 3. Bi is the Biot number. ε_1, ε_2, ε_3, and ε_4 are the dimensionless quantities of the relative heat capacity, density, heat conduction coefficient, and Newtonian heat transfer coefficient, between zone 2 and zone 1 (or zone 3). The reference temperature, T^*, is chosen as the adiabatic temperature for convenience.

We adopted an implicit finite-difference scheme, which could avoid the numerical instability problem. The nonlinear boundary conditions are treated by the traditional Newton-Raphson method.

3. RESULTS AND DISCUSSIONS

Because of the presence of the convective and the radiant heat losses, there exists a maximal converted length for a dense green pellet. For the following parameters, $\beta = 0.5$, $\gamma = 10$, $\theta_0 = 0.093$, $\alpha = 0.001$, Bi = 0.005, $\Omega = 0.001$, the maximal converted length L* for a single pellet under various external heating rate was 1.08.

3.1. Influences of process parameters on the SHS process

Because we are interested in obtaining a longer converted length by a cascade arrangement than the maximal one of a single pellet, it is important to know to what an extent each processing parameter will influence the SHS reaction. We use a criterion to define the 'reactivity' of the SHS process: the time t* for end of the second pellet to achieve at least 90% conversion. We summarize the importance of all the process parameters on the reactivity of the SHS process in Table 1 by using the derivatives of t^* with respect to each parameters.

Table 1
Effects of process and interfacial parameters on the SHS process

	reactivity		reactivity		reactivity
θ_0	-8.265	Bi	-0.00682	ε_3	0.00240
α	12.586	v	-0.00303	ε_4	0.000393
ψ	75.131	ε_1	0.00113	ε_5	4.475
Ω	-80.518	ε_2	0.00113		

Sign of the reactivity shows the benefit of the parameters on the SHS process. If sign of the derivative is negative for a certain parameter, then increase magnitude of this parameter favors the SHS reaction. Next, magnitude of the reactivity represents the importance of parameters. The effect of a certain parameter on the SHS process is more significant if the corresponding derivative has a larger absolute value. It can be readily seen that the convective heat transfer coefficient α, the initial temperature θ_0, the interface thickness e_5, and the radiant heat transfer coefficient from surfaces ψ and from ends Ω are far more important than the rest of parameters.

3.2. Increase of converted length by the cascade arrangement

We would like to see whether the total converted length can exceed the maximal reacted length L^* of a single pellet by cascading two reactant pellets in series. The starting dimensionless parameter values is first selected as (β, γ, θ_0, ϕ, ψ, α, Ω, Bi, ε_1, ε_2, ε_3, ε_4, ε_5, v) equal to (0.5, 10, 0.093, 0.75, 0.3, 0.001, 0.001, 0.005, 0.2, 0.2, 0.2, 0.2, 0.01, 0.2). Under this set of parameters, the maximal converted length can increase from L* = 1.08 of a single pellet to 1.11 of the pellet cascade.

Since the interfacial length ε_5 has a great influence on SHS, effect of ε_5 on the increase of converted length was first studied. It was found that the overall converted length could be raised to 1.14 if ε_5 was reduced to $6.186\text{x}10^{-4}$. Now, we will use this ε_5 value and will pursue further length improvement by varying other interfacial parameters.

First, from Eq. 3, effect of interface heat capacity ε_1 is the same as interfacial density e_2, so it is redundant to study both the effects of ε_1 and ε_2, repeatedly. The converted length remains almost the same as ε_2 varies. This is due to the insignificance of ε_1 or ε_2 on the SHS process given in Table 1. Next, influence of the interfacial heat conduction coefficient ε_3 is considered. Increase of ε_3 from 0.1 to 1.0 could further raise the converted length to 1.18. Next, consideration is given to the weight ratio v. In addition to being the amount of reactant mixtures in the interface, v can also represents an interface modification by adding a small amount of highly exothermic powders, since a large v means larger heat released in the interface. The converted length increase from 1.14 to 1.16 as v varies from 0.2 to more than 5. Table 2. summarizes the progress of converted length increase as interfacial parameters being manipulated.

Table 2
The progress of the maximal converted length as interfacial parameters varied.

	single pellet	pellet cascade	decrease ε_5	increase v or increase ε_3
maximal length	1.08	1.11	1.14	1.16 1.18

In addition to increase the overall converted length by varying the interfacial parameters, effects of the more important processing parameters such as the initial temperature (θ_0), radiant heat transfer coefficients from both the pellet surface (ψ) and pellet end (Ω) are studied. For fair comparisons, when these three parameters are varied, it applies both for a single pellet and for the pellet cascade. The maximal relative ratios of converted length increase are listed in Table 3. It can be seen that more than 10% of length increase can be obtained by adjusting these process parameters.

Table 3
The maximal relative ratios of converted length as the process parameters varied. Intervals following the parameters indicate the range of parameter variations.

	θ_0 (0.1, 0.5)	ψ (0.1, 0.5)	Ω (0.1, 0.5)
maximal increase ratio (%)	12	10.5	12.5

REFERENCES

1. A. G. Merzhanov, in Combustion and Plasma Synthesis of High-Temperature Materials, edited by Z. A. Munir and J. B. Holt ,VCH Publishers, New York, 1, 1990.
2. L. J. Kecskes, R. F. Benck, P. H. Netherwood Jr., J. Am. Ceram. Soc. 73 (1990) 383.
3. J. B. Holt and Z. A. Munir, J. Mater. Sci. 21, 251 (1986).
4. O. Yamada, Y. Miyamoto, and M. Koizumi, Am. Ceram. Soc. Bull. 64 (1985) 319.
5. O. Yamada, Y. Miyamoto, and M. Koizumi, J. Am. Ceram. Soc. 70 (1987) 206.
6. A. Varma, G. Cao, and M. Morbidelli, AIChE J. 36, 1032 (1990).

Materials Science and Engineering Serving Society
S. Sōmiya, R.P.H. Chang, M. Doyama and R. Roy (Editors)

Large-scale combustion synthesis of silicon carbide

C.-C. Chen, Y.-C. Fu and K.-Y. Liao

Department of Chemical Engineering, National Chung Cheng University, Chia-Yi 621, Taiwan

Large-scale synthesis of SiC was conducted by a large and thin reactant sheet (120x120x5 mm), which was made from 80 g of ground reactant powders of silicon and carbon black with 65 ml. of 8 wt.% of PVA solution added, combusted by a custom-built oxy-acetylene torch in the air. The averaged yield was around 95%. The product phase was mainly β-SiC and the morphology of SiC is aggregated particles, or particles sintered together.

1. INTRODUCTION

The self-propagating high-temperature synthesis (SHS), or the combustion synthesis, is emerging as an attractive method to produce powders or densified products of advanced materials [1]. For the advanced materials to be synthesized by the SHS method, two empirical criteria [2] must be satisfied such that the combustion wave could be sustained: $\Delta H/C_p$ at 298K is larger than 2000K and the adiabatic temperature T_{ad} must exceed 1800K. For silicon carbide it is not feasible to use the SHS method under the normal condition (1 atm and 298K), because the heat of formation 69 kJ/mol and the adiabatic temperature Tad 1800K are both too low. Thus, it could be found in the literature that several auxiliary arrangements were adopted to produce SiC by the SHS method. The utilized techniques were preheating the reactant pellet to raise the adiabatic temperature, using a high-pressure nitrogen environment to initiate the Si-N reaction and to trigger the Si-C reaction, increasing the heating surface under high pressure, and adding chemical or oxidative agents to alter the reaction mechanism. The particle sizes of reactant powders of Si and C used in the above studies were all near or much smaller than 1 mm.

Although SiC could be successfully produced by the SHS method using the aforementioned techniques, we found in a previous study [3] that direct combustion synthesis of SiC was possible without any auxiliary assistance. From our previous findings, direct combustion synthesis of SiC without any auxiliary methods was realized. The only limitation is the small longitudinal conversion length. There is no limitation on the transverse direction as long as the green pellet could receive heat from an external heat source. Thus, in order to overcome this drawback and to study the possibility of a large-scale production of SiC, large reactant sheets are used to increase the amount of SiC produced.

2. EXPERIMENTAL PROCEDURES

In this study, powders of Si (99.5% pure, <44 mm, Cerac) and carbon black (99% pure, <2 mm, China Synthetic Rubber) were used. In certain cases, powders were subjected to a fine grinder (DAP-7, Struler) for 60 minutes to reduce particle sizes. The powder reactants in stoichiometric ratio were first mixed, then dispersed in n-hexane by a mechanical stirrer. The resulting slurry was air dried and the dried mixture was cold pressed into a rectangular reactant sheet by an uniaxial single acting press at 5000 psi. The length, width and height of the reactant sheet was 120 mm, 120 mm and 6 mm, respectively. Because the resulted reactant sheet was often too fragile, 65 ml. of 8 wt.% of PVA solution was added into the 80 g of the reactant powders prior to the cold press to obtain a more compact reactant sheet. The thin reactant sheet was air dried for 10 hours. A custom-built comb-shape torch, which had 9 nozzles lined up in series, was applied to the edge of a reactant sheet. Once the ignition started, the torch was continually moving to the unreacted part until the whole reactant sheet was converted. A high-speed CCD (Toshiba, jk-c40) focused on the reactant sheet was used to monitor the progress of the combustion.

The products were analyzed by the XRD (MAC, MXP3) and the SEM (Jeol, JSM-5410). The reaction yield was computed as the weight ratio of w_1/w_0, where w_0 was the weight of the product and w_1 was the weight of product after removing the unreacted Si in 1M NaOH solution for 24 hours and burning the remained carbon in an oven at 600°C for 40 minutes.

3. RESULTS AND DISCUSSIONS

3.1. Cold press of the reactant sheets

Because the reactant sheet is subjected to a torch, a certain degree of structural strength is required for the reactant sheet to endure the torch flame. If the reactant sheet is too fragile, it will be easily blown apart by the torch so that the combustion wave can not be propagated. However, it is not easy to cold press the needed large-thin reactant sheet. Several attempts had been made but without any success. These attempts included: cold pressed the reactant powders repeatedly at a low pressure, and then were subjected to a higher pressure; applied BN as a lubricant to the inner walls of the rectangular die; used springs to increase compaction pressure; mixed paraffin wax with the powders to increase the mobility of powders. The difficulty of forming the compact reactants sheets does not only result from the sample geometry but also from the carbon black powders. Carbon black is not easily to be cold pressed for its poly-chain microstructure. Finally, the successful method to produce compact reactant sheets was to add 65 ml. of 8 wt.% of PVA into 80 g of reactant powders prior to cold press. The added PVA will not contaminate the products, since it will be burned and evaporated at the elevated temperature.

3.2 Combustion of the reactant sheets

It was observed that for a reactant sheet made from Si, carbon black and PVA, the combustion took place only at the burned surface and the propagation length of the combustion wave was less than 1 mm. This result might be caused by the added PVA. Once PVA was burned away, the space originally occupied by PVA became voids for the reactant powders. The reduction of contact areas between Si and C particles prohibited the propagation of combustion wave. In order to increase the reactivity, reactant powders were subjected to grinding to lower the particle sizes. The combustion of the reactant sheet made from ground powders was much better than the previous case without grinding. However, the averaged yield was only around 65%.

In order to achieve a better yield, sample was subjected to preheating. After trials and errors, the following experimental procedures were capable of producing high-purity products. A reactant sheet was first preheated for 1 minute by moving the torch back and forth at a speed of 6 cm/sec. Then, the torch was applied immediately to one edge of the reactant sheet. It took about 10 seconds to ignite the edge of the reactant pellet. The propagation of combustion wave in the longitudinal direction, i.e. the burning direction, easily exceeded 6 mm of the height of the reactant sheet. So, the conversion of the reactants in the longitudinal direction was complete. The combustion wave also traveled transversely (forward) about 15 mm ahead. The torch then moved forward to the unreacted portion of reactant sheet at a speed of 3 mm/sec. The burning process continued until the entire reactant sheet was converted.

3.3 Characterization of products

Because the reactant sheet was relatively large, the reaction may not be uniform throughout the reactant sheet and the product morphology could be position dependent . In order to avoid the possible misinterpretation, samples taken from 6 representative positions (marked in Fig. 1) were all subjected to analysis. From Table 1, yields of the products without preheating were much smaller than those with preheating. Table 1 also showed that the yields at the edges (A, B and D) were smaller than those at the other sample regions. This should due to the heat loss at the sample edge. Also, carbon black powders at the edges were readily oxidized so that the molar ratio between carbon black and silicon was reduced such that the reaction could not be as complete as that in other regions. If the starting molar ratio was increased from 1 to 1.05 and 1.10, it could be seen from Table 1 that yields at six regions were all increased, especially significant at the sample edges.

Table 1
Measured yields (%) at the different locations of products. M.R. is the molar ratio of C and Si. The symbols 'o' and 'x' denote with and without preheating, respectively.

M.R.	1.0, x	1.1, x	1.0, o	1.05, o	1.1, o
A	59.1	67.7	88.7	98.8	93.6
B	61.3	71.4	92.0	90.5	94.3
C	72.0	72.0	97.9	94.4	99.7
D	65.1	72.0	90.2	90.9	97.4
E	66.2	76.0	98.3	99.5	96.8
F	66.7	74.3	94.4	95.7	98.7

Fig. 2 shows the XRD patterns of product powders. Products were mainly β-SiC with a possible small amount of α-SiC. The morphology of the product was mainly aggregated particles as shown in Fig. 3. In some regions, different degrees of sintering were also observed. It can be seen from Fig. 3 that the particle size was small and was similar to that of carbon black, so this suggests a diffusion-controlled reaction mechanism [4]. Formation of SiC involves the diffusion of solid (or liquid) Si into carbon to form the carbide phase.

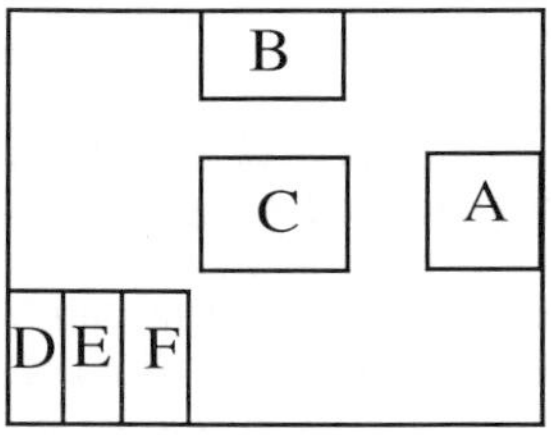

Figure 1. Six regions of product were subjected to analysis.

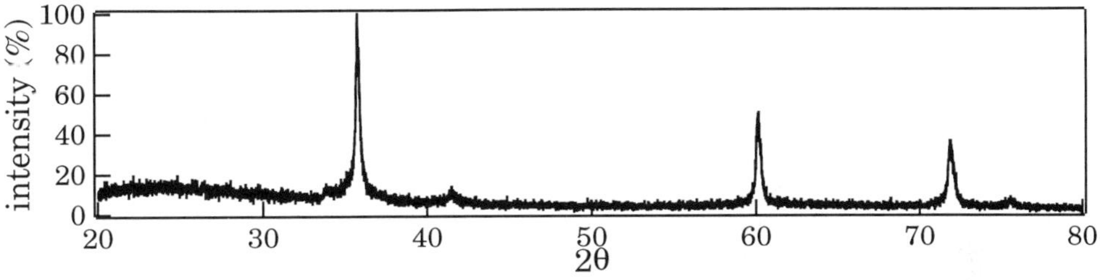

Figure 2. X-ray diffraction pattern of region C of the SiC product prepared from powders with a starting molar ratio between and C 1.1.

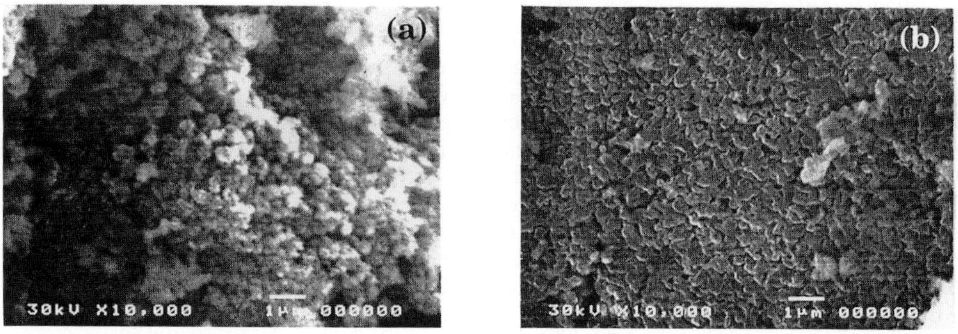

Figure 3. Scanning micrograph of region (a) A and (b) F of the SiC product preparing from powders with a starting molar ratio 1.0.

REFERENCES

1. A. G. Merzhanov, in "Combustion and Plasma Synthesis of High-Temperature Materials", edited by Z. A. Munir and J. B. Holt, VCH Publishers, New York, 1, 1990.
2. Z. A. Munir, Amer. Ceram. Soc. Bull. 67 (1988) 342.
3. Wu andd Chen, "Direct combustion synthesis of silicon carbide", submitted to J. Maters. Sci.
4. R. Pampuch, J. Lis, and L. Stobierski, in "Combustion and Plasma Synthesis of High Temperature Materials", edited by Z. A. Munir and J. B. Holt, VCH Publishers, New York, 211, 1990.

Materials Science and Engineering Serving Society
S. Sōmiya, R.P.H. Chang, M. Doyama and R. Roy (Editors)

A NEW PROCESS TO OBTAIN RE123 BULK SUPERCONDUCTORS WITH ENHANCED SUPERCONDUCTING PROPERTIES IN AIR

M. Kambara[1], Y. Watanabe[1], K. Miyake[2], K. Murata[2], A. Endo[3], Y. Shiohara[3] and T. Umeda[1], [1]Graduate School of Engineering, University of Tokyo, 7-3-1 Hongo, Bunkyo-ku, Tokyo 113, Japan, [2]Shibaura Institute of Technology, 3-9-14 Shibaura, Minato-ku, Tokyo 108, Japan, [3]Superconductivity Research Laboratory - ISTEC, 1-10-13 Shinonome, Koto-ku, Tokyo 135, Japan.

ABSTRACT

$RE_{1+x}Ba_{2-x}Cu_3O_{6+d}$ (RE123; RE = Sm, Nd) crystals with a smaller amount of substitution were produced in air with changing the initial composition from the usual composition of RE123 - $Sm_2Ba_1Cu_1O_5$ / $Nd_4Ba_2Cu_2O_{10}$ (RE211) to the Ba enriched side, which resulted in high T_c values. Another unique and interesting finding includes the reduction of the RE211 particle size with increasing the BaO/CuO ratio in the initial composition. Uniform dispersion of the fine RE211 particles was attained in the RE123 matrix, which led to improve J_c values in relatively low magnetic fields of about 1 T. Therefore, changing the initial composition toward Ba enriched side was found to be a new process to obtain both higher J_c and T_c properties simultaneously for RE123 bulk superconductors produced in air.

I. INTRODUCTION

Higher critical current density (J_c) as well as higher critical superconducting temperature (T_c) properties are required for power applications of $REBa_2Cu_3O_{6+\delta}$ oxide superconductors (RE123). It is well known that introduction of the effective pinning centers for the magnetic flux in the high T_c RE123 matrix is a main factor for the bulk RE123 superconductors with higher J_c values.[1] In other words, the microstructure plays an important role in affecting J_c values. Semi-solid melt growth method, consequently, is considered to be an effective way for obtaining high J_c RE123 bulk materials, since it enables to obtain a highly oriented texture and to introduce the high temperature stable phase ($RE_2Ba_1Cu_1O_5$: RE211) particles as pinning centers through the peritectic solidification of the RE123 phase. For the Y system, the microstructure dispersed with fine particles of the Y211 phase in the Y123 matrix has been successfully obtained by the melt solidification processing, resulting in high J_c values. Similar to the case of the Y system, for the Sm and Nd systems, it is expected that uniform dispersion of the fine particles of the high temperature stable phase, $Sm_2Ba_1Cu_1O_5$ (Sm211) / $Nd_4Ba_2Cu_2O_{10}$ (Nd422), can be attained in the RE123 textures as well. RE123 (RE = Sm, Nd) oxides which have relatively large radii of RE ion, however, are solid solutions due to substitution of RE ions at Ba sites, and have a problem that T_c decreases with increasing the substitution content,x.[2] Therefore it is crucial to control the substitution content for obtaining the bulk RE123 crystals with high T_c and J_c values. Recently, Nd123 and Sm123 crystals fabricated by melt processing under low oxygen partial pressure atmosphere were reported to exhibit higher T_c and critical current density (J_c) values than those of Y123.[3] It was clarified that the substitution of RE ions was suppressed in a reduced oxygen partial pressure atmosphere, which resulted in higher T_c values. On the other hand, it was found from our phase diagram studies of the $NdO_{1.5}$-BaO-CuO system in air, however that the substitution content in the RE123 phase decreases with increasing the BaO/CuO ratio of the tie-line liquid composition.[4] This finding in turn indicates that substitution can be suppressed by selecting the Ba-enriched liquid composition. On the basis of this finding, Nd123 single crystals with a small amounts of substitution could be produced in air and consequently showed higher T_c values.[5]

In this work, aiming at obtaining bulk RE123 with both high T_c and J_c in air, microstructure of the RE123 crystal was investigated with changing the initial composition toward the Ba enriched side.

II. EXPERIMENTAL

Overall compositions were selected on the line from the tie-line composition of Sm123 + 40 mol.% Sm211 (Nd123 + 10 mol.% Nd422) to the Ba-enriched side as depicted in Fig.2. The volume fraction of the Sm211 (Nd422) phases in the initial composition was reported to affect the distribution of RE211 particles in the RE123 crystal so produced. Therefore, the initial composition was selected almost parallel to the metastable liquidus curve of the RE211 + L region at the 123 phase growth temperature in order to maintain the volume fraction of RE211 particles, which arrive at the RE123 growing interface, to be almost constant. The Sm123 bulk crystals were grown by employing the top-seeded melt-texture method combined with the constant undercooling solidification. The mixture powders of Sm_2O_3, $BaCO_3$ and CuO were mixed and calcined were pressed into pellets, and placed on an MgO single crystal substrate in the furnace. The sample was first heated up to 1150 °C and held for 1 h and it was then rapidly cooled down to 1075 °C and kept constant for 1 h. After a tiny cleaved Nd123 single crystal was seeded on the pellet to grow Sm123 crystals, the sample was cooled immediately to 1058 °C below the peritectic temperature. The sample was quenched after holding for 20 h or over. Details of this experiment was found elsewhere.[6]

III. RESULTS AND DISCUSSION

A. Control of the substitution

Bulk Sm123 crystals were produced for the initial compositions with different BaO/CuO ratios from 0.69 to 0.81. Temperature dependences of magnetization were measured for these Sm123 samples as shown in Fig.1. It is found that the T_c value increased slightly from 92 K (BaO/CuO = 0.69) to 93 K (BaO/CuO =0.73, 0.76) and also the transition range became sharper from 6 K to 2 K as the BaO/CuO ratio in the initial composition became higher. This tendency is in good agreement with the results of Nd123 single crystal growth. These Sm123 crystals, however, were almost stoichiometric compounds, and significant differences in substitution content could not be found for the quantitative analysis by the electron probe X-ray microanalysis (EPMA). It has been reported that the substitution range of Sm123 solid solution is smaller that of Nd123 crystals and that T_C values of Sm123 measured by the standard four-probe technique decreased from 93 K (x = 0) to 88 K (x = 0.05) accompanied with the slight change of x value.[2] Therefore, it is considered that compositions of the Sm123 samples in this study are slightly different within the error of EPMA measurement, that is, the substitution content was controlled by the different initial compositions.

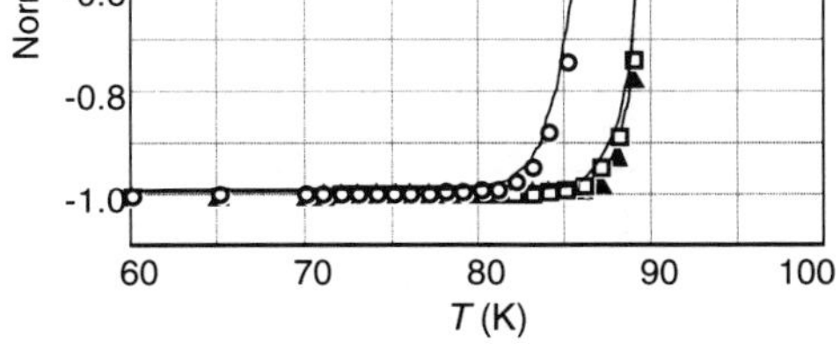

Fig.1. Temperature dependence of DC magnetization for Sm123 crystals grown from different initial compositions.

The fundamental mechanism of the crystal growth of these solid solution including the control of the substitution might be explained in the ternary phase diagram of Fig.2. Diffusion control models proposed for the solidification of Y123 are considered,[7] in which the necessary solute needed for RE123 growth is supplied through the liquid phase from the RE211 particles dispersed in the liquid.

When the sample pellet was heated up to the

temperature where RE211 and liquid phases are stable, different initial composition gives different liquid composition at the RE211/liquid interface at this temperature. Once the sample was cooled down to the growth temperature which is below the peritectic temperature, the liquid changed to that on the metastable liquidus, that is, the liquid composition at the RE211/liquid interface in the undercooled melt for the different initial compositions of 1 and 2 correspond to $C_{L,1}^{211}$ and $C_{L,2}^{211}$, respectively. Since the RE123 peritectic solidification proceeds through liquid diffusion, the 123 phase crystal growth front and the 211 phase grain dissolution front have contacts with the respective liquid. Accordingly, the liquid at the RE123 growing interface ($C_{L,1}^{123}$ and $C_{L,2}^{123}$) can be affected by that at the RE211 dissolving interface due to the diffusion of the RE, Ba and Cu elements satisfying the flux balances. From the tie-line features, the different liquid is equilibrated with the different solid, that is, the Ba-enriched liquid can grow the RE123 solid with which the substitution content is suppressed. Therefore, the composition of the RE123 crystal in turn can be controlled by the initial composition, and the RE123 with a small amounts of substitution can be produced by using the Ba-enriched initial composition.

B. Control of distribution of the high temperature stable phase particles

Another unique and interesting finding includes the reduction of the RE211 particle size with increasing the BaO/CuO ratio of the initial composition. The average Sm211 particle sizes measured for 300 particles in the grown 123 crystal are approximately 2.5 μm, 2.2 μm, 1.5 μm and 1.0 μm in samples with BaO/CuO ratio of 0.69, 0.73, 0.76 and 0.81, respectively. Further the Sm211 particle sizes of the sample with BaO/CuO ratio of 0.76 is as small as that prepared by the MPMG method.[1] It is well known that finer dispersion of high temperature stable phase particles can enhance the J_c values especially in relatively low magnetic fields.[1] Therefore, it is expected that higher J_c values due to the finer 211 particles can be obtained by changing the initial composition toward the Ba enriched side. On the other hand, it was also found that the growth rate of RE123 crystal decreased with increasing the BaO/CuO ratio in the initial composition. Recently, it has been reported that the growth rate of the RE123 crystal and the RE211 phase particle size are important parameters for introduction of the particles at the crystal growth interface and the subsequent spatial distribution of the RE211 particles in the RE123 matrix.[1] This phenomenon has been explained by the prevalent particle pushing/trapping theory:[8] the relationship between critical growth rate (R^*) and particle radius (r^*) for pushing/trapping behavior is approximated as,

$$R^* \propto \frac{\Delta\sigma_0}{\eta \cdot r^{*n}} \qquad (1)$$

where η, n and $\Delta\sigma_0$ are melt viscosity, the power factor ranging from 1 to 2 and the interfacial energy difference among particle, solid and liquid, respectively. According to this theory, smaller particles are pushed out at the interface of the growing crystal at a small growth rate, and the expected amounts of particles from the phase diagram are not introduced in the crystals. However, for the sample with high BaO/CuO ratio of 0.81, it was found that the

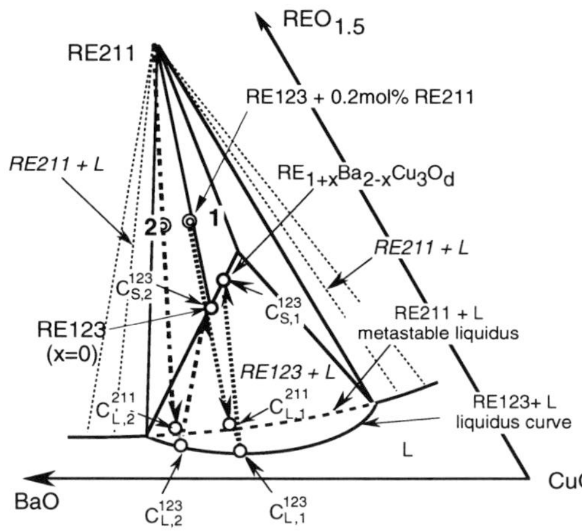

Fig.2. Schematic ternary phase diagram at the RE123 growth temperature, indicating the relationship between the initial composition and RE123 crystal at the growing interface.

Sm211 particles were introduced into the Sm123 crystal grown at larger undercooling, ΔT much more than that at small ΔT. This result indicates that at larger ΔT, i.e. at a higher growth rate, (since growth rate,R, is related with ΔT as a power law equation of $R \propto k\Delta T^n$, where k and n are constant and power factor, respectively), Sm211 particles are trapped in the growing Sm123 at the interface. Therefore, by optimizing the growth conditions, RE123 crystals with uniform dispersion of the finer RE211 particles can be produced with increasing the BaO/CuO ratio in the initial composition.

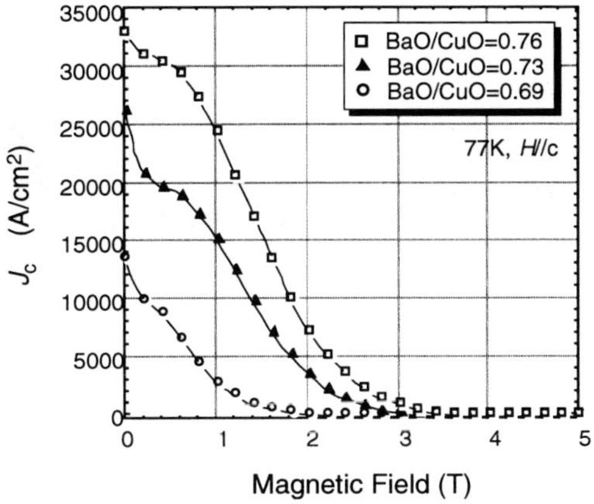

Fig.3. Jc-H curves for Sm123 crystals grown from different initial compositions.

The J_c - H curves for the Sm123 crystals grown from different initial compositions were measured as shown in Fig. 6. The J_c value in relatively low magnetic fields was improved with increasing the BaO/CuO ratio in the initial composition, which was considered to be due to the decrease of the size of Sm211 particles. Further, the J_c value of the sample with the BaO/CuO ratio of 0.76 was higher in comparison with that of Sm123 fabricated by the OCMG method. Therefore, the impact of the BaO/CuO ratio in the initial composition is similar to that of the oxygen partial pressure.

IV. CONCLUSION

Bulk Sm123 crystals with a small amount of substitution were produced in air as the BaO/CuO ratio in the initial composition becomes higher, which exhibited the higher T_c values. It was also found that the RE211 particle size decreased with increasing the BaO/CuO ratio of the initial composition. Accordingly, uniform dispersion of the fine RE211 particles was attained in the grown RE123 crystals, which resulted in the higher J_c values in low magnetic fields. Therefore, the BaO/CuO ratio in the initial composition was found to be an effective parameter for obtaining both higher T_c and higher J_c values for melt processing in air .

Acknowledgment

This work was supported by New Energy and Industrial Technology Development Organization for the R&D of Industrial Science and Technology Frontier Program from.

References

[1] Y. Shiohara and A.Endo: Mat. Sci. Eng., **R29** (1997) 1.
[2] K. Zhang, B. Daborwski, C. U. Segre, D. G. Hinks, Ivan K. Schuller, J. D. Jorgensen and M. Slaski: J. Phys. C: Solid State Phys., **20** (1987) L935.
[3] S. I. Yoo, N. Sakai, H. Takaichi, T. Higuchi and M. Murakami: Appl. Phys. Lett. **65** (1994) 633.
[4] M. Kambara, M. Tagami, X. Yao, E. A. Goodilin, Y. Shiohara and T. Umeda: submitted to J. Am. Ceram. Soc.
[5] X. Yao, M. Kambara, M. Nakamura T. Umeda and Y. Shiohara: Jpn. J. Appl. Phys., **36** (1997) L400.
[6] Y. Watanabe, K. Miyake, A. Endo, K. Murata, Y. Shiohara, T. Umeda: Physica C 280 (1997) 215.
[7] T. Izumi, Y. Nakamura and Y. Shiohara: J. Mater. Res., **8** (1993) 1240.
[8] D.R. Uhlmann, B. Chalmers and K.A. Jackson: J. Appl. Phys., **35** (1964) 2986.

Materials Science and Engineering Serving Society
S. Sōmiya, R.P.H. Chang, M. Doyama and R. Roy (Editors)
1998 Elsevier Science B.V.

Some New Developing Trends For Polymer Matrix Composites

Ren Jie Wu
State Key Laboratory of MMCs,
Institute of Composite Materials,
Shanghai Jiao Tong University, Shanghai 200030, P.R. China

Abstract

Recently, although the progressive rate of polymer matrix composites are very fast, but it still possess many serious problem to be solved, such as high cost, low reliability and difficult to recycle.

Some new developing trends for PMC, include new species, new processing technologies and some recycling measure, which focused on the solving of these problems shown above are reviewed in this paper.

In the mean while , some suggestion of new developing directions are also delivered. They are functional or multifunctional composite, bio-composite and smart or intelligent composite, which may improve the significance of composite in a flash approach.

Introduction

Composite materials are mainly polymer matrix, which have been well developed within fifty years. Previously polymer matrix composites (PMC) play an important role in high techniques, especially in astronautics and aeronautics application or some defense purpose, so it makes PMC developed very fast. But, owing to the "cold war" stopped recently, it forces the PMC change the employable direction, and turn to the general commercial applications. Under this condition PMC is confronted serious challenge with other general materials (metals and plastics). However, people still takes great optimistic expectations, due to PMC has higher performance and designable characters. So that, according to 1995 numerical statement [1], the total production of PMC in the world is about 1.38 million tons, which includes transportation devices (400,000 tons), architecture (250,000 tons), chemical equipment (160,000 tons), ship building (140,000 tons), electronic and electrical devices (120,000 tons), consumption goods (70,000 tons), office appliances (50,000 tons), others (20,000 tons), the less one is the aeronautics and astronautics appliances only 11,000 tons. It means PMC has been stepped in the commercial material ranks. Nevertheless, PMC really has many serious problems to be solved, these problems make some difficulties in the competition with other general materials. First, the cost of PMC is relatively higher compared with other materials. Since the raw materials of PMCs are rather expensive, and processing technologies are not only complicated, but also time expended, so that the cost is hardly to reduce. Secondly, the reliability of PMCs are lower , this is due to many uncertain parameters existed by different components and processing steps, besides the interfacial and interlaminar weak bonding is also influence on the reliability. Thirdly, the sustainable development policy has been carried out in many countries, it requires all materials must be recycling for reuse, on this premise most of the composites are difficult to recycle caused by its multi-components. Obviously, the development trends of PMC is the near future should be led to work around these major problems. On the other hand, some exploration studies on composites are also very important, such as functional or multifunctional composites, bio-composites and biomimetic composites as well as smart or intelligent composites. These new fields will improve the significance of composite and make it to reach a fresh progress, this is a fundamental strategic point of view.

Major Measures and Tentative Plans for the Development of PMCs

1. **Reducing Cost of PMCs**

With the aim to keep vigorous vitality on PMCs, the measures for reducing cost is utmost importance. Especially the expensive high performance fiber reinforcements take a higher percentage in the cost of advanced composites. Recently, some major companies of fiber making tend to extend the production quantity to reach the economic scale, and caused the prime cost decreases gradually. Besides changes the specification of fiber also can deduce the price , fabricates large tow (12K or more) of carbon fiber is obviously cheaper than small tow (3 or 6K).

However, selects an efficient processing technique also be a way to lowering the cost. For instance, resin transfer molding (RTM) method is a reasonable example [2], because it possess many advantages, such as low investment in equipment, use simplified and cheaper mold, short processing period, low energy consumption and less pollution etc. Recently it becomes one of the major process of PMCs, and has combined with vacuum-pressure bag technique to be a powerful processing method. But this technique also had a number of shortcomings, they are a certain limitations for polymer sorts as matrix, and some difficulties to produce large and thin wall structural parts, so it has been paid attention to do further investigations. It still has some other effective process for the cost reduction purpose, e.g. reinforced reaction injection molding (RRIM), pultrusion process and continuous filament winding etc.

Put aside of the traditional processing technique to find some new process to create new PMCs has been attempted in the present time. Various fresh species of composites and their process are suggested, thermoplastics reinforced by fibrils of thermotropic liquid crystalline polymer so called in-situ composite is one of these kinds [3]. When liquid crystalline polymer (LCP) is blended with certain thermoplastics (PP, PET, PC etc.), due to LCP domain within the blend may decrease the melt viscosity during the processing and form well-oriented fibrils as a reinforcing agent. Table I shows the mechanical properties of in-situ composites compared with the matrix polymer and glass

Table I. Mechanical properties of in-situ composites and glass fiber reinforced same matrix composites (containing of LCP and glass fiber are 30%)

Material	Tensile Strength (MPa)	Tensile Modulus (GPa)	Elongation (%)	Bending Strength (MPa)	Bending Modulus (GPa)
PES	63.6	2.50	132	101.9	2.58
PES/LCP_A	125.5	4.90	3.8	125.1	4.11
PES/LCP_B	172.4	8.82	2.6	157.2	6.78
PES/GF	140	4.24	3	190	8.4
PAR	91.0	1.52	135	67.2	1.76
PAR/LCP_A	102	4.27	5.3	108.6	3.32
PAR/LCP_B	51.7	3.28	2.0	90.3	2.94
PAR/GF	135	-	2.5	136	5.8
PC	66.9	2.32	100	93.1	2.47
PC/LCP_A	121	6.72	3.10	132	4.64
PC/LCP_B	154	6.55	4.2	136	5.60
PC/GF	125-145	7.6-8.6	4-3	155-195	6.6-8.0

PES---Poly ether sulphone; PAR---Aromatic polyester; PC---Polycarbonate
LCP_A--- Liquid crystaline polyester; LCP_B---Liquid crystalline polyester amide

fiber reinforced same matrix composites, which indicates the advantage, potential and prosperous future of this new approach. Since the LCP can form the rigid fibrils during the processing, and easy to be orientated in the flowing field, besides, it possess fine dimension and good compatibility in interface, so makes them have a certain improvement in mechanical properties against their matrix polymer. But it must point out, due to the optimal processing parameters have not been handled till now, it may affect on decreasing the aspect ratio of the fibril, it makes the mechanical properties of in-situ composite still not overtakc thc common fiber reinforced plastics. On the other hand, the cost of LCP is relatively higher, it would influence on the development of such in-situ composites. So that in-situ composites have not being a practical material yet, then s further investigations are needed to improve the structure and properties, and reduce price of LCP by extending its production.

Use rod like rigid molecules and have sufficient aspect ratio can use as micro-reinforcements or in-situ composition process. For instance, rod like polybenzoxazole (PBO) and polybenzothiazole (PBT) rigid molecules used to reinforce nylon and polyimide are reported by Du Pont company. Since this kind of "Molecular composites" has microstatic interface, so the effect of interface is unusual and causes composite has excellent performance. Recently, another rigid rod like polyazomethine (PNM) reinforced epoxy resin by in-situ polycondensation process is reported [4]. The mechanical properties of this molecular composite have improved obviously compared with epoxy resin and particle filling expoxy as shown in table II. The suggestion of molecular composite is novel idea, but actually, this system can not realize single molecule as a reinforcement, since molecules will aggregate each other to form micro scale bundles and make aspect ratio lower.

Table II.

Material	Tensile Strength (MPa)	Fracture Toughness (MPa $M^{1/2}$)
Epoxy Resin	50.91	0.83
Epoxy Resin / PNM (5%)	94.25	1.36
Epoxy Resin / SiO_2 (40%)	69.21	0.72
Epoxy Resin / PNM (5%) / SiO_2 (40%)	91.85	1.98

New nano-composite is very attractive, since the reinforcements have nano meter dimension, so their surface area and energy are extremely large, then strengthen the interfacial bonding force and increasing the effective volume fraction of reinforcement. In can be predicted theoretically the modulus of nano composite will be several times larger than its matrix polymer. Actually, due to the aggregation effect causes production of nano composites difficult. There are two methods to form nano-composite, one is to form nano scale reinforcement in the polymer matrix by in-situ reaction, e.g. tetraethoxysilicane added into polyamido-acid solution then nano scale silicate is formed and distributed in polyimide matrix during in-situ imidizine reaction process [5]. The other is using nano scale inorganic particles coated by certain monomer as a surface dispersive agent first to prevent aggregation and then copolymerized with another monomer of desired polymer materials [6].

However, although these attempts have developmental potentials, but they are still in the laboratory state, and have a long way to being commercial products.

2. Improve Reliability of PMCs

The main principles of improving the reliability of PMCs are attempted to prevent defects formed in the composite during processing, doing optimal design, utilize automatic process control and employing effective detection or evaluation on final products. Recently, these dimensional woven and knitted fabrics as reinforcement are used to prevent interlamina debonding, increase impact strength and eliminate other disadvantage of composite effectively [7]. For instance, the addition of through-thickness stitching of two dimensional fabric reinforcement provides up to ten time improvement in the mode I delamination fracture toughness of composite [8]. Concurrently, 3D woven fabrics have made further development into hybrid weaving technique, which can design the hybrid structure according to requirement, and makes it not only to be more reasonable in loading character but also saves the expensive high performance fiber [9]. This new kind of composites have developed very fast, since they can improve reliability and durability really. Recently some companies have produced several computer aided machines to produce 3D woven fabrics, and may fabricate large dimension parts (2.7M) with complicated shape [10]. Besides, 3D woven hybrid fabrics are not only used as reinforcement for structural composite parts but also for functional and multifunctional (heat insulation, antiablation, radiation resistance, electromagnetic wave absorption, armored, corrosion resistance) composite elements by partial combination designing technique [11], so they possess a bright prospects.

Detecting and mentoring technique during the fabrication process are developed for increasing reliability of composites, especially so called smart processing is state of arts technique. It is very attractive, since it can monitoring the certain process parameter and drives the automatic control system to correct it in the mean time, but it is only preliminary attempted in the practical process. Although non-destructive testing or evaluation technique are well developed for detect products of composite, such as acoustic emission, ultra-sonic and laser holography, X-ray, ultra-sonic microwave, edd current, heat radiation imaging etc., Even combination NDE systems are existed, but they still not entirely satisfy the requirement on predicting the complicated damage of composites, so some new method are reported for the NDT of composites purpose. For instance, a generalized modal energy diagnostic method which utilizing vibrating frequency response principle is used to detect composite laminated plates without destructive [12]. It is sensitive and can showing the position and area of the damage. This example illustrates a principle i.e. it needs to create new methods further, in which each method is aimed to a certain kind of damage specially may suitable for the detection on final products of composite to ensure the reliability.

3. **Solving the problems on recycling of PMCs**

Recycling process of composites has been studied now, but still not sufficient. For thermoplastic matrix composites, there is an novel concept of recycling procedure. It is evidence, going with times of recycling, the length of reinforcement becomes shorten. However, in general, a lose of mechanical properties and a cascade principle much be followed, it means use of the composites in progressively less demanding applications.

For thermoset plastic matrix composites, the majority of recycling research works in the literature is concerned with grinding, chipping or flaking the composite past or west into a suitable size, than it may be used as filler in another lower utilizable composites. Although other process such as acid digestion or incineration method is suggested, but it may not accord with environment requirement.

A new approach on recycling of all polymer matrix composites which has advantage in economics and overcome the technical difficulties is so called chemical or tertiary recycling process [13]. This process decomposes polymeric matrix of waste composites into reusable hydrocarbon fractions include monomers, fuels or gases and separates valuable reinforcements from polymers.

The United States Environment Technical Service has pointed out that 850 tons of scrap of CFRP prepreg is generated by industry every year. These discarded scrap prepregs are uncures, preimpregnted composite lamina, so they can be collected, sorted and used to fabricate small structural or non-structural parts. Tensile and flexure testing for reused composite from scrap prepreg are very encouraging, the test results indicate that reasonable mechanical properties can be achieved [14], means that further development of the reuse on scrap preperg is reasonable. The U.S. Environmental Protection Agency indicates that the principle life cycle of composite waste are reduce, reuse, recycle.

Some Strategic Suggestions on the Future Development of PMCs

In order to promote the PMCs can develop continuously, some strategic suggestion should be pointed out, since it will be able to support composites in the competition with other materials from the fundamental point.

1. **Developing functional and multifunctional composites [15].**

Composites due to its composition flexibility, e.g. composibility, connection type and symmetrization can be varied to satisfy the figure of merit for functional materials. Besides, the effects of composite especially the nonlinear multiple effect may create some flesh and excellent functional composites. Although the quantities of functional composite are much smaller compared with structural composites but with higher additional economic value, so that it will be a reasonable way to develop composite.

2. **Carrying out researches on biomimetic composites and biocomposites [16]**

Since almost all natural biomaterials are composites with reasonable structure, such as bamboo, wood, bone and shell etc. They have formed after several millions of years under the natural selection and evolution, so that the utilization of nature raw materials to create composites and produce new composite by means of biomemetic measures are significant. The use of composites in biomedical applications has increased obviously in the past few years, especially the particular interest is used as vivo materials such as hip joints or artificial body organs [17]. This is due to the composites can be designed arbitrarily, so the implant may compatible with the nature condition and match the properties of original organ.

3. **Exploit Smart or Intelligent Composites [15, 18]**

If the material posses the sensation, procession and implementation effects, it can be called as smart or intelligent materials and systems, and generally, they are combination of sensory and actuation materials in an integral form exclude processing device, so they are composites in nature. Smart composites are the primary form, which possess self-diagnosing, self-adapting and self-healing abilities, while intelligent composites are the advanced form, which can not only have the same effects as shown above, but also provide self-decision ability to make the function optimum. Recently, many researching works are progressed, include self-diagnosing structural parts by optical fiber embedded composite and damping device by shape memory alloy and piezoelectric sensor composed with resin and external processing facilities, even some parts of aeroplane are attempted to use.

However, smart and intelligent composites are revolutionary new materials, undoubtedly have excellent progressive future.

References

[1] J.F.Unser, H.Gong, SAMPE J., 32(No.5) 52 (1996).
[2] J.F.Unser, J. of Materials Processing Technology, 37, 363 (1993).
[3] J.S.He, et al, Acta Materiae Compositae Sinaca, (1992), 9, No.4, 53.
[4] Y.Zhang, Ph.D. Thesis of Shanghai Jiao Tong University (1996), to be published in Polymeric Materials Science & Engineering (Chinese).
[5] Y.Chen, et al, Acta Polymerica Sinica.
[6] T.Sun, Annual Report of Chinese State Key Lab.on New Tech. of Composite Fabrication, (1995).
[7] G.Yang, Acta Materiae Compositae Sinica, (1992) 9, No.1, 85.
[8] L.K.Jein, et al, Extended Abs. of 11th Inter. Conf. on Composite Mat., (1997) p2.
[9] T.W.Chou, Progress in Advanced Materials and Mechanics, ed.by T.C.Wang, Peking Univ. Press, Beijing, 1993, p1.
[10] J.C.Zhao, et al , Acta Materiae Compositae Sinica, (1993), 10(No.4), 7.
[11] L.H.Xiao, et al, Acta Materiae Compositae Sinica, (1994), 23(No.2), 11.
[12] L.Hu, et al, Acta Materiae Materiae Compositae Sinica, (1993), 10(No.1), 103.
[13] K.Kemmoshi, et al, Progress in Advanced Materials and Mechanics, ed.by T.C. Wang, Peking Univ.Press, Beijing, 1993, p115.
[14] R.E.Allred, SAMPE J., (1996), 32(No.5), 46.
[15] R.J.Wu, Materials for Mechanical Engineering (Chinese), 1994, 18(No.1), 16.
[16] Q.Y.Zeng, et al, Acta Materiae Compositae Sinica, (1993), 10(No.1), 1
[17] M.Z.Li, Materials Review (Chinese), (1992), No.6, 67.
[18] S.Zhang, Materials Review (Chinese), (1992) No.5, 64.

Materials Science and Engineering Serving Society
S. Sōmiya, R.P.H. Chang, M. Doyama and R. Roy (Editors)

Nanocomposite Synthesis by Electrodeposition in a Disordered Medium

D. Chakravorty [a], S. Banerjee and S. Roy[b]
Indian Association for the Cultivation of Science, Jadavpur, Calcutta 700 032, India
Fax : 91 33 473- 2805 e-mail : mls @ iacs.ernet.in

a **Also affiliated to Jawaharlal Nehru Centre for Advanced Scientific Research Bangalore 560012**

b **Present Address : Datar Switchgear Limited, M.I.D.C., Ambad, Nasik - 422010, India**

Abstract :

Electrodeposition technique has been used to grow nanometer sized silver particles in a melt - quenched silica glass and nanosized copper particles in a silica gel respectively. The particle sizes vary in the range 3.1 to 12.0 nm. depending on the precursor composition and the applied voltage. The growth in both the media have been found to be fractal in nature with the fractal dimension having values in the range 1.75 to 2.0. Generalized diffusion limited aggregation model has been used to explain these results. The electrical resistivity in some of the specimens is controlled by an electron tunneling mechanism. Optical absorption spectra of the copper- silica nanocomposite are explained by Mie theory.

Keywords : Nanocomposites, Electrodeposition, Glass, Gel

1. Introduction

Synthesis and characterization of nanocrystalline materials has become an active field of research in recent years because of novel physical properties expected of such systems [1].Composites of nanometer-sized dispersoids and a host matrix- referred to in this paper as nanocomposites- have been used to understand the behaviour of the nanocrystalline state. A number of synthetic techniques both physical and chemical are available for preparing these materials. An electrodeposition method has been exploited by us recently to prepare nanocomposites involving a metal phase [2]. A brief summary of the salient features is presented in this paper. The principle is based on choosing a suitable amorphous medium containing the metallic ion which needs to be deposited on the cathode and then subjecting the material to an electrical potential between appropriate electrodes. We have used both oxide glasses and gels respectively as the medium. We describe the typical preparation steps in the following sections.

2. Glass-based nanocomposites

Silica- based glasses containing sufficient amount of alkali ions are ion exchanged in a molten bath of silver nitrate at a temperature of 593K for around 50 hours[3]. A typical set of glass compositions used are x Li_2O, 13 CaO, 3 Al_2O_3, 4 B_2O_3,(80-x) SiO_2 (in mol%) with x varying from 20 to 35. The ion exchange step produces a silver-enriched layer with a thickness of the order of a few micrometer within the glass sample. The surfaces are then cleaned using distilled water. Two co-planar silver paint electrodes are applied on the surface of ion-exchanged glass separated by a distance of ~2mm. The sample temperature is then raised typically to 503K so that its resistance attains a value ~10^6 ohms. A dc voltage in the range 1-13 volts is applied for a duration of a few seconds. The current increases to a value of the order of a few hundred milliamperes indicating the formation of metallic percolation channels. The structure of the metal formation is fractal in nature. This has been investigated in detail by optical microscopy[4].

Table 1

Summary of silver particle diameter, geometric standard deviation, fractal dimension and temperature coefficient of resistivity for silver nanocomposites produced by electrodeposition

Glass Sample x	d (nm)	σ	D	TCR (ppm/K)
20	12.0	1.4	1.89	
27	4.0	1.5		1970
30	5.3	1.7	1.81	2420
35	6.0	1.5		2060

Table 1 gives a summary of the mean silver particle diameter d, geometric standard deviation σ(as extracted from the fitting of histogram of particle sizes to log normal distribution function [5]), the fractal dimension D as estimated from optical micrographs and the temperature coefficient of resistivity (TCR) as calculated from the surface resistivity $\rho_{\square}$ vs. temperature data[3].The particle size decreases as the silver ion concentration is increased within the glass. This can be explained on the basis of nucleation and growth mechanism operative in this process[4]· According to this model the critical radius r_c of the metal cluster can be written as

$$r_c = -2\gamma / \Delta G_v \tag{1}$$

where γ is the energy per unit area of the glass-metal interface and ΔG_v is the free energy change per unit volume of the system associated with the formation of the metal phase within the glass. When the silver ion concentration in the glass is increased the value of γ is expected to decrease thereby lowering the value of r_c.

The fractal dimension characteristics have been rationalized[4] using the generalized diffusion-limited aggregation (GDLA) model of Matsushita et al[6]· A simple diffusion-limited aggregation model[7] predicts a value of D equal to 1.67. However, in the present case higher values have been observed According to the GDLA theory the local growth probability at a perimeter site of the growing cluster depends on an exponent parameter η. The value of the latter is related to the number of random walkers (in this case silver atoms) launched from a site at which an initiator walker arrives. The number of such atoms will depend on the compressive stress built up in the glass layer after the ion exchange step. It appears in the present sample series the value of η varies between 1/2 and 1/3 which can explain the higher values of D observed if one uses the relevant expression viz.,

$$D = [D_s^2 + \eta(D_\omega - 1)] / [D_s + \eta (D_\omega - 1)] \tag{2}$$

where D_s is the dimension of space in which aggregation takes place and D_ω the fractal dimension of the random walker trajectory both of which have a value equal to 2 in this case. The decrease in the value of TCR with particle size can be correlated with the lowering of the effective Debye temperature with a reduction in metal particle size [3]. The variation of resistivity has been analysed by using the Ziman equation[9] viz.,

$$\rho = C/\theta_D \; (T/\theta_D)^5 \; J_5 \, (\theta_D/T) \tag{3}$$

where θ_D is the Debye temperature, T the temperature, C a constant and J_5 the Debye integral. The effective Debye temperature θ_D has been estimated by fitting the experimental data to equation (3) . The lowering of the θ_D value leads to an increase in ρ which gives rise to a decrease in the value of TCR β defined as

$$\beta = 1/\rho_o \;\; \partial\rho/\partial T \tag{4}$$

where ρ_o is the room temperature resistivity.

However, in the case of the sample with x=20 the above trend can not be extended. This is due to the disruption of weak links in the metallic chain caused by excessive joule heating and local melting of metal particles. The electrical resistivity for such a sample has been shown to be controlled by an electron tunneling mechanism in the temperature range 140 to 300K. .The activation energy E associated with the electron tunneling is given by

$$E = e^2/4\pi \, \varepsilon \, \varepsilon_0 \, [2/d - 1/(d/2 + s)] \qquad (5)$$

where, e is the electronic charge, d the metal particle diameter, s the inter particle separation, ε the dielectric constant of the matrix and ε_0 the dielectric permittivity of free space. From the electron micrograph of the sample concerned the histogram of inter particle separations has been analyzed and the average value estimated at 6.1 nm. Substituting the relevant values of the other quantities in equation (5) the calculated value of E is obtained as 0.030 eV which is in reasonable agreement with the experimental value of 0.038 eV.
The rather low value of activation energy viz., 0.0035 ev in the temperature range 80 to 130K can be explained on the basis of a quantum size effect [3]. In the latter case it appears that there are few continuous chains formed by ultra fine metal particles within the glass matrix for these metal particles the energy levels become discrete in accordance with Kubo's theory[7] with the separation being of the order of a few meV for particle size of the order of 3 nm[3]

3. Gel-based nanocomposites

A typical set of gels with precursor compositions represented by y $Cu(NO_3)_2 \cdot 3H_2O$, (100-y) Si $(0C_2H_5)_4$ (in mole %) with y having values of 0.1 and 1.0 respectively are first of all synthesized. The sols are prepared by dissolving a weighed amount of the copper salt in ethyl alcohol and water and then pouring the required amount of $Si(0C_2H_5)_4$ on to it. The pH of solution is maintained at ~ 5 by a suitable addition of nitric acid. After pouring the resultant solution into a flat bottomed plastic container it is allowed to gel under ordinary atmosphere for a period of two weeks. For electrodeposition a gel piece of approximate dimension 10 mm x 5mm x 0.5 mm is placed on a lucite block. A sharp edged copper plate is used as anode and a sharp tip copper rod acts as the cathode both of them being clamped on the specimen surface by spring loading. The electrodes are separated by a distance of ~5 mm. The electrodeposition voltage is in the range 8- 35 volts applied for a period of around 20 min. The growth of metallic copper phase is fractal in nature. Detailed microstructure has been investigated both by optical and transmission electron microscopy[11]. The fractal dimensions for the samples show a semi oscillatory behaviour as a function of applied voltage with the D values having a range 1.75 to 2.0. A generalized diffusion limited aggregation and a dielectric breakdown mechanism seem to be operative and can explain such an unusual behaviour [8] Computer simulation studies on electrochemical deposition under coupled diffusion fields of ion concentration and electric potential have shown that the concentration gradient is raised as the applied voltage is increased. This leads to an increase in the effective value of η as discussed earlier in relation to equation (2) and as a result the fractal dimension D decreases as a function of voltage. At a critical voltage however the concentration gradient becomes large enough to cause a dielectric breakdown at the double layer concerned. The narrow discharge channels have a fractal structure and as the potential gradient goes down due to the discharge the fractal dimension should show an increase from ~ 1.75 to ~2.0. The cycle of the above two mechanisms is then repeated which brings about a semi-oscillatory behaviour of the fractal dimension as a function of voltage.The metal particle diameters can be varied from 6.2 to 20.8 nm by a suitable choice of gel composition and applied voltage.The values are summarized in table 2.

Table 2

Summary of copper diameters in silica gel by electrodeposition

Sample y	Applied Voltage (volts)	d (nm)
0.1	15.0	15.0
	28.0	9.2
1.0	10.0	20.8
	15.0	10.1
	18.0	6.2

In a variant of the above method gels with precursor compositions z Cu $(NO_3)_2 \cdot 3H_2O$, (100-z) $Si(0C_2H_5)_4$ (in mole %) with z having values 5, 10 and 15 respectively are synthesized by the steps indicated as above. Gel pieces are crushed to a size ~ 5μm and a thick paste is made using distilled water. A copper cathode is painted with the paste and a copper anode is placed on top of it. Electrodeposition is carried out by applying voltages on to the cell. The resultant gel powder with metal phase is scraped off the cathode plate. These are dispersed in a solution of polystyrene dissolved in tetrahydrofuran. Corning glass slides are dip coated with the solution and then dried. Optical absorption spectra are studied in the wavelength range 300 to 600 nm.

Metal particles are found to have sizes in the range 3.1 to 11.4. Optical absorption maxima are found in the wavelength range 380 to 470 nm with the position shifting to higher wavelengths as the particle size is increased. The results can be explained on the basis of Mie theory [13,14.] According to the latter the absorption coefficient α of a composite is given by

$$\alpha = [18\pi \, f \varepsilon_m^{3/2} \varepsilon_2]/\lambda \, [(\varepsilon_1 + 2\varepsilon_m)^2 + \varepsilon_2^2] \qquad (6)$$

where, f is the volume fraction of the particles embedded in the medium, ε_m is the dielectric constant of the latter, λ is the wavelength of radiation used, $\varepsilon = \varepsilon_1 + i\varepsilon_2$ is the dielectric constant of the metal phase. A size dependent dielectric constant ε for the metal particles has been used which is as follows[15]

$$\varepsilon = (1-\omega^2_p/\omega^2) + i \, [\omega^2_p/\omega^3 \, (\sigma/\varepsilon_o + 2V_f/d)] \qquad (7)$$

where, ωp is the Drude plasma frequency, ω the angular frequency, σ the dc conductivity of the metal phase V_f the fermi velocity of bulk metal. By fitting the experimental optical absorption data to equation (7) the values of the parameters f and σ/ε_0 for different samples have been extracted. These are summarized in table 3. The electrical conductivity as estimated by this analysis are in satisfactory agreement with results reported earlier for copper nanoparticles in glass ceramic systems[16].

Table 3

Summary of f and $\sigma/\varepsilon o$ values extracted from optical absorption data for different copper-silica nanocomposites

Sample z	Applied voltage volts	Median diameter d (nm)	f	σ/ε_0 (x 10^{16}) Sec^{-1}
5.0	5.0	11.4	0.0031	1.2
	10.0	9.8	0.003	1.2
	15.0	8.0	0.0029	1.2
10.0	5.0	10.5	0.003	1.2
	10.0	8.8	0.0029	1.2
	15.0	7.5	0.0028	1.3
15.0	5.0	6.3	0.0027	1.4
	10.0	4.7	0.0026	1.5
	15.0	3.1	0.0022	1.5

4. Conclusions

Nanoparticles of metal can be grown within an oxide glass or a gel by electrodeposition method. In the present case, silver in silica glass and copper in silica gel have been synthesized. The metal growth is fractal in nature. A GDLA mechanism can be used to explain the results. Electron tunneling mechanism is operative in composites with metal particles in a non-percolative configuration. The optical absorption spectra of nanocomposites of copper in a polymer medium can be explained by Mie theory.

Acknowledgment

D. Chakravorty thanks DST, Govt. of India for support of this research. S. Banerjee thanks C.S.I.R, New Delhi for a senior Research Fellowship.

References :

1. D. Chakravorty and A.K. Giri in Chemistry of Advanced Materials, ed C.N.R. Rao (Blackwell Scientific Publications, Oxford, 1992) p. 217

2. S. Roy and D. Chakravorty, Appl. Phys. Lett.59, (1991) 1415

3. B. Roy, S. Roy and D. Chakravorty, J. Mater. Res. 9,(1994) 2677

4. S. Roy and D. Chakravorty, Phys. Rev. B 47, (1993) 3089

5. B.Roy and D. Chakravorty, J. Phys. Condens Matter 2, (1990) 9323

6. M. Matsushita, K. Honda, H. Toyoki, Y. Hayakawa and H. Kondo, J. Phys. Soc. Jpn 55, (1986) 2618

7. T. A. Witten and L. M. Sander, Phys. Rev. Lett. 47, (1981) 1400

8. J. M. Ziman, Electrons and Phonons (Claredon, Oxford, 1960) p. 364.

9. P.A. Tick and F.P. Fehlner, J. Appl. Phys. 43, (1962) 362.

10. R. Kubo, J. Phys. Soc. Jpn. 17, (1962) 975

11. S. Bandyopadhyay, B. Roy and D. Chakravorty, Solid State Communications 99, (1996) 835

12. L. Niemeyer, L. Pietronero and H.J.Wiesman, Phys. Rev. Lett. 52, (1984) 1033

13. G. Mie, Ann. Physik 25, (1908) 377

14. S. Bandyopadhyay and D. Chakravorty, J. Mater. Res (1997) (In print)

15. U. Kreibig, J. Phys.F Met. Phys 4, (1974) 999.

16. B.Roy and D. Chakravorty, J. Appl Phys, 74, (1993) 4190

Materials Science and Engineering Serving Society
S. Sōmiya, R.P.H. Chang, M. Doyama and R. Roy (Editors)

Intercalation of n-Alkylamines into Copper Molybdenum Bronze

Nobukazu Kinomura, Katsuya Mizumoto and Nobuhiro Kumada

Institute of Inorganic Synthesis, Faculty of Engineering, Yamanashi University
Miyamae-cho 7, Kofu, 400 Japan

ABSTRACT

Copper molybdenum bronze, $Cu_{0.21}MoO_3$, with a layered structure was reacted with n-alkylamines in the acetonitrile or aqueous solution and neat liquid of n-alkylamines in the temperature range from room temperature to 70 °C for up to 3 weeks. Alkylamines with the number of carbon atoms from 3 to 16 were found to form the intercalation compounds with the bronze. The basal spacing of the intercalation compounds increases linearly with the length of alkylamine. As copper ions in the bronze remained intact in the interlayer space on the reaction and protons were not included in the bronze, the intercalation reaction is considered to be not an ion exchange reaction nor an acid-base reaction.

1. INTRODUCTION

Layered compounds are well known to undergo soft chemical reactions such as intercalation and ion-exchange reaction and various kinds of organic molecules and ions are taken up into the interlayer spaces. Many organic bases can be introduced into the interlayer spaces of inorganic layered compounds either by the acid-base reaction with protonated host compounds or ion-exchange reaction. The thermal molecular redox reaction also makes intercalation of organic bases possible. The acid-base reaction between organic Lewis bases and protonated host compounds is the most popular reaction for the intercalation of organic molecules into layered oxides. Some metal oxychlorides have been reported to form sorption-type intercalation compounds [1-3]. However not many oxides intercalate directly organic molecules and produce the sorption-type compounds.

Molybdenum bronzes with layered structures are considered to be typical low dimensional materials and attract much attention by their highly anisotropic electrical conductivities [4]. In spite of their layered structures, no soft chemical reactions such as ion exchange, intercalation and deintercalation involving this phase have been reported so far. Recently we have reported that molybdenum blue bronze intercalated n-alkylamines after the hydrogen insertion by the spillover technique [5]. On the course of this study, copper molybdenum bronze, Cu_xMoO_3, was found to intercalate directly n-alkylamines. The copper molybdenum bronze

This work supported by a Grant-in-Aid for Scientific Research from the Ministry of Education, Science and Culture, Japan was prepared by Tian et al. [6] and its layered structure was determined by Steiner et al. [7]. In this paper we report the intercalation of n-alkylamines into the copper molybdenum bronze.

2. EXPERIMENTAL

The copper molybdenum bronze (CMB) was prepared as described in ref. [6]; a mixture of powders of Cu (metal) and MoO_3 with a molar ratio of 0.21:1 was sealed in an evacuated silica tube and heated at 500 °C for 120 h. X-ray powder diffraction (XRD) pattern of product showed only that of CMB. Intercalation of n-alkylamines, $C_nH_{2n+1}NH_2$ where n ranges from 1 to 16, to CMB was carried out using aqueous solution for n=1 and 2 and about 0.8 M acetonitrile solution for n=3-16. Powder of the host compounds was added to a solution and stirred vigorously in the temperature range from room temperature to 50 °C for up to 3 weeks. Neat liquid of n-alkylamines (n=3-16) was also reacted with CMB from room temperature to 70 °C for up to 3 weeks. The product was washed well with acetonitrile and dried in air.

XRD patterns were recorded with Cu Kα radiation on a diffractometer. Thermogravimetric and differential thermal analysis (TG-DTA) was measured with a heating rate of 10 °C/min in air and Ar. Species evolved on pyrolysis in a He stream were analyzed by mass spectrometry. As TG-Mass analysis indicated that water molecules were not included in the intercalation compounds, amount of intercalated n-alkylamines was calculated from the weight loss observed in TG measurement in Ar up to about 500 °C. IR spectrum was obtained with a FT-IR spectrometer by KBr method. Electric conductivity was measured by two probe method on a pressed body.

3. RESULTS AND DISCUSSION

The basal spacings and compositions of the intercalation products of CMB in the acetonitrile solutions are listed in Table 1. When n≤2, CMB was dissolved. Upon the intercalation copper ion was analyzed to remain unchanged in the interlayer space and almost no water molecules were found to be included in the products after drying. Therefore the chemical composition of intercalation compounds can be represented as

Table 1. Basal spacing and amount of n-alkylamine of the products, $(C_nH_{2n+1}NH_2)_xCu_{0.21}MoO_3$

n	basal spacing(Å)	x
3	13.15	1.85
4	14.87	1.74
5	17.47	2.18
6	19.61	1.80
7	21.94	1.39
8	22.88	2.39
9	26.23	1.68
10	26.97	1.70
11	29.04	1.71
12 *	31.12	2.79
13 *	33.04	2.00
14 *	35.21	2.19
15 *	37.57	2.07
16 *	39.42	2.03

*; two phases

$(C_nH_{2n+1}NH_2)_xCu_{0.21}MoO_3$. As single phase of intercalated compounds was not obtained when n≥12, the largest basal spacing and average composition of a product are given in Table 1. As a typical example XRD pattern and TG curve of product are shown in Figs. 1 and 2. The butylamine derivative of CMB was practically an insulator with an electric resistivity of about $1.6x10^9$ Ω·cm at room temperature whereas CMB itself was a semiconductor with an electric resistivity of about $1.0x10^3$ Ω·cm at room temperature.

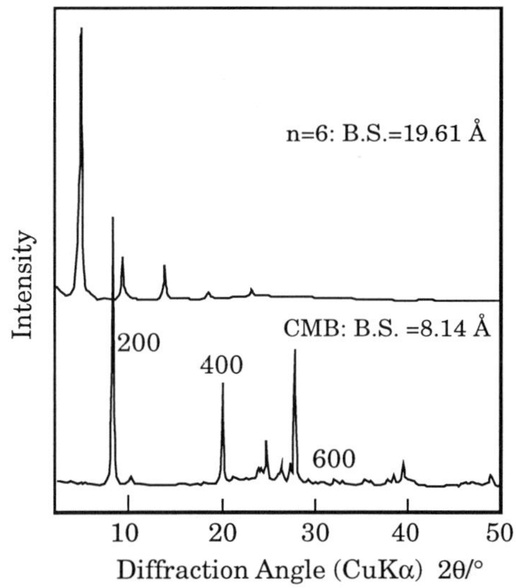

Fig. 1. XRD patterns of CMB and its n-hexylamine derivative.

The basal spacings given in Table 1 are plotted vs. carbon number in the alkyl chains in Fig. 3. The basal spacings of the product in neat liquids were almost coincided with those of products in the acetonitrile solutions when n≤9, but basal spacings slightly larger than those of products in the acetonitrile solutions were observed in the case of n≥10, although single phase was not obtained even by using neat liquids. As seen in this figure, the basal spacing increases linearly with number of carbon atoms and there seems to be two kinds of slopes; for n≤9 and n≥10. The increase in basal spacing with increase of n to n+1 in $C_nH_{2n+1}NH_2$, corresponding to addition of one methylene unit to alkyl chain, is plotted against n in Fig. 4. The odd-even alternative change of basal spacing which is considered to reflect a regular arrangement of all-trans alkyl chains in the interlayer space is clearly seen in the region of n≤9 [8]. On the other hand almost constant increase in basal spacing is observed when n≥10. From this fact one can conclude that n-alkylamines with n≤9 are arranged regularly with the all-trans conformation in the interlayer space of CMB, while in the case of

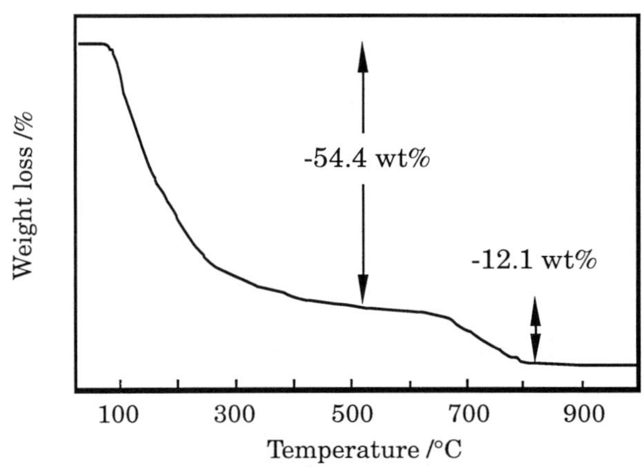

Fig. 2. TG curve of octylamine derivative

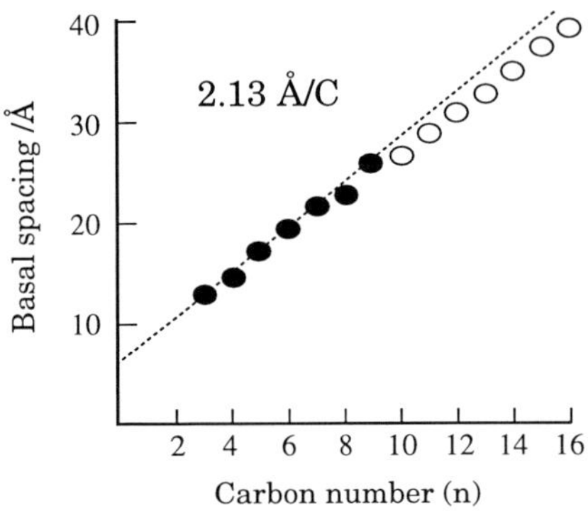

Fig. 3. Basal spasing vs. n

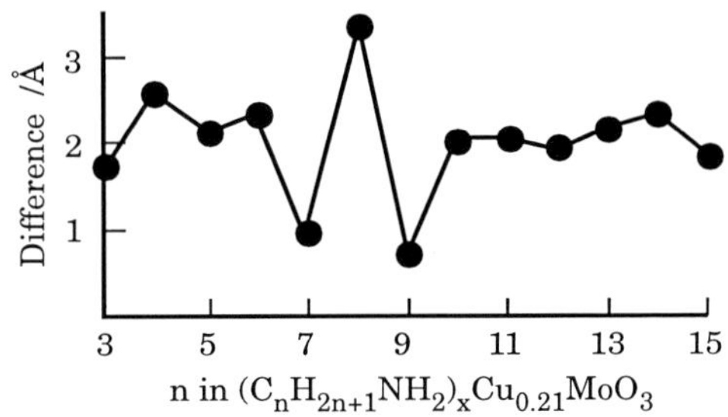

Fig. 4. Difference in basal spacings between $(C_nH_{2n+1}NH_2)_xCu_{0.21}MoO_3$ and $(C_{n+1}H_{2(n+1)+1}NH_2)_xCu_{0.21}MoO_3$

$n \geq 10$ there would exist kink-block or gauche-block in the alkyl chain in the interlayer space. From the slope of the line (2.13 Å/C) for the intercalated compounds with $n \leq 9$, alkyl chains in the interlayer space is considered to form bilayer arrangement of all-trans alkyl chain with the tilt angle of 67° to the MoO_3 layer.

CMB reacts readily with n-alkylamines, while in the case of blue potassium molybdenum bronze hydrogen insertion and successive hydration are necessary for the intercalation of n-alkylamines [5]. This noticeable difference in the intercalation behavior between the two bronzes may attribute to the difference in the cation in the interlayer space as well as in the structure of layer. Probably capability of formation of amminecomplexe of copper ion plays an important role in the sorption-type intercalation of n-alkylamines into CMB.

REFERENCES

1. J.-H. Choy, D.-Y. Noh, J.-C. Park, S.-H. Chang, C. Delmas and P. Hagenmuller, Mat. Res. Bull., 23 (1988) 73.
2. J.-H. Choy, J.-W Uh, J.-K. Kang, A. Weiss and M. Rey-Lafon, J. Solid Sate Chem., 77(1988) 60.
3. D. A. Odink, K. Sang and S. M. Kauzlarich, Chem. Mat., 4 (1992) 906.
4. M. Greenblatt, Chem. Rev., 88 (1988) 31.
5. N. Kinomura, K. Mizumoto and N. Kumada, J. Solid State Chem., 128(1997)256.
6. S.-B. Tian, F. Zhou, J.-C. Yang and Z.-C. Li, Solid State Ionics, 57(1992) 109.
7. U. Steiner, T. Morgenstern, W. Reichelt, H. Borrmann and A. Simon, Z. anorg. allg. Chem., 620 (1994) 1905.
8. K. Beneke and G. Lagaly, Clay Miner., 17 (1982) 175.

N. Kinomura is a professor in the Dept. of Engineering at Yamanashi University. He obtained his D. Sc. from Osaka University in 1972.
K. Mizumoto is a graduate student at Yamanshi University.
N. Kumada is an associate professor in the Dept. of Engineering at Yamanshi University. He obtained his D. Eng. from Tohoku University in 1991.

Materials Science and Engineering Serving Society
S. Sōmiya, R.P.H. Chang, M. Doyama and R. Roy (Editors)

Distinguishing ceramic interfaces through their bonding: a new approach

Hui Gu

Japan Science and Technology Corporation, "Ceramics Superplasticity" project, JFCC 2F, 2-4-1 Mutsuno, Atsuta-ku, Nagoya 456, Japan

Interfaces in ceramic materials often have distinguishing bonding characters as compared to the matrix surrounding them. Such characters can be reflected in the Electron Energy-Loss Near Edge Structures (ELNES) of a suitable element, such as silicon $L_{2,3}$ edge in silicon-based ceramics. Employing an ELNES subtraction procedure based upon the principle of spectrum orthogonization in a vector space, spectral contributions from and outside an interface regions can be separated from each other. The resulted interface ELNES can tell us not only how and how much the bonding of interfaces are different in from their matrix, and between different types of interfaces, but they can also lead to further quantification of the chemical width, the composition and the density of the interfacial regions. These unprecedented knowledge offer new criteria to distinguish ceramic interfaces, and they will gain further insight into the role these interfaces playing in influencing the properties of these materials.

1. INTRODUCTION

Controlling the interfacial chemistry is crucial to achieve the desired macroscopic properties in many high performance ceramic materials. Electron Energy-Loss Spectroscopy (EELS) is sensitive to the local structural and chemical changes at/near the interfaces since it probes the coordination, thus the local structure in 3D, and the electronic states, thus the chemical bonding of the interested elements, reflected in the corresponding energy-loss near-edges structures (ELNES). As an analytical tool, EELS gives also underlining compositional changes at the interfaces with the same measurements. This method is particularly useful in studying the ceramic interfaces where local structure and/or bonding usually have substantial changes and where segregation and secondary phases often occur. It has been shown that the boundaries between two Si_3N_4 grains there is always covered with a thin (~1 nm) amorphous film made of silicate [1]. Previous works has revealed that the bonding state of the GB film is substantially different from any of the related bulk phases [2]. In this paper, I will first describe the principles to separate interfacial ELNES from the matrix, and later give a few examples of interfaces and their distinct features in Si_3N_4-related materials.

The microscope involved in this work is a VG HB501 dedicated STEM at Max-Planck-Institut für Metallforschung in Stuttgart, attached with a Gatan PEELS spectrometer. This system can deliver a fine electron probe of 0.5 nm for analysis, and an energy resolution better than 0.7 eV. The later is crucial to reveal ELNES details of interests.

2. Vector algebra picture for separating interfacial ELNES

ELNES for silicon $L_{2,3}$ core-loss edge in a series of compounds exhibit systematic changes as depicted in Fig. 1. A prominent peak at or near Si-$L_{2,3}$ edge onset shifts towards higher energy following the decrease of covalence, and subsequently the length, of Si-ligand bonds in the order from Si-C_4 to Si-O_4. There are also weaker features reflecting other electronic structures near the probed Si atoms. The detailed interpretation of these features requires extensive calculations, such as the molecular-orbital method [3].

In spatially-resolved studies of interfaces, these ELNES features can help to remove the unwanted matrix signals included in the probed areas. The separation of the matrix and the interface ELNES is achieved by: a) considering ELNES as vectors in a multi-dimension space (energy pixels); b) using a particular feature from the matrix ELNES as a new coordinate which must be orthogonal to another unknown coordinate associated with the interface ELNES; c) projecting the as-measured ELNES onto such matrix feature to know how much the matrix ELNES should be removed. This essentially is a vector orthorgonization process as represented in Fig. 2. A complete description of this picture and operation is given in Ref. [4].

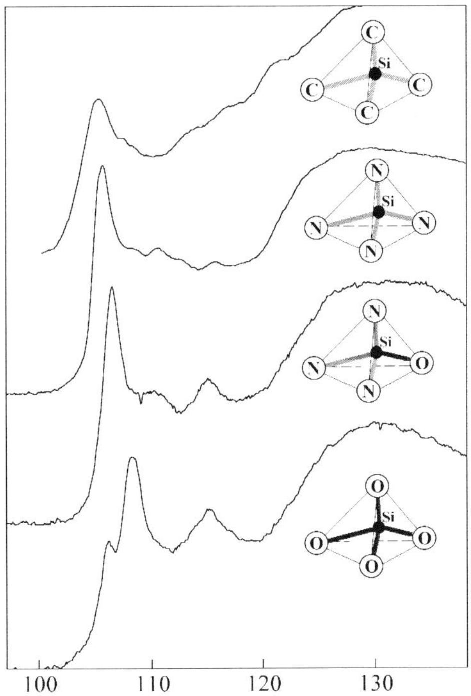

Figure 1 Silicon $L_{2,3}$ ELNES from crystalline SiC, Si_3N_4, Si_2N_2O and amorphous SiO_2. The ligand atoms are schematically depicted.

3. ELNES from typical interfaces in Si_3N_4-based ceramics

Using or repeatedly using the principles stated in the previous section, faithful ELNES of different interfacial regions can be achieved from measured spectra. Si-$L_{2,3}$ ELNES from a 12 Å thick grain boundary (g. b.) film in HIPed high-purity undoped Si_3N_4 [5] is shown in Fig. 3 (b) with ELNES of the matrix (a). The g. b. ELNES reveals the film an oxide-based phase by the appearance of a new peak at 115 eV, which is the signature of Si-O bonds (see also Si_2N_2O and SiO_2 ELNES in Fig. 1). The amorphous nature of g. b. film is

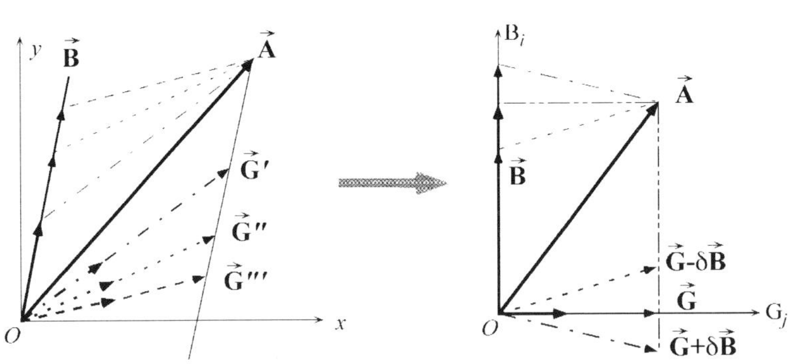

Figure 2 Vector description of orthogonizing the matrix (**B**) and interface (**G**) ELNES in order to separate them in as-measured ELNES (**A**). x-y are two spectral pixels; B_i and G_j are ELNES features. See text and Ref. [4] for details.

indicated by the absence of any feature as sharp as those in the crystalline phases, reflecting the lack of structural ordering in the film. However, this film is also substantially different from amorphous SiO_2 (d) by the relative height of this peak and by the energy position of the main peak (near the onset) which is sensitive to the bonding nature and is related to the ratio of N:O ratio among the ligand atoms surrounding Si (Fig. 1). The broadness of this peak suggests that the bonding and the composition are not uniform across g. b. film; its center of weight is located rather close to the main peak in Si_2N_2O ELNES, indicating a significant amount of nitrogen inside the film.

This new method opens the possibility to compare the bonding and structure of different interfaces. ELNES of the interface between Si_3N_4 grain and SiO_2 glass (c in Fig. 3) shows similar overall shape as the g. b. film in Si_3N_4, except a double-peak feature at the place of the broad peak near the onset (b). This new feature suggests that the spatial transition from Si-N_4 to Si-O_4 configurations, or from mainly covalent bonding to dominant ionic bonding, takes place through two transient layers of different degrees of covalent/ionic nature and of different N:O ratios. This distinct interfacial structure disapproves the model that g. b. films in Si_3N_4 are made of two such interfaces. The width of such interface is also wider than the films [6].

The chemical additives can also change the bonding of interfaces. Si-$L_{2,3}$ ELNES from a g. b. film (b) and from an interface (c) between Si_3N_4 grain (a) and La-Si-Al-N-O glass (d) in a same sample with Si_3N_4 grains grown from the super-cooled oxynitride glass [7] are compiled in Fig. 4. La has segregated to both g. b. film and to the grain/glass interface. Its signal overlaps with Si-$L_{2,3}$ ELNES, and has been removed from the spectra (b), (c) and (d) in Fig. 4. The main peaks at the edge onset of (b) and (c) are both sharper than the undoped interfaces in Fig. 3, indicating the bonding within the interfacial regions are more uniformly distributed at the presence of rare-earth cations. Another interesting phenomenon is that ELNES of g. b. film is more different from Si_3N_4 grain than ELNES of the interface (c). The main peak in (b) is clearly shifted to higher energy, signifying increased ionicity of Si atoms inside the film. It remains at almost the same position in (c) as compared to (a), suggesting at the interface there is no significant bonding change from Si_3N_4 grains, at least at Si sites. The peak related to Si-O

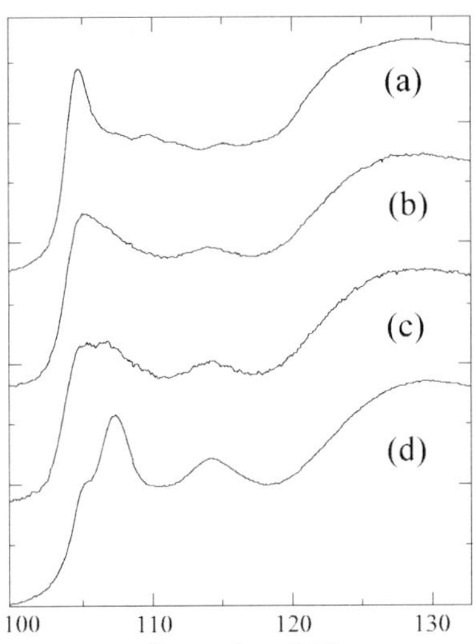

Figure 3 Si-$L_{2,3}$ ELNES from g. b. film (b) between two Si_3N_4 grains (a), and from the interface (c) between Si_3N_4 and SiO_2 (d).

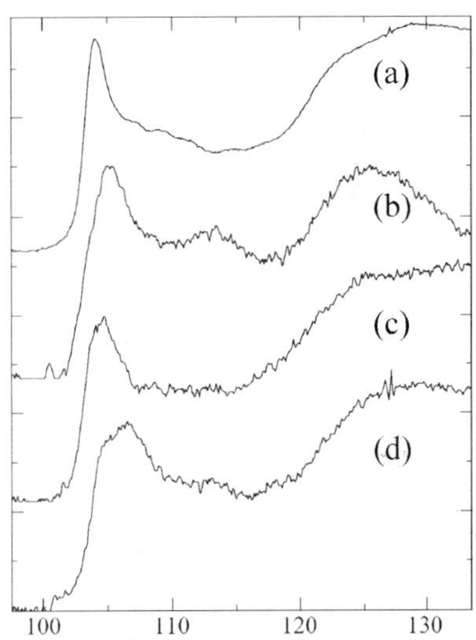

Figure 4 Si-$L_{2,3}$ ELNES from g. b. film (b) in La-doped Si_3N_4 (a), and from the interface (c) between Si_3N_4 and La-rich glass (d).

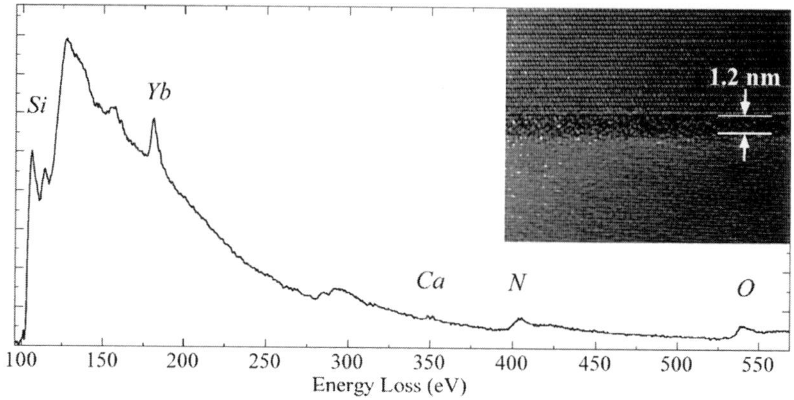

Figure 5 Spectrum of a g. b. film (b) (after ELNES separation) reveals the film composition.

bond at 115 eV is even absent in (c), further supporting this picture. It is probable that most of the oxygen are bonded directly to La or Al at the interface to the La-rich glass while some of them are bonded to Si in the g. b. film. It is also possible that the anisotropy of the Si_3N_4 grain surfaces plays an important role in the difference of bonding between both types of interfaces.

4. Other related quantities for distinguishing interfaces

From the interfacial ELNES, other parameters can be deduced to further explore the differences between the interfaces. Due to the direct links between an interfacial ELNES and the number of atoms it represents, the space which these atoms occupy, and the number of the other elements presented in the same space, it is possible to quantify the density, the chemical width and the composition of an interfacial region after successful separation of its ELNES from the matrix [4, 6]. An example is given in Fig. 5: a wider range spectrum of a g. b. film in a Yb-doped Si_3N_4 sample reveals the presence of all elements, which permits the direct quantification of the film composition. The image of this g. b. film is also shown in the inset. The ratio of N:O is directly measurable from such spectra, which allows a quantitative relation to the bonding change in the interfacial regions. For the measurement of the density and the chemical width of an interfacial region, the size of the probed area must be known. After quantifying the partitioned intensities belonging separately to the interface and the matrix, the two quantities can be calculated. However, it is beyond the scope of this paper to describe more details about these quantities. They are derivatives of this ELNES approach. [8]

References

1. D. R. Clarke, Ultramicroscopy, 4 (1979) 33.
2. H. Gu, M. Ceh, S. Stemmer, H. Müllejans and M. Rühle, Ultramicroscopy, 59 (1995) 215.
3. I. Tanaka and H. Adachi, Phil. Mag., B72 (1995) 459.
4. H. Gu, Ultramicroscopy, (1997) submitted.
5. I. Tanaka, G. Pezzotti, T. Okamoto, Y. Miyamoto and M. Koizumi, J. Am. Ceram. Soc., 72 (1989) 1656.
6. H. Gu, MRS Symp. Proc., 453 (1997) 703.
7. M. Krämer, M. J. Hoffmann, G. Petzow, J. Am. Ceram. Soc., 76 (1993) 2778.
8. The author thanks Prof. I. Tanaka and M. J. Hoffmann for providing the samples.

Materials Science and Engineering Serving Society
S. Sōmiya, R.P.H. Chang, M. Doyama and R. Roy (Editors)

INTERFACE MORPHOLOGICAL STABILITY OF RE123 SUPERCONDUCTOR VIA PERITECTIC REACTION

Masaki SUMIDA[1], Yuh SHIOHARA* and Takateru UMEDA, Department of Metallurgy, Graduate School of Engineering, the University of Tokyo, 7-3-1, Hongo, Bunkyo-ku, Tokyo, 113 Japan, * Superconductivity Research Laboratory, 1-10-13, Shinonome, Koto-ku, Tokyo, 135 Japan, [1] Research Fellow of the Japan Society for the Promotion of Science

Abstract

Interface stability of RE123 superconductors is analyzed in accordance with the constitutional supercooling criterion. As single crystal growth of the 123 phase is largely dependent on the interface stability, quantitative analysis has been greatly required. Considering, at a certain supercooled situation from the peritectic temperature, a unidirectional peritectic solidification of the 123 phase from the 211+liquid mixture by liquid diffusion limitation, solute concentration distribution around the 123 and an isolated 211 particles is derived precisely under the bispherical coordinate. Interface stability is determined with refer to competition of the liquidus temperature gradient and the imposed temperature gradient at the 123/L interface along the 123 growth direction.

From the analysis above, it is clarified that the constitutional supercooling must exist when a 211 particle comes close to the 123 growth interface. It can also predict that larger 211 particle, larger growth rate, or smaller imposed temperature gradient causes easy occurrence of constitutional supercooling ahead of the 123 growth interface. This tendency agrees with the microstructure observation and it is concluded the interface stability of the 123 phase is explainable quantitatively.

1. Introduction

From equilibrium phase relationships of RE-Ba-Cu-O system, it is generally recognized that the RE123 phase forms via a peritectic reaction between $RE_2Ba_1Cu_1O_x$ (RE211) and BaO-CuO liquid phases. When a unidirectional solidification of RE123 from RE211+L mixture is carried out, the growth morphologies of the solidified RE123 vary from faceted planar, cellular, and equiaxed blocky due to the increase of the growth rate, *R*, or decrease of the temperature gradient at the solidification interface, G_S [1], as shown in Fig. 1. The equiaxed blocky morphology is formed by multiple nucleation and growth of the RE123 phase ahead of the 123 growth front, as shown in Fig. 2. The growth interface stability has to be considered carefully to achieve the bulk single crystal growth of RE123 superconductors, however, it has not been reported a quantitative analysis about the growth interface morphological stability of RE123 from the RE211+L mixture via peritectic reaction. In this paper, a simple model is proposed with use of the constitutional supercooling criterion.

2. Assumptions

Consider a hypothetical situation of an isolated 211 particle and infinite plane of the 123 phase at a certain supercooled situation as schematically depicted in Fig. 3, 123 phase crystal grows upwards unidirectionally under a positive imposed temperature gradient being limited by the solute transport via liquid phase due to dissolution of the 211 particle [2- 4]. The 211 particle is assumed to be large enough to neglect the curvature effect into the analysis. The local equilibrium can be assumed if we neglect the interface kinetics at the 123/L and L/211 interfaces. The temperature gradient, G_S, around the diffusion field is considered to be linear, as the distance between the 123/L and the L/211 interfaces are so small. Thus, the temperature difference exists between 123/L and L/211 interfaces which is expressed as ΔT_G. Assuming the 123/L and L/211 interfaces are uniform composition and isothermal during growth, the solute concentrations, the interface temperatures and supercoolings at these interfaces are depicted in the equilibrium phase diagram, Fig. 4. The solute concentration at the 123/L and the L/211 interfaces are expressed as C_L^{123} and C_L^{211}, respectively. As the growth rate of the 123 phase is generally small [1] compared to the liquid diffusion rate, the

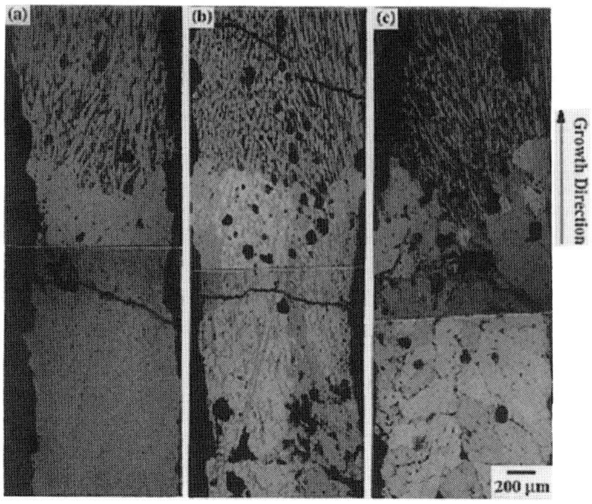

Fig. 1 Polarized microphotographs of the longitudinal section around the solidification interfaces. Initial composition of the precursor is Sm123, G_S = 50 K/cm (a) R = 0.000083 cm/s, (b) R = 0.00019 cm/s, and (c) R = 0.00028 cm/s. Partial molten regions are on the upper part, white regions are Sm211 particles and dark regions are quenched liquid phase. Solidified Sm123 locates on the lower part of the pictures.

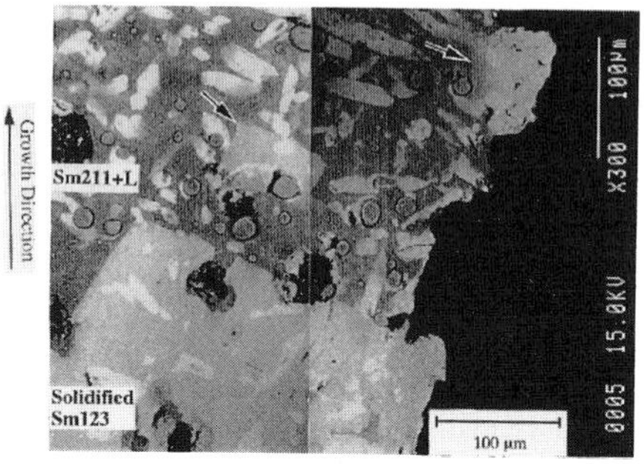

Fig. 2 SEM image of the longitudinal section around the Sm123 growth front. New and gray Sm123 grains forms at the inner wall of the sample rod or in the partial molten mixture ahead of the growth interface.

solute concentration field around the 123, and 211 phases can be considered to be stational, approximately.

3. Solute concentration field

The Laplace equation is solved to obtain the solute concentration distribution around the 123/L/211 phases precisely under the bispherical coordinate [5]. Under the boundary conditions as follows, B.C. 1 at $\mu = \mu_0$, $C = C_L^{211} - C_L^{123}$, and B.C. 2 at $\mu = 0$, $C = 0$, the concentration distribution can be expressed as Eq. (1).

$$\frac{C - C_L^{123}}{C_L^{211} - C_L^{123}} = \sqrt{2}\sqrt{\cosh\mu - \cos\eta}\sum_{n=0}^{\infty}\left[\frac{\exp\left\{-\left(n+\frac{1}{2}\right)\mu_0\right\}}{\sinh\left(n+\frac{1}{2}\right)\mu_0}\sinh\left\{\left(n+\frac{1}{2}\right)\mu\right\}P_n(\cos\eta)\right] \tag{1}$$

4. Constitutional supercooling

From the assumption of the local equilibrium, relationship between the C_L^{211} and the C_L^{123} can be obtained from the equilibrium phase diagram. Taking into account the solute flux balance at the 123/L and L/211 interfaces, C_L^{123}, C_L^{211} and the differential equation to determine the decomposition rate of the spherical 211 particle can be obtained. Here, the necessary condition not to exist the constitutional supercooling ahead of the growing 123 phase is given as $G_S \geq m_L^{123}\, \partial C/\partial z\,|_{z=0}$. If this equation is not satisfied, the constitutional supercooling exists. Differentiating Eq. (1) at $z = 0$ to take the z-component, and from the orthogonality of the zonal harmonics functions, it transforms to Eq. (2),

$$G_S \geq m_L^{123}\left(C_L^{211} - C_L^{123}\right)\sqrt{2}\frac{(1-\cos\eta)^{\frac{3}{2}}}{\sqrt{b^2 - r_{211}^{\ 2}}}\sum_{n=0}^{\infty}\frac{(2n+1)P_n(\cos\eta)}{\left(b/r_{211} + \sqrt{(b/r_{211})^2 - 1}\right)^{2n+1} - 1} \tag{2}$$

where,

$$C_L^{211} - C_L^{123} = \left(1 - m_L^{123}/m_L^{211}\right)\left(C_P - \frac{(Rr_{211}/D)\left(8\sqrt{2}\pi/9f_L^{211}\right)^{\frac{2}{3}}C_S^{123} - \left\{G_S b/m_L^{211} + \left(1 - m_L^{123}/m_L^{211}\right)C_P\right\}\sum_{n=0}^{\infty}\frac{\sqrt{(b/r_{211})^2 - 1}}{\left(b/r_{211} + \sqrt{(b/r_{211})^2 - 1}\right)^{2n+1} - 1}}{(Rr_{211}/D)\left(8\sqrt{2}\pi/9f_L^{211}\right)^{\frac{2}{3}} - \left(1 - m_L^{123}/m_L^{211}\right)\sum_{n=0}^{\infty}\frac{\sqrt{(b/r_{211})^2 - 1}}{\left(b/r_{211} + \sqrt{(b/r_{211})^2 - 1}\right)^{2n+1} - 1}}\right) + \frac{G_S b}{m_L^{211}}$$

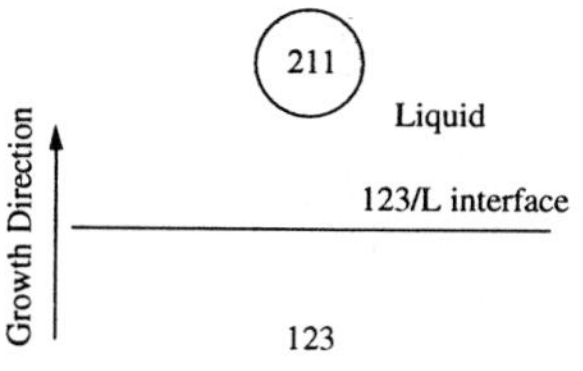

Fig.3 Schematic illustration around the solidificaiton interface. A spherical 211 particle in the liquid decomposes for solute supply to the planar 123 interface during the peritectic solidification.

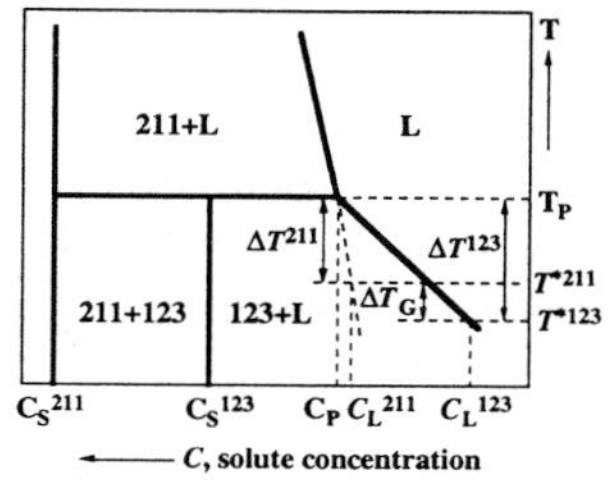

Fig. 4 Schematic illustration of the quasi-binary phase diagram in the 211-$Ba_3Cu_5O_8$ system and the difinition of the symbols.

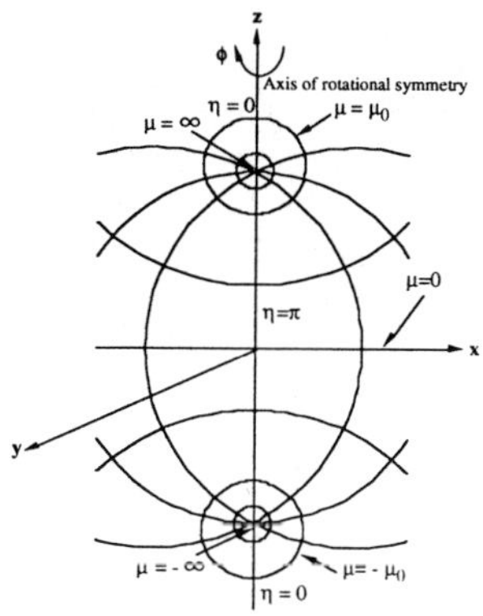

Fig. 5 Bispherical coordinate

Where b is the distance between the 123/L interface and the center of the 211 particle along the z-axis, $b = a \coth \mu_0$. The spherical 211 particles are assumed to be configurated to be isotropically close packed in the liquid of which volume fraction is f_L^{211}, and radius is r_{211} ($= a^*|\text{csch}\,\mu_0|$).

Fig. 6 shows the normalized concentration distribution around the 123/L/211 phases (C_L^{123} = 0, C_L^{211} = 1, and r_{211} = 1 in Eq. (1)) under several different configurations. The decrement of the r_{211} corresponding to the growth of the 123 phase is not involved in these calculations. These figures show that the solute concentration increases along the z direction and becomes steep with decrease of the b/r_{211} value. Physical properties and typical values used in these calculations are summarized in Table I [6]. Figs. 7 show $G_S - m_L^{123} \partial C/\partial z\,|_{z=0}$ at $\eta = \pi$ for several different values of R and r_{211}. These figures show that Eq. (2) may be satisfied when b/r_{211} is large enough. As b/r_{211} becomes small, however, $G_S - m_L^{123} \partial C/\partial z\,|_{z=0}$ comes to minus infinity, that is, the constitutional supercooling must exist when the 123 crystal interface plane comes close to the 211 particle. And these figures also show that the larger R, or the larger r_{211} induces easier occurrence of the constitutional supercooling. It is noted that the r_{211} and f_L^{211} are the determining parameters of the microstructure formation during this kind of peritectic solidification.

5. Growth interface morphological stability of solidified Sm123

Fig. 1 shows the unidirectionally solidified Sm123 phase can be grown with a faceted planar interface below a certain critical growth rate, meaning that only the constitutional supercooling can not predict the stability of the 123 growth interface. Microstructure observation indicates that the 123 growth morphological transition occurred with the nucleation and the growth of multiple 123 grains ahead of the growth front. This model does not include the interface kinetics of 123/L or L/211 interfaces. When the constitutional supercooling is small enough, however, the planar 123 interface can maintain unless it overcomes the activation energy for the nucleation and the growth of a new 123 phase ahead of the interface. That is one of the reasons why the 123 phase can grow with planar interface, even if there must exist the constitutional supercooling.

6. Conclusions

From diffusion analysis using the bispherical coordinate, it was clarified that the

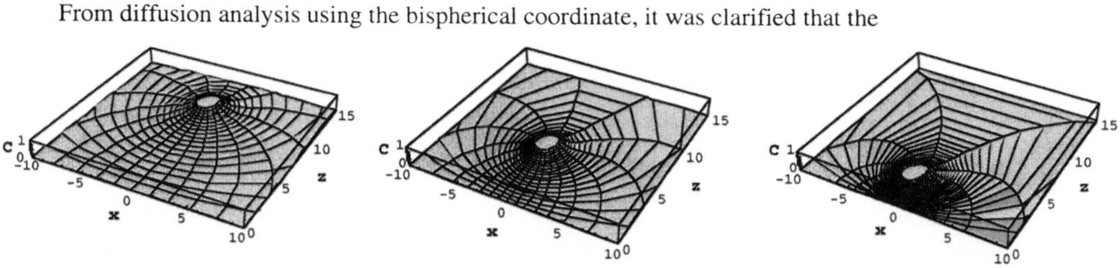

(a) $b/r_{211} = 10$ (b) $b/r_{211} = 5$ (c) $b/r_{211} = 2$

Fig. 6 Concentration distribution in the liquid around a spherical 211 particle (circle in the middle) and the planar 123 phase (z = 0) under several diffrent configrations, (a) b/r_{211} = 10, (b) b/r_{211} = 5, (c) b/r_{211} = 2. Concentration gradient around the origin (x = 0, z = 0) becomes steep as b/r_{211} decreases. Radius of a 211 sphere and the boundary conditions were normarized to simplify the calculation, that is, $C_L^{123} = 0$, $C_L^{211} = 1$, and $r_{211} = 1$.

Table I Physical properties and the typical values of the operating parameters

Value	Property	Value	Property
$D = 0.00005\ cm^2/s$	Diffusion coefficient in liquid	$C_P = 0.018$	Solute concentration at the peritectic point
$m_L^{211} = 6720$ K/mol	Liquidus slope of 211 phase at T_P	$f_L^{211} = 0.3083$	Volume fraction of 211 phase in liquid
$m_L^{123} = 3330$ K/mol	Liquidus slope of 123 phas at T_P	$r_{211} = 0.0001 - 0.001$ cm	Radius of a 211 particle
$C_S^{211} = 0.500$	Solute concentration of 211 solid phase	$R = 0.0000278 - 0.000278$ cm/s	Growth rate
$C_S^{123} = 0.16666$	Solute concentration of 123 solid phase	$G_S = 50 - 1000$ K/cm	Temperature gradient at the 123/L interface

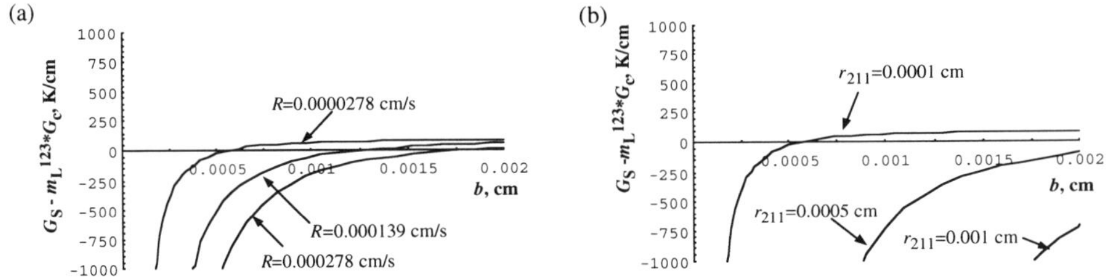

Fig. 7 (a) Growth rate, R, and b dependence of $G_S - m_L^{123}\partial C/\partial z|_{z=0}$ at $\eta=\pi$ (at the origin). As R increases, the constitutional supercooling occurs farther ahead of the 123/L interface. $r_{211} = 0.0001$ cm, $R = 0.0000278, 0.000139, 0.000278$ cm/s. (b) 211 particle radius, r_{211}, and b dependence of $G_S - m_L^{123}\partial C/\partial z|_{z=0}$ at $\eta=\pi$. As r_{211} increases, the constitutional supercooling occurs farther ahead of the 123/L interface. $r_{211} = 0.0001, 0.0005, 0.001$ cm, $R = 0.0000278$ cm/s. $f_L^{211} = 0.3083$, $G_S = 100$ K/cm.

constitutional supercooling must exist ahead of the 123 phase. It exists more easily when under larger R, smaller Gs, and distribution of larger 211 particles in the mixture.

7. Acknowledgment

This work was supported by NEDO as a part of Industrial Science and Technology Frontier Program.

References

1. M. Sumida, Y. Nakamura, Y. Shiohara, and T. Umeda, J. Mater. Res., Vol. 12, No. 8, Aug., 1979-1989 (1997)
2. M. J. Cima, M. C. Flemings, A. M. Figueredo, M. Nakade, H. Ishii, H. D. Brody, and J. S. Haggerty, J. Appl. Phys. 72 (1), 1 July, 179-190 (1992)
3. T. Izumi, Y. Nakamura, and Y. Shiohara, J. Cryst. Growth 128, 757-761 (1993)
4. T. Izumi, Y. Nakamura, and Y. Shiohara, J. Mat. Res. Vol. 7, No. 7, July, 1621-1628 (1992)
5. P. M. Morse, H. Feshbach, THEORETICAL METHODS OF MODERN PHYSICS Part 1 (McGraw-Hill Publishing, Tokyo), pp 1298
6. M. Sumida, M. Tagami, Ch. Krauns, Y. Shiohara, and T. Umeda, Physica C 249, 47-52 (1995)

Materials Science and Engineering Serving Society
S. Sōmiya, R.P.H. Chang, M. Doyama and R. Roy (Editors)
1998 Elsevier Science B.V.

Preparation and thermoelectric properties of p-type Si-Ge thermoelectric materials by mechanical alloying

H. Okamura[a], M. Niino[a], Y. H. Park[c], M. Miyajima[b] and R. Watanabe[a]

[a]Department of Materials Processing, Faculty of Engineering, Tohoku University, Sendai 980-77, Japan

[b]Daikin Industries Ltd. MEC Laboratory, 3 Miyukigaoka, Tsukuba, Ibaraki 305, Japan

[c]Tohoku National Industrial Research Institute, Sendai 983, Japan

We report on the powder metallurgical preparation and the thermoelectric properties of heavily doped p-type Si-Ge solid solution. The solid solution powders were prepared by mechanical alloying(MA) from elemental silicon, germanium and boron as a dopant. We aimed at large reduction of the thermal conductivity via very fine microstructure obtained by MA process.

Thermoelectric properties were measured in the temperature range from 300K to 1073K, and the thermal conductivity reduction of more than 20% compared with non MA samples were achieved. The variations in the amount of boron also appear to have a effect on microstructure, hence on the thermal conductivity.

1. INTRODUCTION

The vaccum melting / chill casting / grinding / hot pressing sequence is currently used to prepare Si-Ge alloys for high temperature thermoelectric conversion applications[1]. The figure of merit, Z, is a measure of how good a material is for thermoelectric conversion and is given by

$$Z=\alpha^2\sigma / \kappa \tag{1}$$

where, α is the Seebeck coefficient, σ is the electrical conductivity and κ is the thermal conductivity. The figure of merit also depends on the carrier concentration and there is some optimum carrier concentration at which the figure of merit achieves a maximum[1, 2].

In this study, to obtain low thermal conductivity thermoelectric material, i.e., high figure of merit, p-type $Si_{80}Ge_{20}$ alloys with very fine microstructure were prepared by mechanical alloying(MA) and subsequent hot pressing. This method is also expected to save industrial production cost. Also we studied the feasibility of carrier concentration controling based on this method.

2. EXPERIMENTAL PROCEDURE

The powder of silicon(5N), germanium(5N), and p-type dopant boron(2N) were used as starting materials, and mixed to the nominal composition of $Si_{80}Ge_{20}$-xat%B(x=0.25, 0.5, 1.0). The mixed powder were mechanically alloyed using stainless vibratory mill and steel milling balls in Ar atmosphere for milling time of 5, 10, 20, 40, 75, 100, 150, and 200h. The weight ratio of filled powder to the milling balls was 1/75.

The milled powder was sintered by hot pressing in a graphite die under a pressure of 60MPa at the temperature of 1523K for sintering time of 3600s. Argon gas flow was used as the sintering atmosphere.

The alloying process was observed from the change of the X-ray diffraction patterns of milled powder. The homogeneity of milled powder was estimated by SEM-EDX. The microstructures of the sintered compacts were observed by TEM.

The thermal conductivity of the sintered Si-Ge compacts was measured by laser flash method from 300K to 1073K. Seebeck coefficient was determined from 300K to 1073K as the ratio of the voltage developed between two Ta electrodes in contact with the sample, one electrode 10K warmer than the other, and the temperature difference between the Ta electrodes as measured by thermocouples placed very near the points of contact of the electrodes with the sample. The electrical conductivity was measured from 300K to 1073K by 2-probe method using the same arrangement as Seebeck coefficient measurement. The DC current flow was added between the Ta electrodes. The Hall coefficient, from which the carrier concentration and the Hall mobility were calculated, were measured by van der Pauw method at room temperature.

3. RESULTS AND DISCUSSION

3.1. Phase analysis and microstructures

The X-ray diffraction patterns of vibratory ball milled Si-Ge powder are shown in Fig.1. The peaks from Ge disappeared after 20h of processing, and the peaks from Si shifted to lower diffraction angle and broadened with increase of milling time. The change of d_{111}, interplaner spacing of (111), of Si and Ge in milled powder which indicate the shift in composition stopped after 100h. This shows that alloying of elemental Si and Ge completed after 100h. The composition analysis by SEM-EDX indicated homogeneity of Si and Ge element. From X-ray fluorescence analysis, the impurities from the milling vessel and balls were 600ppm of Fe, but this is a negligible amount compared with the amount of added dopant boron from the view of the carrier concentration control.

The SEM micrographs of milled powder after 200h milling, and the TEM micrographs of sintered compacts are shown in Fig.2. Mechanically alloyed powder consisted of particles of

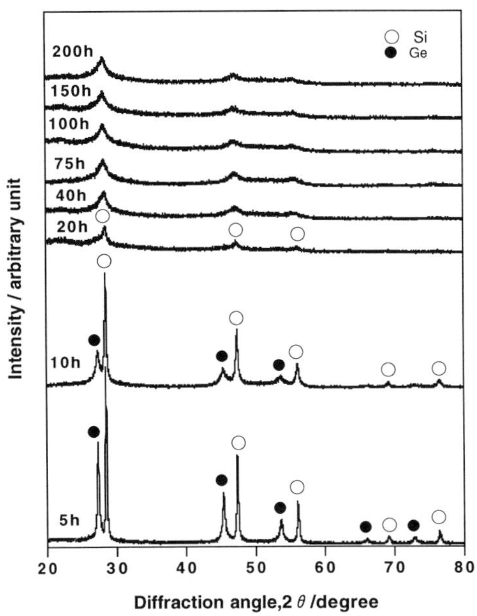

Fig.1 The X-ray diffraction patterns of vibratory ball milled Si-Ge powder with milling time.

Fig.2 SEM and TEM micrographs of p-type Si-Ge sample.((a):powder, (b):as sintered)

about 0.2μm in diameter, and sintered compacts kept submicron size grains. Twinning patterns which are likely to be introduced by mechanical alloying process are detected on micrograph. The difference in grain size with the amount of boron was not observed.

3.2. Thermoelectric properties

The Seebeck coefficient, the electrical conductivity, the thermal conductivity, and the figure of merit of p-type Si-Ge sintered compacts as a function of temperature are shown in Fig.3. Solid lines represent the measurement for hot pressed sample using melt solidified powder[3]. The considerable thermal conductivity reduction of more than 20% compared with melt solidified sample were achieved. This is because of small grain size and a large number of lattice defects, however, they were accompanied by the decrease in electrical conductivity as shown in Fig.3(b). As a result, the figure of merit were almost equal to melt solidified sample at lower temperature as shown in Fig.3(d). However, the figure of merit of MA samples were enhanced to 125% of melt solidified one at 1073K. A better understanding of the relationship between dopant concentration and thermoelectric properties may provide the basis for further significant improvements.

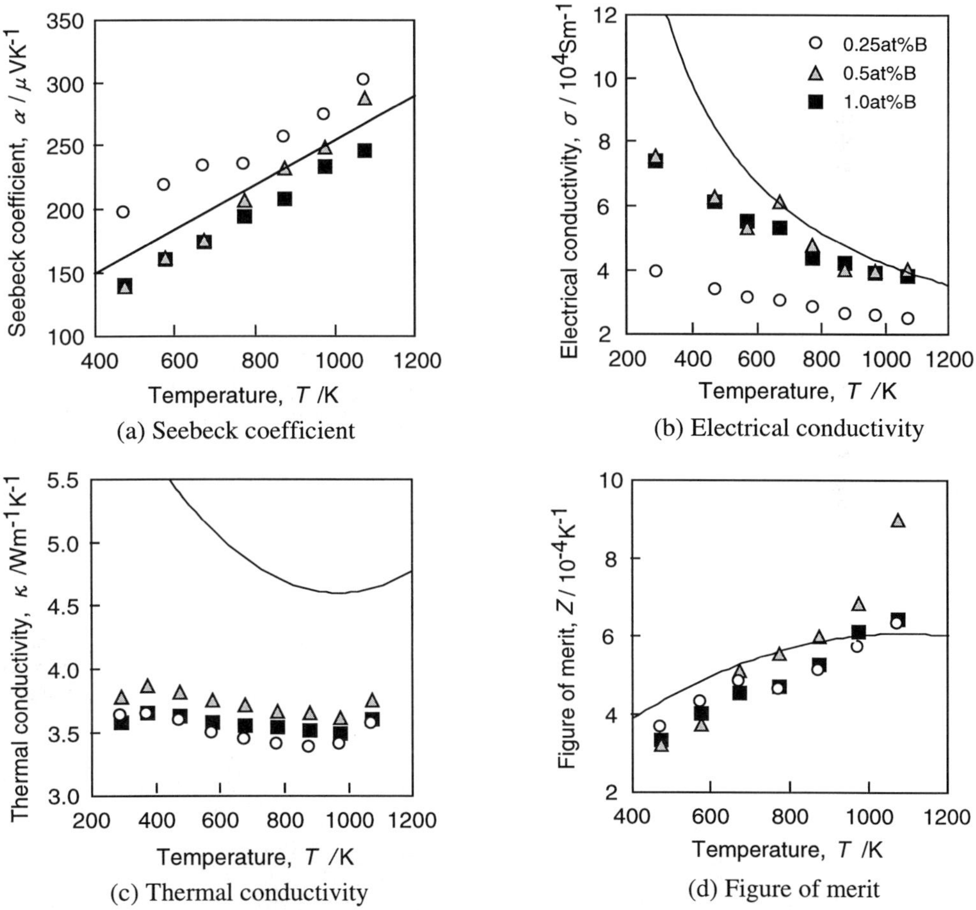

Fig.3 Thermoelectric properties as a function of temperature for p-type Si-Ge sintered compacts. (Solid line : Ref. [3])

The Hall mobility, the carrier concentration, and the theoretical carrier concentration of p-type Si-Ge sintered compacts with various amounts of boron at room temperature are shown in Table 1. The theoretical carrier concentration means the carrier concentration which is calculated assuming that each of added boron atoms produce one hole. The carrier concentration increased with increase in the amount of boron, and appear to be saturated at 1.0at%B. As the Seebeck coefficient and the electrical conductivity depend on the carrier concentration[2], the saturation of them in Fig.3(a) and (b) are consistent with the change of n in Table1. The amount of dopant which do not produce holes, n_{th}-n, increased with increasing the amount of boron. There is a large possibility that these boron act as phonon scattering centers[4] which cause decrease in the lattice thermal conductivity κ_{ph} which is given by

$$\kappa_{ph}=\kappa-L\sigma T \qquad (2)$$

where, L is the Lorentz number and T is the absolute temperature. The experimentally measured thermal conductivity κ and the electrical conductivity σ obtained from 300K to 1073K were used to calculate κ_{ph}. The result is shown in Fig.4. As expected, the lattice thermal conductivity decrease with increasing the amount of boron. We will consider in detail the effect of excess boron in other place.

Table 1 Hall mobility, carrier concentration, and theoretical carrier concentration of p-type Si-Ge sintered compacts with various amount of boron at room temperature.

Boron (at%)	$\mu_H \times 10^3$ ($m^2V^{-1}s^{-1}$)	$n \times 10^{-26}$ (m^{-3})	$n_{th} \times 10^{-26}$ (m^{-3})	$(n_{th} - n) \times 10^{-26}$ (m^{-3})
0.25	2.57	0.844	1.230	0.386
0.5	2.70	1.363	2.459	1.096
1.0	2.62	1.501	4.918	3.417

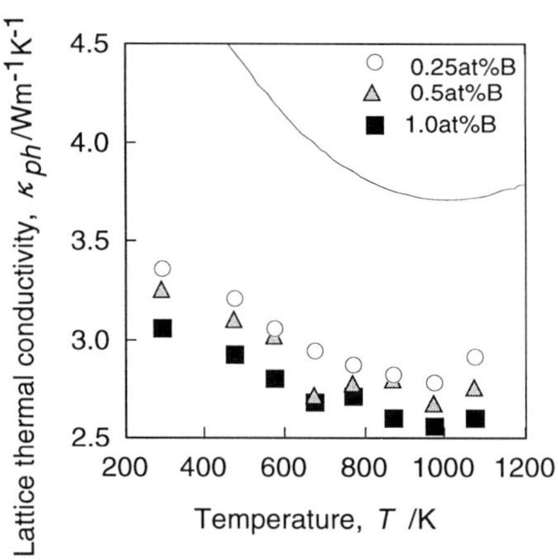

Fig.4 Lattice thermal conductivity as a function of temperature for p-type Si-Ge sintered compacts. (Solid line : Ref. [3])

4. SUMMARY

(1) Si-Ge-B solid solution powder were successfully prepared by mechanical alloying and consisted of particles of about 0.2μm in diameter. The sintered compacts kept submicron size microstructure.
(2) The thermoelectric properties were measured in the temperature range from 300K to 1073K. The thermal conductivity reduction of more than 20% compared with non MA samples were achieved. As a result, the figure of merits were enhanced to 125% of non MA samples at 1073K.
(3) The carrier concentration increased with the amount of dopant. On the other hand, the lattice thermal conductivity which is related to the microstructure decreased with the amount of dopant B. The latter results seemed to be caused by the difference in the amount of dopant which do not produce holes and act only as phonon scattering centers.

REFERENCES

1. K. Uemura and I. Nishida, Thermoelectric Semiconductors and Their Applications, Nikkan-Kougyou Shinbun-Sha, Tokyo, 1988. (in Japanese)
2. C. Wood, Energy Convers. Mgmt., 24(1984)317.
3. V. Raag, Proc. 2nd ICT, (1978)5.
4. J. P. Fleurial, Proc. 12th ICT, (1993)1.

Materials Science and Engineering Serving Society
S. Sōmiya, R.P.H. Chang, M. Doyama and R. Roy (Editors)

Interlayer interactions and magnetic properties of ferromagnetic multilayers

Y. Suezawa

Department of Information Engineering, Teikyo Heisei University,
2289-23, Uruido, Ichihara-Shi, Chiba 290-01, Japan

The interaction between ferromagnetic layers across a thin insulator layer is a fundamental topic where the spin-polarized electron tunnelling is considered to play an essential role. In the present work, the interlayer interaction of Ni/Al_2O_3/$Ni_{80}Fe_{20}$ trilayers was investigated together with their magnetic hysteresis curve and spin-polarized tunnel effect. Ferromagnetic resonance study showed an additional peak for trilayers with Al_2O_3 layers thinner than 400 Å, indicating an interlayer interaction. The spin-polarized tunnel effect was observed for Ni/Al_2O_3/$Ni_{80}Fe_{20}$ junctions with thicker Al_2O_3 layers. The interlayer interaction is discussed in relation to the tunnel effect and the micro-structure of the Al_2O_3 layer.

1. INTRODUCTION

The interlayer interaction in ferromagnetic multilayers has been an attractive subject for many years both from fundamental and practical viewpoints [1, 2]. The interest has been again stimulated by the discovery of the giant magnetoresistance in Fe/Cr multilayers [3], which has led to the search for MR reading head materials.

The simplest form of magnetic multilayers is bilayers, where two magnetic layers are in direct contact. The exchange interaction at the interface is expressed as

$$E = -J\cos(\phi) \tag{1}$$

where E is the areal density of the coupling energy, J is the exchange constant, and ϕ is the angle between the magnetizations of the layers on both sides of the interface. Previously, we reported the magnetic properties of bilayers which was composed of a soft $Ni_{80}Fe_{20}$ and a hard Fe layer. Owing to the exchange interaction, the magnetization of the soft $Ni_{80}Fe_{20}$ layer was pinned to that of the hard Fe layer at the interface, and reversible magnetization changes and reversible magetoresitance changes were observed [4]. The magnetic behavior of such soft/hard bilayers depends on their thickness, depending how deep the pinning effects the soft magnetic layer.

The exchange interaction exists also in metallic multilayers where ferromagnetic layers are separated by a nonmagnetic spacer layer. The interaction is due to metal electrons in the layer, oscillatory and depends on the interlayer thickness. Such

RKKY-type coupling has been observed in Fe/Cr, Co/Cr [5], Fe/Au, Fe/Cu [6] and in other systems.

The exchange type interaction is expected even in trilayers which consist of ferromagnetic layers separated by a thin insulator. Slonczewski theoretically showed that spin-polarized tunneling electrons caused an coupling expressed also by the form of Equation 1, and that the exchange constant J was proportional to the mean conductance G_0 across the insulator layer[7]. Such interaction due to the spin-polarized tunnelling electrons was observed in Fe/C/Fe trilayers [8].

It should be noted that the tunnel conductance depends on the angle ϕ between the magnetizations of two ferromagnetic electrode layers as

$$G = G_0(1 + \varepsilon \cos \phi) \tag{2}$$

where ε is the product of spin polarizations of those ferromagnetic layers and a factor due to the insulator barrier height [7]. This so called magnetic valve effect was observed in Fe/Ge/Co at 4.2 K[9], Ni/NiO/Co at 4.2 K and room temperature [10,11] and Ni/Al_2O_3/Co at room temperature[12]. Recently, relative resistance change of 18 % was observed in Fe/Al_2O_3/Fe junctions at room temperature [13].

In the present work, the magnetic properties of $Ni_{80}Fe_{20}$/Al_2O_3/Ni trilayers was studies in relation to the interlayer interaction, by observing magnetization curves, spin-dependent tunnelling and ferromagnetic resonance.

2. EXPERIMENTAL METHOD

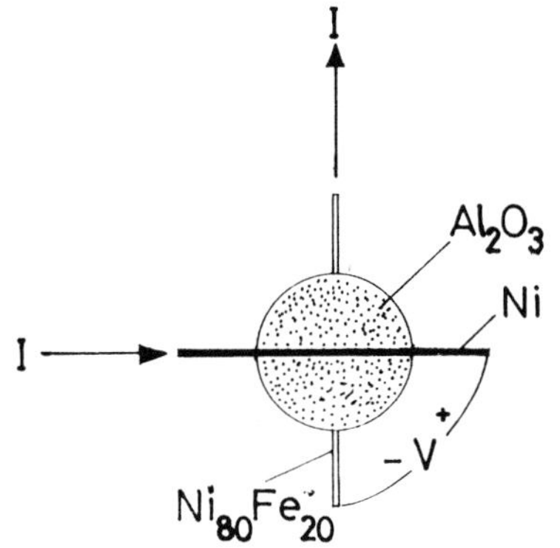

Figure 1. Structure of Ni/Al_2O_3/$Ni_{80}Fe_{20}$ junction and measurement of the tunnel characteristics.

To study the interlayer interactions, $Ni_{80}Fe_{20}$/Al_2O_3/Ni trilayers were prepared as follows. First, a Ni film was evaporated on to a glass substrate. Next, a Al film of 100, 400 and 700 Å thick were evaporated onto the Ni film. These Ni and Al films were evaporated sucseccively without breaking vacuum. The Al/Ni film is then exposed to air for 48 hours to form an Al_2O_3 layer. Finally, a $Ni_{80}Fe_{20}$ film was evaporated onto the Al_2O_3/Ni film. The thickness of the Ni and $Ni_{80}Fe_{20}$ film is 700 and 1200 Å, respectively. These evaporations were done in vacuum range of 10^{-6} Torr. The substrate temperature was 150 °C during the evaporation of the Ni film but 50°C during the Al and $Ni_{80}Fe_{20}$ film to avoid mutual diffusion at their interface. These samples were disk-shaped with its diameter of 5 mm. Ni/Al_2O_3/$Ni_{80}Fe_{20}$ junctions were prepared in a similar manner. Its structure is schematically shown in Figure 1. Both electrodes are 10 mm long and 0.2 mm wide. The Al_2O_3 insulator layer is disk-shaped with its diameter of 5 mm. The vacuum was broken between the evaporations to exchange masks and to oxidized the Al layer. During the evaporations of Ni and $Ni_{80}Fe_{20}$ film, a static magnetic field of 5 Oe was applied along their long axes so that the easy axes for magnetization were formed along these directions. Disk-shaped trilayers were similutaneously prepared, to measure the magnetic characteristics.

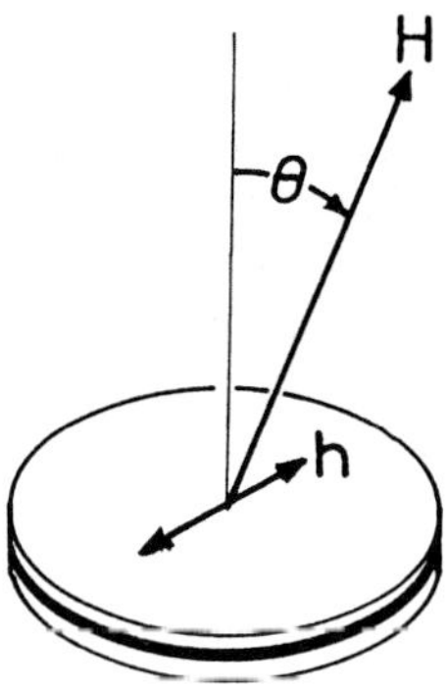

Figure 2. Static magnetic field, *H*, and microwave magnetic field, *h*, in ferro-magnetic resonance measurement.

The magnetic hysteresis loops (M-H loop) were observed by using a sample magnetometer (VSM). The magnetoresistance of the $Ni/Al_2O_3/Ni_{80}Fe_{20}$ junctions were measured as indicated by Figure 1. The constant current I (0.1 mA) was supplied from the one end of the Ni electrode to the $Ni_{80}Fe_{20}$ counter electrode via the Al_2O_3 layer. The voltage, V, was measured between the Ni and $Ni_{80}Fe_{20}$ electrodes at their opposite ends. Thus, the resistance across the Al_2O_3 layer can be obtained as $R = V / I$. Static magnetic fields were applied parallel to the junction surface and at various angles relative to the Ni electrode.

The interlayer interactions were studied by means of ferromagnetic resonance (FMR). A microwave magnetic field h (9 GHz) was applied in the film plane and a static magnetic field H (0 ~ 15 kOe) was applied perpendicularly to the microwave field (Figure 2). The absorption of microwave power was measured as a function of static magnetic field for various field directions. Two absorption peaks can be usually observed for the multilayers since the resonance field is different for each ferromagnetic layer. These resonance fields shift to lower values as the static magnetic field is leaned to the in-plane direction.

3. RESULTS AND DISCUSSION

A typical magnetic hysteresis curve (M-H curve) of $Ni/Al_2O_3/Ni_{80}Fe_{20}$ film is shown in Figure 3. The thickness of Ni, Al_2O_3 and $Ni_{80}Fe_{20}$ layer was 700, 700, and 1200 Å, respectively. The magnetic moment of the film is shown as a function of magnetic field, which was applied along the easy axis of Ni layer and along the hard axis of the $Ni_{80}Fe_{20}$ layer. There are abrupt changes at the coercive force of the Ni layer, H_c(Ni), and the anisotropy field of $Ni_{80}Fe_{20}$ layer, H_k(NiFe), as indicated in Figure 3. This hysteresis curve can be considered as the superposition of the open easy-axis-loop of Ni layer and the closed hard-axis-loop of the $Ni_{80}Fe_{20}$ layer. For field values such that $|H| > H_c(\mathrm{Ni}) > H_k(\mathrm{NiFe})$, the magnetizations of Ni and $Ni_{80}Fe_{20}$ layer are saturated in the field direction and parallel to each other. For field values such that $H_k(\mathrm{NiFe}) < |H| < H_c(\mathrm{Ni})$ only the magnetization of the $Ni_{80}Fe_{20}$ layer is reversed to the field direction but that of the Ni layer remains, thus they are oppositely oriented. In this way, the angle between the magnetizations of those layers could be changed with external magnetic field.

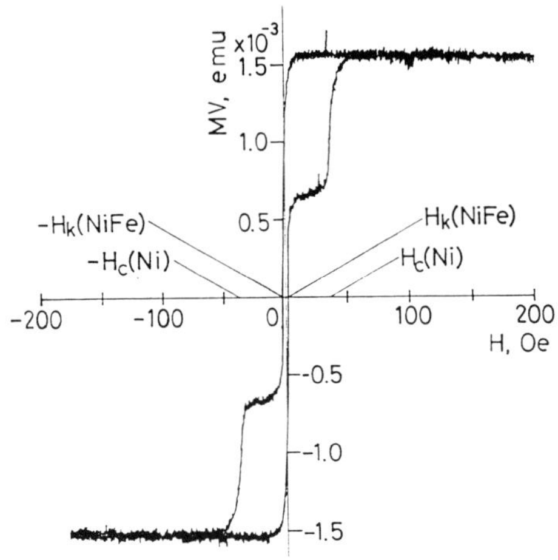

Figur 3. magnetic hysteresis curve for $Ni/Al_2O_3/Ni_{80}Fe_{20}$ film .

The magnetoresistance effect of a $Ni/Al_2O_3/Ni_{80}Fe_{20}$ junction is shown in Figure 4. The resistance across the Al_2O_3 layer is shown as a function of the magnetic field applied along the Ni electrode, i.e. its easy axis. The resistance increased for field values between H_c(Ni) and H_k(NiFe), where the magnetizations of these electrodes are oppositely oriented as mentioned above. The resistance change was in accordance with that of spin-dependent tunnelling indicated by Equation (1). The minimum value of the resistance was 3.217 Ω and its relative change was about 1 %. This result is not unreasonable for a tunnel resistance. The thickness of the Al_2O_3 layer seems too large for tunnelling to occur. However, we have reported that the Al_2O_3 layer formed by air oxidation of an evaporated Al film was composed of aluminium grains each covered by a thin layer of Al_2O_3, and that tunnel barriers were formed between those grains[14]. For those junctions with a Al_2O_3 layer thinner than 400 Å, the I-V characteristics showed that their elctrodes were shorted, indicating formation of micro-bridges.

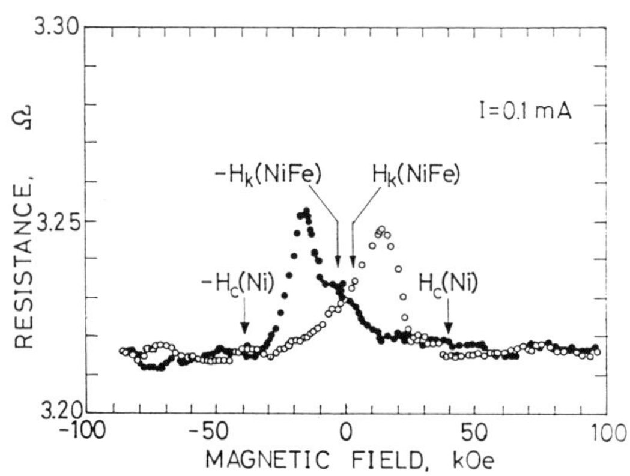

Figure 4. Resistance versus magnetic field curve for $Ni/Al_2O_3/Ni_{80}Fe_{20}$ junction.

Typical FMR spectra are shown in Figure 5 for a $Ni_{80}Fe_{20}$(1200 Å)/Al_2O_3 (100 Å)/Ni(700 Å) film. In this figure, the power absorption is shown as a function of static magnetic field, and the angle between the field and the film normal is shown as a parameter. Two main peaks were observed for each field angle. These peaks indicated by the marks ▼ is due to the Ni layer and those indicated by the marks ▽ is due to the $Ni_{80}Fe_{20}$ layer. These main peaks shifted towards lower fields as the field direction was changed from the film normal to the film plane. Subsidiary small peaks were observed at the lower field side of the $Ni_{80}Fe_{20}$ peak in the case of the field along the film normal. These are

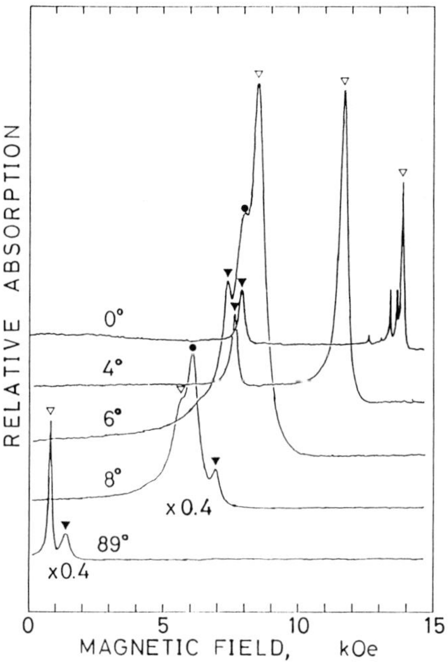

Figure 5. Ferromagnetic resonance spectra of $Ni_{80}Fe_{20}/Al_2O_3/Ni$ film. The angle between the static magnetic field and the film normal is shown as a parameter. The marks ▽ and ▼ indicate the peak due to the $Ni_{80}Fe_{20}$ and the Ni layer respectively.

due to spin-wave but excluded from the present discussion. It is noticed that the third peak appeared when the field angle was 6°, as indicated by the mark ●. Such additional peak appeared for the field angle between 6 ~ 10°. This newly appeared peak is considered due to either the formation of alloy layers at the interfaces or the magnetic coupling between the Ni and $Ni_{80}Fe_{20}$ layer. To examine the possibility of such alloy layer formation, Ni/Al_2O_3 and $Al_2O_3/Ni_{80}Fe_{20}$ film were prepared similutaneously with the above trilayers. These films showed FMR spectra with a single peak only, excluding such possibility.

It should be noted that the angle between the Ni and $Ni_{80}Fe_{20}$ magnetization is larger in the above field directions and that the exchange coupling produces stronger torque between them. Thus, the third peak is considered due to the magnetic coupling.

To see the effect of Al_2O_3 layer thickness on the interlayer interaction, FMR spectra were observed for $Ni_{80}Fe_{20}$(1200 Å)/ Al_2O_3(400 Å)/Ni(700 Å) and $Ni_{80}Fe_{20}$(1200 Å)/ Al_2O_3(700 Å)/Ni(700Å) trilayers, which were prepared simultaneously with the above trilayers as mentioned before. A weaker indication of interlayer coupling was observed also for the samples with Al_2O_3 layer of 400 Å thick. For the trilayers with Al_2O_3 of 700 Å thick, no additional peak was observed, indicating no inter layer interaction. Since the tunnel effect study suggested formation of micro-bridges in those trilayers with thinner Al_2O_3 layer as mentioned above, the interaction is possibly due to such micro-bridge formation. However, it should be mentioned that both micro-bride and spin-dependent tunnel electrons are expected to cause exchange type coupling. Therefore, it is not clear so far if the above inter layerinteraction is totally due to the micro bridge formation or in part due to the spin-dependent tunnelling.

Acknowledgement: The author wish to thank our graduate students, D. Aoki, S. Iwata and H. Hasegawa for help with experiments.

REFERENCES

1. A. Yelon, Physics of Thin Films, Academic Press, New York, 1971, p.205.

2. P. Grünberg and F. Sourenbach, MRS int'l Mtg. on Adv. Mats 10 (1989) 255.
3. M.N. Bibich, J.M. Broto, A. Fert, F. Nguyen Van Dau, F. Petroff, P. Eitenne, G. Creuzet, A. Friedrich and J. Chazelas, Phys. Rev. Lett. 61 (1988) 2472.
4. S.S.P. Parkin, N. More and K.P. Roche, Phys. Rev. Lett. 64 (1990) 2304.
5. Q. Leng, V. Cros, R. Schäfer, A. Fuss, P. Grünberg and W. Zinn, J. Magn. Magn. Mat. 126 (1993) 367.
6. Y. Gondō and Y. Suezawa, Thin Solid Films 281-282 (1996) 496.
7. J.C. Slonczewski, Phys. Rev. B39 (1989) 6995.
8. M. Pomerants, J.C. Slonczewski and E. Spiller, J. Appl. Phys. 61 (1987) 3747.
9. M. Julliere, Phys. Lett. 54A (1975) 225.
10. S. Maekawa and U. Gäfvert, IEEE Trans. on Magn. MAG-18 (1982) 707.
11. Y. Suezawa and Y. Gōndo, Proc. Int'l. Symp. Phys. Magn. Mat., Sendai, 1987 (World Scientific, Singapore, 1987), p.303.
12. Y. Suezawa, F. Takahashi and Y. Gondō, Jpn. J. Appl. Phys. 31 (1992) L1415.
13. T. Miyazaki and N. Tezuka, J. Magn. Magn. Mat. 139 (1995) L231.
14. Y. Suezawa an Y. Gondō, J. Magn. Magn. Mat. 126 (1993) 524.

Materials Science and Engineering Serving Society
S. Sōmiya, R.P.H. Chang, M. Doyama and R. Roy (Editors)

Preparation of CdS nanoparticles at the monolayer of a positively charged surfactant

W. Huang and W. -L. Yu

Institute of Materials Research and Engineering, National University of Singapore, Singapore 119260

Semiconducting nano-particles and their composites with functional polymers are potential electroluminescent (EL) materials. Nanosized particles have been prepared by mechanical reduction of larger structures, by direct chemical synthesis, and by colloid chemical techniques using controlled mixing of precursors under well-adjusted conditions [1]. An interesting approach was based on membrane mimetic systems, which provided chemical and spatial control for the *in situ* generation and stabilisation of ultrafine semiconducting and magnetic particles [2]. Many membrane mimetic systems of organised surfactant aggregates have been used in this field. In particular, CdS and related semiconductor particles have been *in situ* generated in reversed micelles [3,4], surfactant vesicles [5,6], bilayer lipid membranes (BLMs) [7], and between the headgroups of Langmuir-Blodgett (L-B) films [8]. It also has been found that monolayers are particularly useful as templates for the *in situ* generation of nano-particles and even nanocrystalline particulate films with the advantages of stabilisation and convenient transfer to substrates [9], which is essential for the fabrication of devices. Recently CdS semiconductor particles have been prepared under monolayers of anionic surfactants [10]. However, the effort on preparing CdS under monolayers of cationic surfactants was failed maybe due to the unsuccessful absorption of Cd^{2+} onto a positively charged monolayer [10].

We report here the preparation of CdS semiconductor nanoparticles under a positively charged monolayer prepared from N-methyl-p-(p-tetradecyloxystyryl)pyridinium iodine (TSPMI). The amphiphiles containing hemicyanine groups have received considerable attention due to the pronounced optical nonlinearity of the chromophore groups [11,12]. Therefore, we also expect some interesting electro-optical properties arising from the interactions between the semiconductor nano-particles and the chromophore groups based on this preparation.

TSPMI was synthesised from 4-tetradecyloxybenzaldehyde (obtained by reacting 4-hydroxybenzaldehyde with 1-bromotetradecane in the presence of K_2CO_3) and N-methyl-4-methylpyridinium iodine through an aldol-condensation reaction [12]. After recrystallizing five times from methanol, a yellow needle crystal was obtained (Anal. Calcd. for $C_{28}H_{42}ONI$: C, 62.79; H, 7.90; N, 2.62. Found: C, 62.71; H, 7.58; N, 2.97). The surface pressure – area isotherm at 25 °C for TSPMI (using chloroform as spreading solvent) was measured on a KRÜSS film balance (Model FB-1) as Fig. 1 with a precision of ± 0.2 mN m^{-1} for the surface pressure and of ± 0.5 Å^2/molecule for the molecular area. The breakdown pressure is about 52 mN m^{-1}, and the area required at this point is 0.25 nm^2/molecule. This large area requirement implies that the packing within the monolayer is controlled mainly by the head group, not by a tight packing of the hydrocarbon tails of the amphiphile.

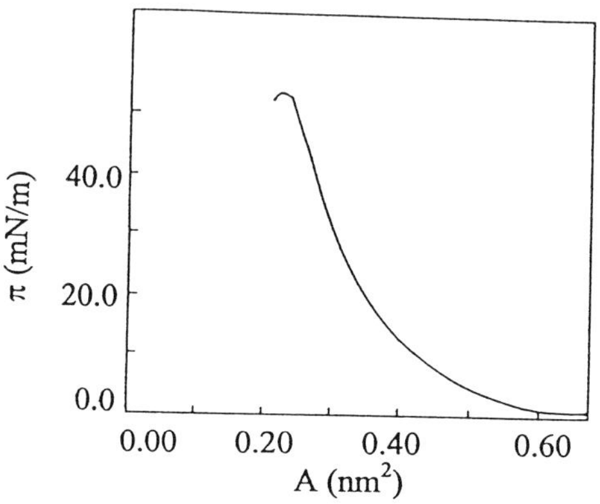

Figure 1. Surface pressure – area (π – A) isotherm of TSPMI on deionic water at 25 °C.

A similar experimental set-up as described by Zhao *et al.* [9b] was used to generate CdS particles on the monolayer. A circular glass trough (coated with paraffin) (1.5 cm deep, 5.5 cm in diameter) was placed on a clean flat glass plate and covered by a circular glass jar (5.0 cm high, 8.5 cm in diameter) whose flat frosted bottom provided a gas-tight contact. The aqueous solutions of $CdCl_2$ and Na_2EDTA with equal molar ratio constituted subphase. In these aqueous solutions, Cd^{2+} was converted to $CdEDTA^{2-}$, so the solutions are named as $CdEDTA^{2-}$ solutions below. An appropriate amount of the spreading solution of 1.6×10^{-3} M TSPMI in chloroform was carefully injected onto the clean subphase (cleaned by sucking the surface) to give a coverage of 0.25 nm^2 per molecule of TSPMI. Subsequent to 20 min of incubation, a certain amount of H_2S was introduced to the atmosphere covering the monolayer by a syringe *via* a rubber septum. During the generation of CdS particles, the reaction system was protected from light. After a period, the glass jar was taken off. The resultant particles along with the supported monolayer were transferred onto the solid substrates (quartz plates or carbon-coated Cu grids) by horizontal lifting. The substrate with the CdS particles was dried under dynamic vacuum at room temperature for 8 h before characterisation.

Transmission electron micrography (TEM) (taken on a JEOL CX100 instrument) confirmed the formation of CdS nanoparticles at the monolayer of TSPMI. Some TEM pictures are shown in Fig. 2. Fig. 2a and 2d reveal the globular shape of the generated particle clusters, and it can be identified from Fig. 2d that the globular clusters are constituted by smaller particles. The density of the particles increases with the increase of the concentration of $CdEDTA^{2-}$ in subphase. The particles generated on the subphase of 5.0×10^{-5} M $CdEDTA^{2-}$ solution are well separated (Fig. 2a), whilst the particles generated on the subphases of 1.0×10^{-4} and 1.0×10^{-3} M $CdEDTA^{2-}$ solutions connect to

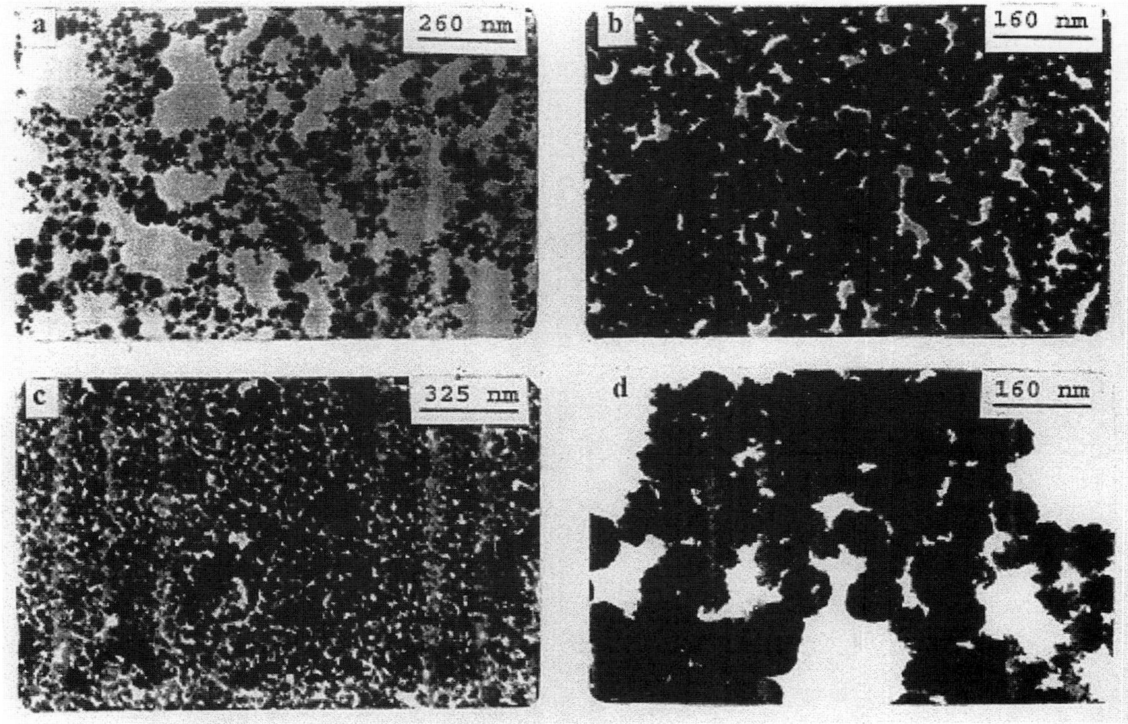

Figure 2. Transmission electron micrographs of CdS particles generated at the monolayer of TSPMI on the subphase of aqueous solutions of $CdEDTA^{2-}$ with the concentration of 5.0×10^{-5} (a), 1.0×10^{-4} (b), 1.0×10^{-3} (c), and 1.0×10^{-4} M (d) by exposure to H_2S for 10 min (a), 5 min (b, c) and 1 h (d). The amount of injecting H_2S was constant to be 90 μl. The particles were subsequently transferred onto carbon coated copper grids.

a film — particulate film (Fig. 2b and 2c). On the other hand, the size of the particles increases with the concentration of $CdEDTA^{2-}$ in the subphase, *e.g.*, the mean diameters of the particles in Fig. 2b and Fig. 2c are around 15, and 30 nm respectively. Increasing the reaction time results in the growth of the particles. Comparison between Fig. 2b and Fig. 2d reveals that the mean diameter of the particles increases from 15 nm at 5 min to about 100 nm at 1 h.

In summary, the nanoparticles of CdS semiconductor have been prepared at the monolayer of a cationic functional surfactant — TSPMI — by prior converting Cd^{2+} to $CdEDTA^{2-}$. The particles can be transferred onto substrates at different stages of the generation and growth of the particles. The density of the particles on the monolayer — from separated particles to film, and the size of the particles — from 15 nm to around 100 nm, can be controlled by adjusting the concentration of $CdEDTA^{2-}$ in subphase and the reaction time.

References

1. E. Matijevic, *Pure and Appl. Chem.,* 60 (1988) 1479.
2. (a) J. H. Fendler, *Membrane Mimetic Chemistry,* Wiley-Interscience, New York, 1982. (b) J. H. Fendler, *Chem Rev.,* 87 (1987) 877.
3. (a) P. Lianos and J. K. Thomas, *Chem. Phys. Lett.,* 125 (1986) 299. (b) P. Lianos and J. K. Thomas, *J. Colloid Interface Sci.,* 117 (1987) 505.
4. L. Motte, C. Petit, L. Boulanger, P. Lixon, M. P. Pileni, *Langmuir,* 8 (1992) 1049.
5. (a) Y. M. Tricot and J. H. Fendler, *J. Phys. Chem.,* 90 (1986) 3369. (b) H. J. Watzke and J. H. Fendler, *J. Phys. Chem.,* 91 (1987) 854.
6. (a) G. X. Zhao and W. L. Yu, *Chinese Chem. Lett.,* 5 (1994) 1071. (b) W. L. Yu, J. Pei, W. Huang, G. X. Zhao, *Materials Chem. Phys.,* 49 (1997) 87.
7. X. K. Zhao, S. Baral, R. Rolandi, J. H. Fendler, *J. Am. Chem. Soc.,* 110 (1988) 1012.
8. K. C. Yi and J. H. Fendler, *Langmuir,* 6 (1990) 1519.
9. (a) X. K. Zhao, Y. X. Yuan, J. H. Fendler, *J. Chem. Soc., Chem. Commun.,* (1990) 1248. (b) X. K. Zhao, S. Xu, J. H. Fendler, *Langmuir,* 7 (1991) 520.
10. J. H. Fendler, in *Anvances in the Applications of Membrane Mimetic Chem.,* T. F. Yen, R. D. Rilbert and J. H. Fendler Ed., Plenum Press, New York, 1994.
11. I. R. Girling, N. A. Cade, P. V. Kolinsky, R. J. Jones, I. R. Peterson, M. M. Ahmad, D. B. Neal, M. C. Petty, G. G. Robert, W. J. Feast, *J. Opt. Soc. Am.,* B4 (1987) 950.
12. D. Lupo, W. Press, U. Scheumann, A. Laschewsky, H. Ringsdorf, I. Ledoux, *J. Opt. Soc. Am.,* B5 (1988) 300.

Materials Science and Engineering Serving Society
S. Sōmiya, R.P.H. Chang, M. Doyama and R. Roy (Editors)

Distorted chalcopyrite-type structure of $AgGaS_2$ under high pressure

H. Kitahara[a], N. Ishizawa[a], F. Marumo[b] and Y. Noda[c]

[a]Materials and Structures Laboratory, Tokyo Institute of Technology,
4259 Nagatsuta, Midori-ku, Yokohama 226, Japan

[b]Department of Geosciences, Nihon University,
3-25-40 Sakurajosui, Setagaya-ku, Tokyo 156, Japan

[c]Department of Materials Science, Faculty of Engineering, Tohoku University,
Aramaki, Aoba-ku, Sendai 980-77, Japan

Single-crystal X-ray diffraction study under high pressure has been carried out to obtain crystallographic information on a high-pressure phase of $AgGaS_2$ above 4.2 GPa employing a diamond-anvil cell. The high-pressure phase has monoclinic symmetry with space group Cc, a = 8.065(4) Å, b = 8.039(1) Å, c = 6.226(3) Å and β = 128.50(3)° at 5.6 GPa. The monoclinic cell can be approximated as a pseudotetragonal cell of $a_{pt} = b_{pt}$ = 5.694(2) Å, c_{pt} = 9.750(6) Å, $\alpha_{pt} = \beta_{pt}$ = 88.70(3)° and γ_{pt} = 89.82(3)° on the basis of the conventional chalcopyrite-type tetragonal cell. The principal response of the crystal structure to compression is a volume reduction of S_6 octahedral interstices. AgS_4 tetrahedra are more distorted than GaS_4 tetrahedra in the high-pressure phase.

1. INTRODUCTION

Carlone *et al.*[1] carried out a Raman spectro-scopic study on $AgGaS_2$ under high pressures, and observed two pressure-induced phase transitions at 4.2 GPa and 11.6 GPa. Werner *et al.*[2] observed three phase transitions of $AgGaS_2$ with powder X-ray diffraction study, the first near 5 GPa, the second near 12 GPa, and the third near 15 GPa.

In the present paper, the authors report more detailed crystallographic data on the phase transition of $AgGaS_2$ taking place at 4.2 GPa.

2. EXPERIMENTAL

A piece of crystal with dimensions of about 100-150 μm in length and width and 50-80 μm in thickness was mounted in a diamond-anvil cell designed by Miyake *et al*[3]. The diamond-anvil cell was mounted on a precession camera or a four-circle diffractometer (Rigaku AFC-5) with LiF-monochromated Ag Kα to examine the symmetry and cell dimensions of high-pressure phase structure. The pressure range surveyed was from ambient to about 6 GPa.

Diffraction intensity measurements were performed at the experimental station BL-14A at the Photon Factory, National Laboratory for High Energy Physics. The synchrotron radiation was monochromatized to be 0.5 Å in wavelength. Data were measured at 5.5 GPa. All diffraction studies were carried out at room temperature.

3. RESULTS AND DISCUSSION

3.1. Unit cell of the high-pressure phase

Using 24 reflections measured in the sinθ/λ range between 0.294 and 0.446 Å^{-1}, the cell dimensions at 5.6 GPa were determined as a = 5.690(5) Å, b = 5.693(1) Å, c = 9.751(1) Å, α = 88.69(2)°, β = 88.66(4)°, and γ = 89.86(5)° without placing any constraints on the parameters.

Assuming that $a = b$ and $\alpha = \beta$, the pseudotetragonal cell can be reduced into a monoclinic one by taking $\boldsymbol{b_m}$ unique as:

$$\begin{pmatrix} a_m \\ b_m \\ c_m \end{pmatrix} = \begin{pmatrix} 1 & 1 & 0 \\ -1 & 1 & 0 \\ -0.5 & -0.5 & 0.5 \end{pmatrix} \begin{pmatrix} a_{pt} \\ b_{pt} \\ c_{pt} \end{pmatrix},$$

where subscripts m and pt indicate the monoclinic and pseudotetragonal cells, respectively. The geometrical relation between the monoclinic and pseudotetragonal unit cells is shown in Fig. 1. The cell dimensions were determined as a_m = 8.065(4) Å, b_m = 8.039(1) Å, c_m = 6.226(3) Å, and β_m = 128.50(3)° using the same 24 reflections. Crystal data for the high-pressure phase are given in Table 1 with those at ambient-pressure for comparison. In the table the dimensions of pseudotetragonal cell were calculated from that of the monoclinic cell backwardly.

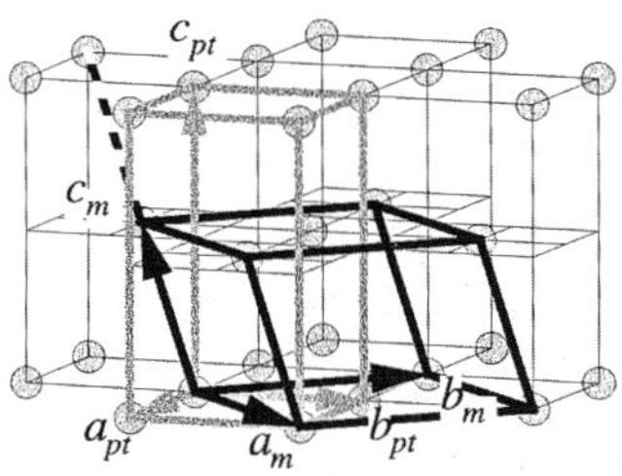

Figure 1. Relation between the pseudotetragonal cell defined by $\boldsymbol{a_{pt}}$, $\boldsymbol{b_{pt}}$, and $\boldsymbol{c_{pt}}$, and the monoclinic cell defined by $\boldsymbol{a_m}$, $\boldsymbol{b_m}$, and $\boldsymbol{c_m}$ [4]. Circles show the body-centered pseudotetragonal lattice points.

3.2. Possible space group for the high-pressure phase

In tetragonal $I\bar{4}2d$, diamond-glide planes parallel to {110} provide a reflection condition $2h+l=4n$ for hhl and its equivalent reflections. The diffraction experiments revealed that the set of diamond-glide planes parallel to (110) is lost in the high-pressure phase while the set of diamond-glide planes parallel to $(1\bar{1}0)$ is preserved. With this symmetry reduction, the $\bar{4}$ axes and all the rotation and screw diads disappeared automatically in the high-pressure phase. Therefore it was plausible that the high-pressure phase has the space group Cc, which is a *translationengleiche* subgroup of $I\bar{4}2d$.

All the crystals above the transition point suffered from twinning. The orientation relation of the twin components is schematically shown in Fig. 2. The twin components are quadruplets related by the twofold axes parallel to $\boldsymbol{a_t}$, $\boldsymbol{b_t}$ and $\boldsymbol{c_t}$ of the tetragonal $I\bar{4}2d$ cell. They can also relate each other by the $\bar{4}$ axis of $I\bar{4}2d$. Each component has $\boldsymbol{c_{pt}}$ parallel or antiparallel to $\boldsymbol{c_t}$.

Table 1
Crystal data for the ambient and high-pressure phase of $AgGaS_2$

	Ambient-pressure (at 0.1 MPa)	High-pressure (at 5.6 GPa)	
Crystal system	Tetragonal	Pseudotetragonal	Monoclinic
Space group	$I\bar{4}2d$		Cc
Z	4	4	4
Cell dimensions			
a (Å)	5.7626(5)	5.694(2)	8.065(4)
b (Å)	5.7626(5)	5.694(2)	8.039(1)
c (Å)	10.3128(9)	9.750(6)	6.226(3)
α (deg)	90	88.70(3)	90
β (deg)	90	88.70(3)	128.50(3)
γ (deg)	90	89.82(3)	90
Cell volume (Å^3)	342.47(7)		315.9(2)
Density (g cm^{-3})	4.69		5.083
Temperature (K)	298		298

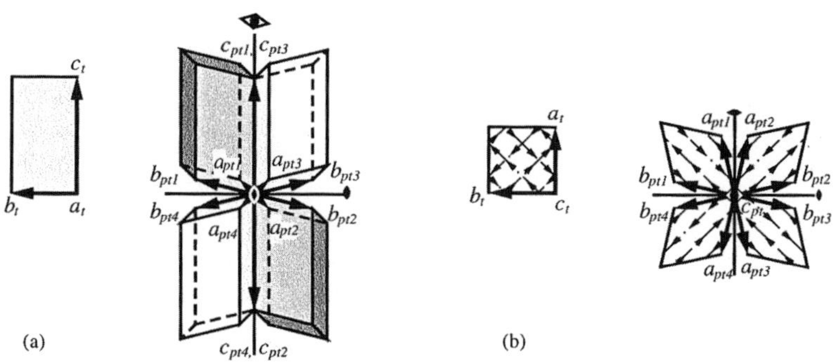

Figure 2. The geometrical relation among the four twin components formed above the transition point and the orientation of the ambient-pressure phase viewed along (a) $\boldsymbol{a_t}$ and (b) $\boldsymbol{c_t}$[4]. The subscripts *pt* and *t* indicate pseudotetragonal and tetragonal cells respectively. The deformation of the high-pressure cell is exaggerated.

3.3. Structure determination of the high-pressure phase

The structure was solved by the trial-and-error technique starting from the ambient-pressure model. Least-squares refinement of the structure parameters was carried out using a program LINKT[5]. The atomic positional parameters were refined together with anisotropic temperature factors of Ag, Ga and isotropic temperature factors of S(1), S(2). The R factor for the high-pressure phase at 5.5 GPa became 0.059 in the final cycle.

3.4. Description of high-pressure structure at 5.5 GPa

The high pressure phase has two crystallographically independent octahedral interstice $S_6(1)$ and $S_6(2)$, each of which consists of three S(1) and three S(2) atoms. Volumes and the distortion indices[6] of coordination polyhedra and volumes of selected interstices are summarized in Table 2. The volumes of the octahedral interstices $S_6(1)$ and $S_6(2)$ are 0.97 and 0.86 times larger than those at ambient pressure, respectively. Such large volume reduction of $S_6(2)$ should be a primary response to compression from structural point of view.

The S-S distances and the S-Ag-S angles of AgS_4 tetrahedra have a variety in the high-pressure phase compared to those of GaS_4. Under high pressure, the GaS_4 tetrahedra keep relatively regular shapes compared to AgS_4. The distortion of GaS_4 tetrahedra can be approximated as a kind of rotation rather than deformation. The average bond distances for Ag-S and Ga-S decrease by 2.6 % and 1.3% respectively, when the pressure varies from 0.1 MPa to 5.5 GPa. All these characteristics highlight the difference of the bonding nature between the Ag-S and Ga-S, namely, the Ag-S bonds are more ionic and have smaller force constants than the Ga-S bonds[7].

This study was supported by Grant-in-Aid for Scientific Research (B), No. 08455298.

Table 2
Polyhedral volumes and distortion indices of $AgGaS_2$

	0.1 MPa Tetragonal	5.5 GPa Monoclinic	
AgS_4 tetrahedron			
Volume($Å^3$)	8.385(5)	7.4(3)	
Quad. elong.	1.0157	1.04	
Angle var.	57.81	157	
GaS_4 tetrahedron			
Volume($Å^3$)	6.072(4)	5.8(3)	
Quad. elong.	1.0004	1.00	
Angle var.	1.627	17.5	
S_4 tetrahedral interstices		$S_4(1)$	$S_4(2)$
Volume($Å^3$)	7.041(2)	6.8(2)	6.2(3)
S_6 octahedral interstices		$S_6(1)$	$S_6(2)$
Volume($Å^3$)	28.539(6)	27.7(5)	24.6(8)

REFERENCES

1. C. Carlone, D. Olego, A. Jayaraman and M. Cardona, Phys. Rev. B 22 (1980) 3877.
2. A. Werner, H. D. Hochheimer and A. Jayaraman, Phys. Rev. B 23 (1981) 3836.
3. M. Miyake, T. Fujii, H. Ishii and F. Marumo, Rep. RLEM, Tokyo Inst. Tech. 7 (1982) 1.
4. H. Kitahara, N. Ishizawa, F. Marumo and Y. Noda, Phys. Rev. B 55[5] (1997) 2690.
5. K. Tanaka and F. Marumo, Acta Crystallogr. A39 (1983) 631.
6. K. Robinson, G. V. Gibbs, and P. H. Ribbe, Science, 172 (1971) 567.
7. H. Kitahara, N. Ishizawa, F. Marumo and Y. Noda, Acta Crystallogr. to be submitted.

Materials Science and Engineering Serving Society
S. Sōmiya, R.P.H. Chang, M. Doyama and R. Roy (Editors)
1998 Elsevier Science B.V.

AN ANAEROBIC FIXED BED REACTOR WITH A POROUS CERAMIC CARRIER

Masayuki Kaneno
Corporate *Research and Development Group, NGK Insulators, Ltd., 2-56 Suda-cho, Mizuho-ku, Nagoya 467*

ABSTRACT

From the point of preservation of the environment, waste water from the industry should be treated before exhausting outside. An anaerobic treatment is one of the eminent technology. A large number of studies on anaerobic fixed bed reactors have been reported, but there have been few studies on the application of this technology to thermophilic anaerobic treatment. The study reported in this paper was conducted to examine the thermophilic treatment performance of an anaerobic fixed bed reactor containing a porous ceramic carrier ('microbe immobilized ceramic', MIC) developed for the immobilization of anaerobic organisms. This porous ceramics with adequate pore diameter is quite favorable as MIC material. When boiled soybean wastewater (55,000 mg/l COD) was treated anaerobically at a high temperature (54℃) in a reactor where 31% of the volume was filled with the MIC carrier, it was found that highly loaded operation with up to 65kg COD/ ▪ /d was possible. The COD load per unit area of carrier surface under these operational conditions reached 0.397kg COD/m²/d, and the performance of the reactor was excellent.

INTRODUCTION

From the point of environmental problem, wastewater from the industry should be treated before exhausting outside. Wastewater from the food processing industry is often highly concentrated compared to ones from other industries, and thus disposal of this wastewater requires special measures to prevent environmental pollution. In order to reduce the organic components of the wastewater, and also to produce methane as a source of usable energy, such high strength wastewaters have been treated using anaerobic processes in recent years. There are many references concerning anaerobic fixed bed reactor treatment of high strength wastewaters.

There are also a number of reports on carriers for immobilization of organisms. One of these reports showed that in the treatment of bean blanching wastewater using anaerobic fixed beds of red draintile clay, grey potter's clay, needle-punched polyester and polyvinylchloride, the reactors filled with red draintile clay and needle-punched polyester had superior treatment capacity, and the methane generation per unit surface area of these carriers was 0.045 - 0.055 ▪ (STP)/m²/d.
A comparative study has been made of the development of methanogenic fixed films on polyvinylchloride and baked clay carriers, using acetic acid as the substrate.
It was found that baked clay was an excellent carrier. A comparison was also made between the needle-punched polyester and the red draintile clay in thermophilic treatment.

The author has examined methanogenic micro-organisms attached to carriers. He reported that porous ceramic carriers are superior to polymers and that ceramic carriers with macropores larger than 10 μm were particularly effective in immobilizing micro-organisms. "Miso" is a traditional Japanese food which is made by fermenting boiled soybeans. The wastewater which is produced when boiling the soybeans is discharged from miso factories. This wastewater is usually hot and highly concentrated. After dilution, the wastewater is currently treated in an aeration tank for a considerable time, using a large amount of energy. In our preliminary experiment the boiled soybean wastewater was treated in an anaerobic fixed bed with a microbe immobilized ceramic (MIC) carrier. It was found that thermophilic treatment (at 54℃) can treat a load about 1.7 times that of mesophilic treatment.

This paper reports work undertaken to examine the thermophilic treatment of boiled soybean wastewater using an anaerobic fixed bed reactor with a MIC carrier and confirms the effect on treatment performance of using carriers with immobilized micro-organisms. The work also confirms the treatment performance stability of the carrier anaerobic fixed bed reactor (CAFBR) during fluctuating loading rates.

EXPERIMENTAL

Effect of Carrier on Treatment Performance

Carrier. A porous ceramic carrier was used for the immobilization of micro-organisms. This microbe immobilized ceramic (MIC) carrier contains many pores with diameters of 10 μm or greater, which are thought to be effective in the immobilization of anaerobic micro-organisms. The MIC carrier has a high porosity, with a pore volume of 0.815 cm^3/g, making it a suitable material for immobilization of micro-organisms. Figure 1 shows the pore size distribution of the carrier. Despite the high porosity, this MIC carrier has a high mechanical strength with a bending strength of about 133.3 $kg\text{-}f/cm^2$.

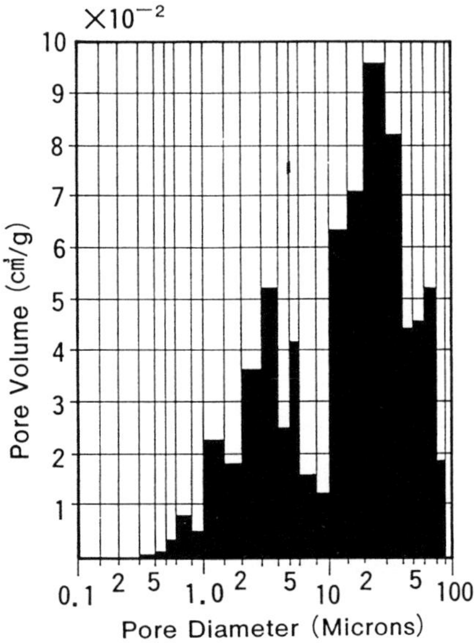

Fig. 1. Pore volume distribution of MIC carrier

Reactor. Four sets of experimental reactors were prepared. As shown I Table 1, the amount of carrier in the reactor (carrier fill ratio) was varied from 0 to 34.5% to confirm the effect of the carrier on reactor performance, and to determine the maximum BOD loading rate. The CAFBR is an upflow anaerobic fixed bed reactor (see Fig. 2), and the substrate is fed by a peristaltic pump into the bottom of the reactor. The reactor has a water jacket structure in which hot water is circulated in the outer part and the interior of the reactor is kept at 54℃.

TABLE 1 Specifications for the CAFBR

Item	Reactor N	Reactor S	Reactor M	Reactor L
Carrier	-	MIC	MIC	MIC
Carrier fill ratio, %	0	12.5	31.0	34.5
Carrier surface area, m^2	0	0.13	0.36	0.41
Working volume, 1	3.2	2.8	2.2	2.1
Reactor volume, 1	3.2	3.2	3.2	3.2

For the purpose of comparison, Reactor N was operated without the carrier. The rate of circulation was 2 1/hr, which is equivalent to a velocity of 15 cm/hr inside the reactor.

Seed. The experiment was initiated by filling the reactor with effluent from a fermentation tank where anaerobic micro-organisms had been cultivated with the boiled soybean wastewater as a substrate. The original anaerobic (thermophilic) micro-organisms in the fermentation tank were screened from sludge sampled from a hot spring.

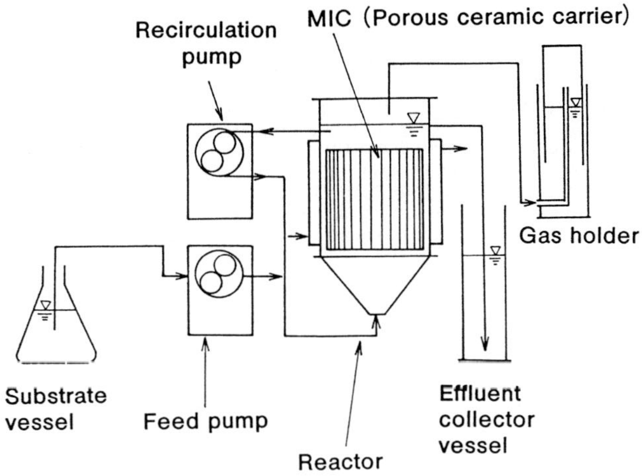

Fig.2. Upflow anaerobic fixed bed reactor with microbe immobilized ceramic carrier (CAFBR)

Substrate. Table 2 shows the composition of the boiled soybean wastewater used in the experiment. The concentration of the wastewater was quite high, as shown by the COD of about 55,000 mg/l and the BOD of about 34,000 mg/l. The wastewater was light yellow in color and transparent. It contained a moderate amount of nitrogen and phosphorus. The wastewater had been discharged at 80-90℃, was collected in a 20 l tank, and was stored in a low temperature chamber (4℃), before being used for the experiment. There was no addition of nutrients or pH adjustment. A typical Japanese miso factory would discharge several tens of m^3/d of boiled soybean wastewater.

TABLE 2 Typical Composition of Boiled Soybean Wastewater

Item	Range	Average
Suspended solids (SS), mg/l	100 -700	240
Total chemical oxygen demand (TCOD), mg/l	32,000 - 66,000	55,100
Total biochemical oxygen demand (TBOD), mg/l	31,000 - 49,000	34,000
Total organic carbon (TOC), mg/l	20,000 - 27,000	23,400
Total Kjeldahl nitrogen (TKN), mg/l	800 - 1,500	1,060
Total phosphate (TP), mg/l	200 - 330	240
Soluble sulfate, mg/l	250 - 700	310

Procedure. After filling Reactors N, S, M, and L with the fermentation tank effluent mentioned previously, the boiled soybean wastewater was fed into the reactors.
The BOD loading rate was increased step by step until an undesirable phenomenon such as a sudden decrease in gas generation rate, a sudden drop in pH, or a sudden decrease in the BOD removal rate, was observed. These phenomena indicate a reactor over-load. The maximum BOD loading rate was defined as the highest BOD loading rate tested where undesirable phenomena due to over-loading were not observed and stable treatment performance was maintained. Since the minimum supply rate of the peristaltic pump used as the feed pump was 120 ml/d, the experiment was initiated on the 17.5th day of the hydraulic residence time (HRT).

Effect of Carrier on Treatment Efficiency

Maximum loading rate. Figures 3 and 4 show the results of experiments using Reactor N without the carrier and Reactor M with a carrier fill ratio of 31%.
The figures show the variations in pH and soluble BOD removal rate of the effluent, gas generation rate, and changes in HRT and BOD loading rate of the reactor. In order to understand accurately the status of fermentation from the amount of gas generated, this is shown as the amount relative to carbon fed to the reactor. When process conditions are favorable, a constant value for gas generation should be found, irrespective of the BOD load. If all the carbon used is converted to CH_4 and CO_2, 1 g of C should theoretically generate about 1.87 l (STP) of gas. However, since part of the carbon could remain untreated or is converted into micro-organism cells, the amount of gas generated in the case of normal methane fermentation is empirically 1.2 - 1.5 l/g C utilized.

In the case of Reactor N without the ceramic carrier, flocs of suspended micro-organisms tended to be washed out with the effluent and as a result, the treatment performance decreased due to the loss of micro-organisms. After repeated washout of micro-organisms, the fermentation process became very unstable, as indicated by changes in pH and gas generation. While the performance changed to some extent due to changes in the concentration of micro-organisms in the reactor, the maximum BOD loading rate was 6.5kg/m^3/d and the minimum treatment time was 6 days for Reactor N.

On the other hand, Reactor M, with a ceramic carrier fill ratio of 31%, showed very stable fermentation, as can be seen in Fig. 4. Up to 23 weeks, the process conditions remained very stable with a pH of 7.2 - 7.3, gas generation of 1.3 - 1.4 l/g of carbon used, and a BOD removal rate greater than 90%. However, in the 24th week , pH, BOD removal rate, and gas generation rate decreased rapidly as the BOD loading rate was increased up to 58kg/m^3/d.
Therefore, Reactor M had a minimum BOD loading rate of 40kg/m^3/d and a maximum treatment time of 1.0 day.

Similar experiments were carried out for Reactors S and L, to determine the maximum BOD loading rate and minimum treatment time. The results are summarized in Table 3 together with the results of Reactors N and M.

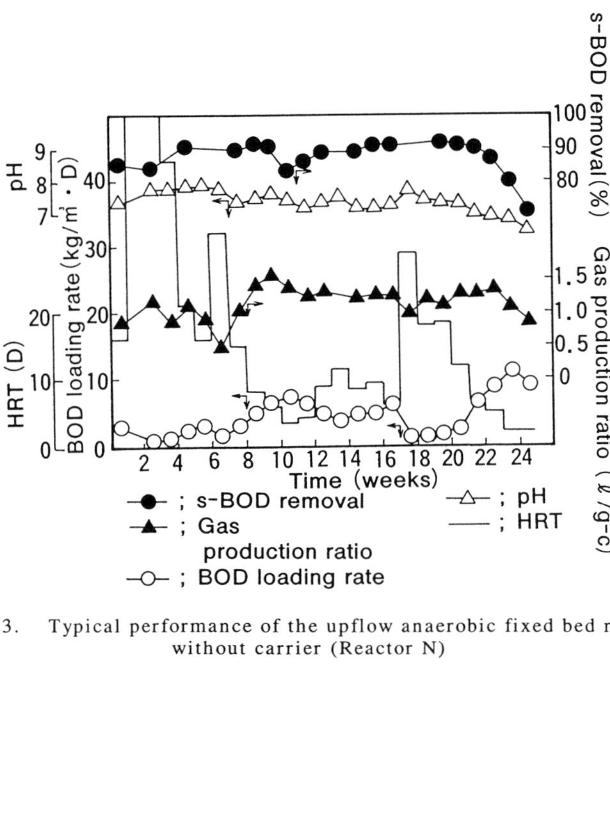

Fig. 3. Typical performance of the upflow anaerobic fixed bed reactor without carrier (Reactor N)

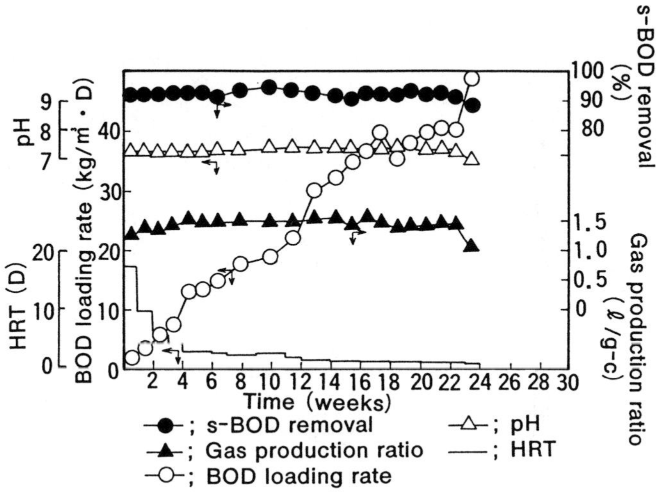

Fig. 4. Typical performance of the upflow anaerobic fixed bed reactor without carrier (Reactor M)

TABLE 3 Maximum BOD Loading Rate for CAFBR

Item	Reactor N	Reactor S	Reactor M	Reactor L
Carrier fill ratio, %	0	12.5	31.0	34.5
Maximum amount of BOD treated, g/d	21	50.4	88	98.7
Maximum BOD volumetric loading rate*, kg/m³/d	6.5	18	40	47
Maximum BOD volumetric loading rate**, kg/m³/d	6.5	15.8	27.5	30.8
Minimum treatment time*, d	6	2.2	1.0	0.85

*relative to working volume; **relative to reactor volume.

As shown in Table 3, the maximum BOD loading rate increased with increasing carrier fill ratio. For Reactor N without the carrier it was 6.5kg/m³/d, and it increased up to 47kg/m³/d for Reactor L with a carrier fill ratio of 34.5%.

Figure 5 shows the relationships between the maximum BOD loading rate and the surface area of the carrier for each reactor. The abscissa of Fig. 5 is the total surface area (apparent) of the carrier, and the ordinate is the maximum BOD loading rate per day for each reactor. When these data are plotted for Reactors L, M, S, and N, the observed values for Reactor S deviate slightly upward from the line connecting Reactors L and M and the origin. At the maximum BOD loading rate, Reactor S was operating at a relatively long HRT of 2.2 days. It is thought that suspended thermophilic micro-organisms increase and are not washed out, hence, suspended micro-organisms contribute to treatment performance. If one assumes that the contribution of suspended micro-organisms in Reactor S is comparable to that in Reactor N, the remainder of the total amount of BOD treated each day after the BOD treated by the suspended micro-organisms is deducted is the amount of BOD treated by the attached micro-organisms in Reactor S, i.e., 32g/d (value plotted in parentheses in Fig. 5). This value is on the line linking the origin and points L and M. Consequently, the slope of this line is the BOD loading rate for the biofilm, which has a value of 0.245kg BOD/m²/d.

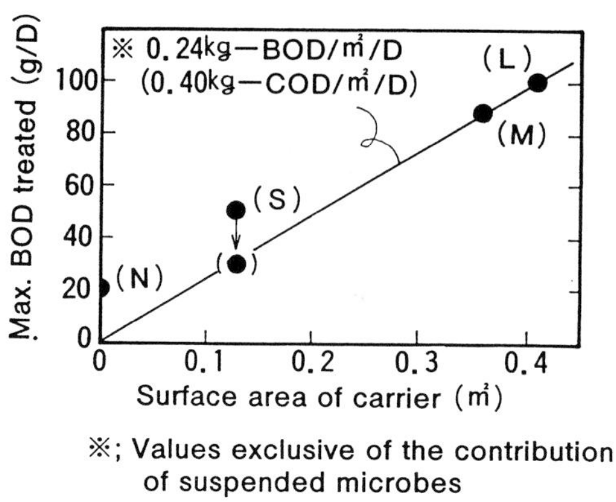

Fig. 5. Relationship between maximum BOD loading rate and surface area of carrier

The reason for this high value may be that a large number of thermophilic anaerobic micro-organisms were attached to the carrier (MIC), the pore size of which is of a particular size designed to immobilize micro-organisms. The number of micro-organisms attached to the carrier in Reactor L is shown in Table 5. 90g of micro-organisms (as volatile solids) were attached to 0.41m² of apparent surface of the carrier, i.e., 220g/m² carrier, which corresponds to 43,000mg/l of working volume. The amount of micro-organisms as suspended solids was measured as 17g, which corresponds to 8,000mg/l of working volume. However, the suspended micro-organisms are not thought to contribute greatly to methane fermentation due to the relatively short HRT of 0.85 day.

TABLE 4 Amount of Immobilized Micro-organisms in Reactor L

Distribution	*Weight	**Concentration	note
On ceramic carrier	90g	43,000mg/l	220g/m²***
In liquid	17g	8,000mg/l	
Total	107g	51,000mg/l	

* in reactor ** Versus working volume
*** Total surface area 0.41m² of carrier

Gas Production

The composition of the gas generated throughout the experiment (excluding the over-loaded period) was: methane 55 - 65%; carbon dioxide 35 - 40%. Therefore, the rate of methane gas generation per unit surface area of carrier was as high as 0.137m³ (STP) CH_4/m²/d.

CONCLUSIONS

An anaerobic fixed bed filled with a porous ceramic carrier (designed for improved immobilization of micro-organisms) was used to investigate thermophilic methane fermentation of boiled soybean wastewater. The following results were obtained:

1. Porous ceramic material having several ten microns of pore size was proved to be an excellent carrier for immobilizing anaerobic microbe.

2. Boiled soybean wastewater can be treated by a CAFBR at a high loading rate and with thermophilic methane fermentation.

3. The treatment capacity of the CAFBR was proportional to the surface area of the carrier, with a COD film loading rate of 0.397kg COD/m²/d.

4. The CAFBR treatment process was found to be stable under fluctuations in the loading rate of 26 to 51kg COD/m³/d.

Materials Science and Engineering Serving Society
S. Sōmiya, R.P.H. Chang, M. Doyama and R. Roy (Editors)

A proposal of predicting formulae for influence of stress on magnetostriction in grain oriented silicon steel

H.Masui

Teikyo University, Department of Materials Science and Engineering
1-1,Toyosatodai, Utsunomiya-shi, 320, Japan

Generalization of the formulation for the influence of the stress on the magnetostriction in grain oriented silicon steel,which is known to be one of main causes for noise problem of transformer, has been investigated. This enabled computing design to evaluate an occurrence of the magnetostriction under various practical stress conditions, proposing generalized predictive formulae for the occurrence.

1. INTRODUCTION

Recently,noise generated by transformers has become important from an environmental point of view. The magnetostriction in grain oriented silicon steel is recognized as a main cause of the noise. As extensive data show, the magnetostriction is remarkably influenced by various stress conditions through manufacturing and application process of the steel not only until a construction of transformer but all time during its operation. The reason why the magnetostriction is remarkably influenced by stress condition, such as the deterioration by compressive stress,has been elucidated by many researchers[1-9] recently. Even with these studies,however,it was not always easy to estimate the influence of superimposed stresses on the magnetostriction. This study,first of all, aimes to derive a predictive equation of stress influence on the occurrence of the magnetostriction in grain oriented silicon steel by both hypothetical theory and elaborate experimental procedure,and introduce an application method of the equation with an exemplified procedure even upon practical superimposed stress condition.

2. EXPERIMENTAL PROCEDURE

3% grain oriented silicon steel (Hi-B, 0.23mm, B8:1.93T) was prepared for experiments. Two kinds of surface coating stress conditions were made in such a way as one (sample A) with high tension insulation coating film on mill glass and another (sample B) by dissolving off the high tension coating film from mill glass by alkali solution. Magnetostriction in longitudinal (rolling) direction was measured in a range of 0.5~1.9T at 50Hz by using a Doppler type magnetostriction tester simultaneously applying compressive stress from zero to 3 MPa along the rolling direction on 60×300 mm specimens.

Observation of magnetic domain without compulsory magnetic field was done by a 200KV scanning electron microscope enclosing a special device , which applies compressive stress from zero to 8 MPa to the rolling direction on 15 ×70 mm specimen.

3. EXPERIMENTAL RESULTS

3.1. Magnetostriction

Two kinds of the different surface coating stress samples were prepared. Sample A had a coating stress of 14.7MPa, while sample B showed 3.9 MPa. Magnetostriction(0-Peak, a difference from magnetostriction at maximum flux density T_{max} to that at 0 flux density T_0) at 1.7 T is shown in Figure 1. With even small increase of compressive stress, elongation by magnetostriction in sample B became remarkably large, while extremely small increase of the elongation was perceived even on 3 MPa compressive stress in sample A.

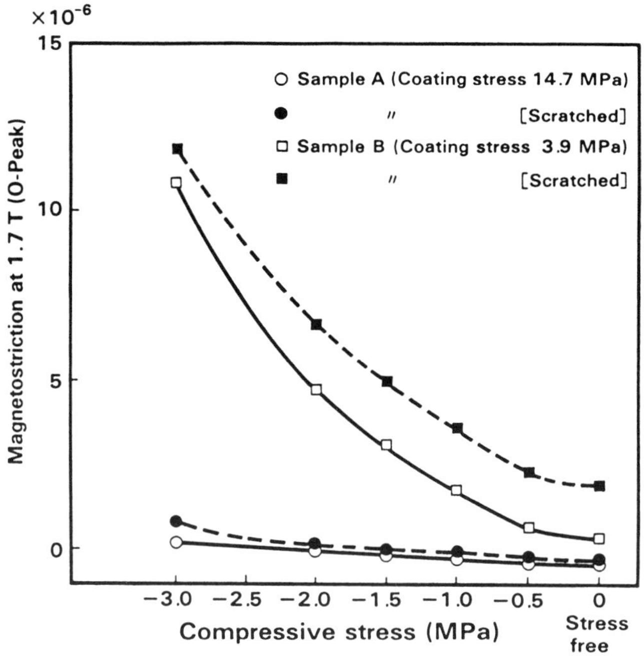

Figure 1. Effect of compressive stress to specimen on magnetostriction (1.7T, 50Hz,0-Peak) in different coating stress sample A and B.

3.2. Observation of domain by scanning electron microscope

Observation of magnetic domain by scanning electron microscope under the stress condition in sample A was exemplified in Figure 2. In stress free condition of sample A, exclusively main domains (180° main domains) were observed(2-(a)) except around the scratch, while moderate amount of the supplemetary domains (90° supplementary domains) which bear 90° domain wall component and ,in this case,are normal to rolling direction,under 4 MPa compressive stress (2-(b)) and much more the supplementary domains under 6 MPa compressive stress (2-(c)) were observed. On the other hand, in sample B, the critical compressive stress for the occurrence of the supplementary domains was just near 1 MPa and mostly the finer supplementary domains were observed even under 2 MPa compressive stress. The difference in observation between sample A and B coincides substantially with that in the magnetostriction in Figure 1. Scratched line on surface perpendicular to rolling direction in the test specimens contributes also as a marker to recognize the direction and scale of the magnetic domains in scanning electron microscopic observation.

4. A PROPOSAL OF THEORY FOR STRESS ON THE MAGNETOTRICTION

It is already known that elongation by magnetostriction of steel is caused by movement of 90° domain walls according to the equation $\triangle L/L=(3/2)\lambda(1-\cos^2\phi)$ which has maximum elongation at $\phi=90^o$,where ϕ is the angle between the axis of easy magnetization and the direction of magnetic fields[1]][2].

Therefore,in the case of grain oriented silicon steel,the important factor for estimating magnetostriction is how to evaluate the conditions for the formation of the 90° domain walls. Magnetoelastic energy theory has been frequently applied even for explaining the magnetostriction in grain oriented silicon steel[1-9]. Magnetostriction in grain oriented silicon steel is,however, far susceptible to influence of stress condition and, therefore ,

(a) Stress free (b) Compressive stress 4MPa (c) Compressive stress 6MPa

1mm

Figure 2. Appearance of the supplementary domains under applied compressive stress to sample A (Coating stress 14.7 MPa).

might be rather related directly to stress dynamics or strain elastic energy.

As was made clear by Honda and Kaya[3][4] , iron has positive magnetostrictive elongation in <100> easy magnetization axis , and in the nature of thing, as experimented by Williams[5][6], magnetization inside the domains turns toward tensile direction and forms clear domains parallel to the direction if tensile stress is applied to <100> of the iron crystal. The fact that generally the elongation in iron crystal is dominated by strain elastic energy, needs a precise calculation for the formation of magnetic domains in various stress conditions by using anisotropic elastic coefficients.

As indicated in the authors' observation on magnetic domain under stress field without compulsory external magnetic field in scanning electron microscope, the domains drastically changed their direction by stress condition. This suggests that strain elastic energy which dominates elastic deformation may change a direction of domains, resulting in a large influence on the magnetostriction. This yields a hypothesis that the largest strain elastic energy of those for the three possible directions of magnetization($x_{[100]}$, $y_{[010]}$ and $z_{[001]}$) should detemine the direction of the domains under the particular stress condition. The Goss orientation texture usually has main (180°) domains parallel to $x_{[100]}$ axis of cubic crystal. Then, more physical condition for the formation of the 90° supplementary domains is expressed by

$$E^{el}x_{[100]} \leqq E^{el}y_{[010]}\ (z_{[001]}) \qquad \cdots(1)$$

(balance of strain elastic energy in easy magnetizaion axes)

As a geometrical symmetry of Goss orientation to the steel sheet, usually $E^{el}y_{[010]}$ may be equal to $E^{el}z_{[001]}$ in the grain oriented silicon steel. Chikazumi[1][2] pointed out already that magnetoelastic energy associated with each axis of easy magnetization comes to differ from each other according to the application of stress,and this tends to cause the formation of a 90° domain wall. By using σ_{X-Y-Z}:stress tensor in specimen axes, σ_{x-y-z}: stress tensor in easy magnetization axes, ξ_{x-y-z}:strain tensor in easy magnetization axes, S:anisotropic elastic coefficient tensor, and T:tensor for coordinates transformation ,the authors, by a mathematical procedure for instance by Masui[10],derived strain elastic energy tensor E^{el}_{x-y-z} .

5. EXAMPLES OF APPLICATION OF THE THEORY

Figure 3 indicates a critical compressive stress for the formation of the supplementary domains based on the condition (1),being related with an occurrence of the magnetostriction, in different coating stress sample A and B.

This figure shows that the critical compressive stress based on the condition (1)is 3.5 MPa (sample A) and 0.8 MPa (sample B),while in observation it was near or slightly less than 4 MPa(A) and 1MPa(B) respectively. This coincidence enables substantially forecasting magnetostrictive elongation as shown in Figure 1.

Under preferable normal condition of $E^{el}x_{[100]} > E^{el}y_{[010]}\ (z_{[001]})$ in grain

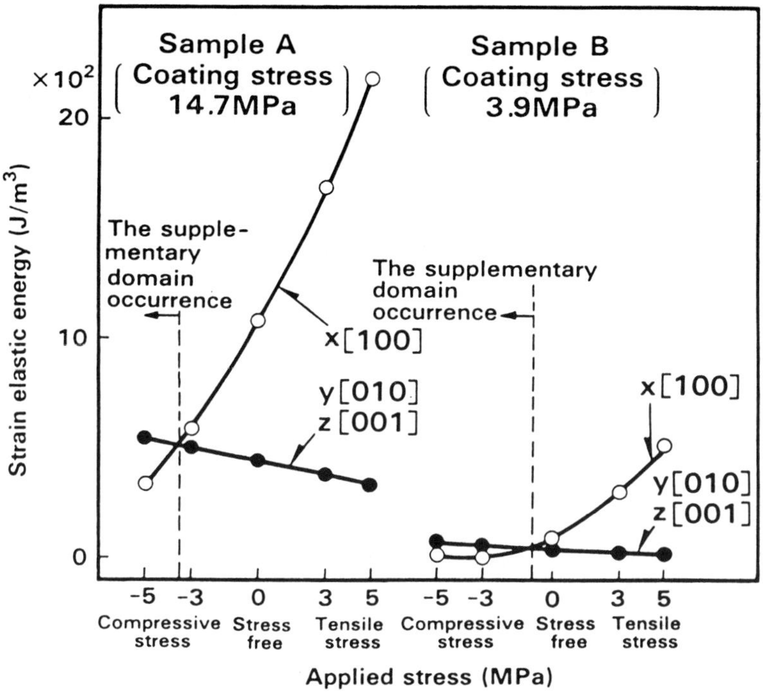

Figure 3. Effect of coating stress on strain elastic energy in easy magnetization axes x[100], y[010] and z[001] under various applied stress conditions.

oriented silicon steel, magnetostrictive elongation by magnetization in the rolling direction, if the specimen being set parallel to the rolling direction as usual, shall no longer take place basically, since main domains have already full elongation. On the other hand, under undesirable condition of $E^{el}x_{[100]} \leqq E^{el}y_{[010]}\,(z_{[001]})$, the supplementary domains or the triangular closure domains most deviated from the rolling direction shall inevitably rotate to the magnetic field applied in the rolling direction, being accompanied with a movement of 90° domain walls, definitely bringing about elongation proportional to the magnetic flux density.

Moses et al[8][9] studied the magnetostriction in the rolling direction under bi-axial longitudinal and transverse stress condition in grain oriented silicon steel and got important experimental data. Calculation by the equation (1) was carried out in similar complex stress condition as that by the experiments by Moses and the region $\Delta E^{el}=E^{el}x_{[100]}-E^{el}y_{[010]}\,(z_{[001]}) \leqq 0$ indicates the formation of the supplementary domains, which fairly explained the experimental results by Moses et al.

These proove a determinative effect and practical importance of the strain elastic energy on the formation of the supplementary domains in grain oriented silicon steel, as supported by observation in scanning electron microscope under the stress conditions.

6. CONCLUSION

Causes of noise emitted from transformer core are more or less attributed to magnetostriction of the core materials. The magnetostriction in grain oriented silicon steel is fatally influenced by stress condition on the products. A hypothesis on a critical condition for the formation of the 90° supplementary domains which determinatively cause the magnetostriction in grain oriented silicon steel, as expressed by $E^{el}x_{[100]} \leqq E^{el}y_{[010]}\ (z_{[001]})$ (balance of strain elastic energy in easy magnetizaion axes) was ascertained by elaborate experiments as well as the magnetostriction data under bi-axial longitudinal and transverse stress to specimen by Moses et al.

REFERENCES

1. S.Chikazumi: Physics of Magnetism,(John Wiley & Sons,Inc.),(1964)152.
2. S.Chikazumi:Physics of Ferromagnetism,Vol.II(Japanese),(SYOKABO,Japan),(1984)106.
3. S.Kaya and H.Takaki:Sci.Repts. Tohoku Univ.,(1936)314.
4. K.Honda and S.Kaya: Sci.Repts. Tohoku Univ. vol.15,(1926)721.
5. H.J.Williams,R.M.Bozorth and W.Shockley:Phys.Rev. 75(1949)115.
6. H.J.Williams and W.Shockley: Phys.Rev. 75(1949)178.
7. R.M.Bozorth:Ferromagnetism,(IEEE Press),(1993)595.
8. A.J.Moses:IEEE Trans.Magn.,MAG 15(1979)1575.
9. A.J.Moses and P.S.Phillips:IEEE Trans.Magn.,MAG 14(1978)353.
10. H.Masui:Kimitsu Research Report(Japanese),(NIPPON STEEL),(1975)56.

Materials Science and Engineering Serving Society
S. Sōmiya, R.P.H. Chang, M. Doyama and R. Roy (Editors)

Insulating Metal; Al/Al_2O_3 Nano-Structure-Controlled Materials

Yasuhiro Sugaya and Osamu Inoue

Device Engineering Development Center, Matsushita Electric Ind.Co.,Ltd.
Kadoma, Osaka 571, Japan

We have studied the structure and properties of our newly developed "Insulating Al/Al_2O_3 materials". They show the high thermal conductivity of metal and high electrical resistivity of oxide. The values of thermal conductivity and dielectric constant could be controlled with the consisting elements and the relative density.

1. INTRODUCTION

Recently, many studies have been conducted on composite ceramics. The main purpose of these studies is improvement of mechanical properties. There are few studies about controlling the electrical properties, such as barrier layer condenser[1)]. The authors have developed series of nano-structure-controlled materials, which are artificially controlled bulk materials in an aim to develop unprecedented electrical properties with a new concept. "Fe-based metal/insulator material" is one of the nano-structure-controlled materials as insulating soft magnetic metal, which have the high saturation magnetic flux density of the core magnetic metal and the high electrical resistivity of oxide film [2-4)].

Similar hybrid properties, such as high thermal conductivity as metal and high electrical resistivity as oxide, are strongly required for electric components. These functional materials are limited to such as SiC-BeO[5)] and AlN[6)], which are now facing with low productivity due to their low sinterability. We solve these demands by the newly developed "Insulating Al/Al_2O_3 materials". The materials have the structure, in which aluminum grains are separated completely from each other by thin insulation layers on the order of nano-meter-thick. In this study, we report the synthetic process, thermal conductive and electrical properties.

2. Experimental Procedure

The fabrication process of the Al/Al_2O_3 materials is shown in Figure 1. The starting material was gas atomized powder of aluminum with average particle size of 20 μ m. The insulation film that consisted of Al_2O_3 was formed on the surface of each particle by an oxidation process. The thickness of

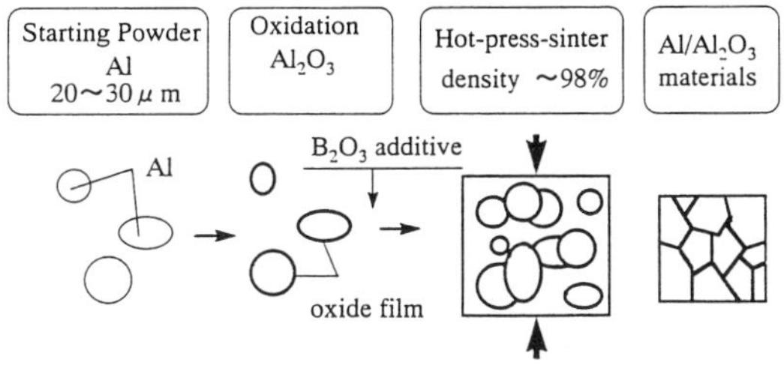

Figure 1. Fabrication process of Al/ Al_2O_3 materials

Al_2O_3 layers, 50nm～200nm, were precisely controlled by the oxidation conditions at 600～650℃×50hr. The oxidized powders were molded and hot-press-sintered to various density of 92-98% at temperature, 673-873K, and at pressure, 30～50MPa. Some samples were hot-press-sintered with addition of B_2O_3. Density of sample was measured by the Archimedes method. Thermal conductivity was introduced by thermal resistivity which calculated from temperature difference caused by a heater.

3. Results and Discussion

Figure 2 shows the typical fractured surface of Al/Al_2O_3 material. The spherical particles were deformed to a polygonal shape and were densified to nearly 100%. Al_2O_3 film was not clearly observed because it was very thin.

Figure 3 shows the relationships with the volume fraction of oxide and electrical resistivity (ρ). Densities of the samples were approximately 92%. Open circle shows ρ of the additive free sample and closed circle shows the sample with B_2O_3 additive. The solid line shows a calculated ρ based on a simple model. The measured ρ values were in a good agreement with the calculated ones when the oxide contents are higher than 4 vol.%. The measured ρ gradually decreased at a volume fraction of less than 4 vol.%. The decrease was assumed to be due to the following process. Before the hot-press-sintering, each spherical particle was perfectly coated by the insulated oxide film. During the densification process, spherical particle deformed into polygonal shape as shown in Figure 2, and the surface area of the particle became larger than that of the original one. Therefore, cracks were formed on the insulating oxide film, and direct contacts of the metal

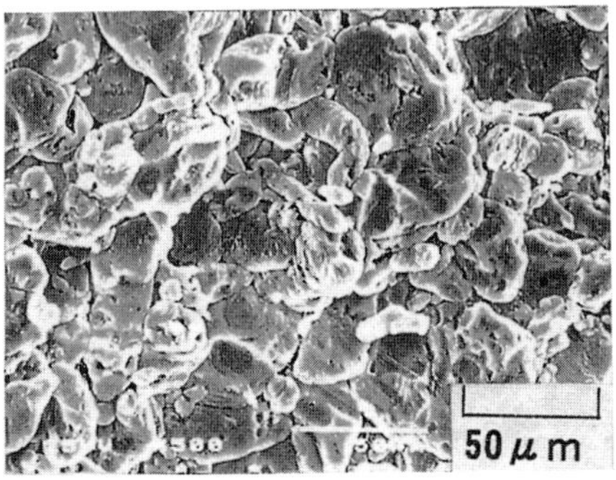

Figure 2. Fracture surface of Al/ Al_2O_3 materials

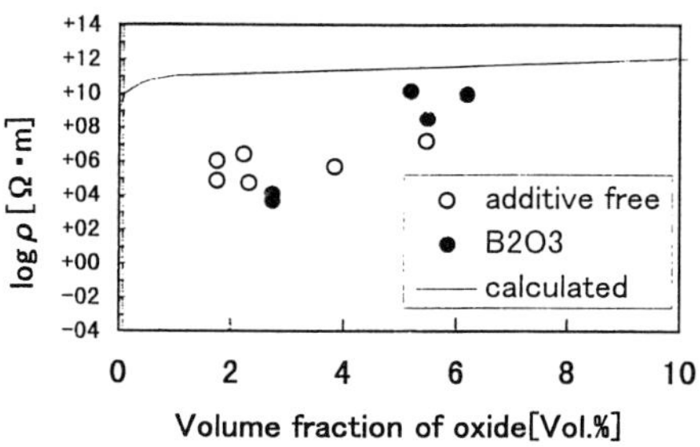

Figure 3. Volume fraction of oxide vs. electrical resistivity (ρ)

Table 1 Properties of typical Al/Al_2O_3 samples

No.	Al_2O_3 film δ [nm]	B_2O_3 [Vol.%]	relative density [%]	ρ [Ω·m]	thermal conductivity [W/m·K]	εr at 1MHz mes.	εr at 1MHz cal. model 1	εr at 1MHz cal. model 2	tan δ 1MHz
A	77	0	92	5×10^{4}	—	—	—	—	—
B	173	0	98	1×10^{-2}	—	—	—	—	—
C	123	2.5	98	8×10^{9}	64	2132	2430	1620	0.009
D	144	8	92	8×10^{9}	17	83	1916	90	0.013

particles were occurred through the cracks. The contact naturally caused the breakdown of ρ. The contacts could be easily occurred when the film thickness was thin and the density was high. Table 1 shows several properties of four typical Al/Al_2O_3 samples (A~D). If the sintered body densified more than 98% such as sample B, the sample exhibited low ρ of 1×10^{-2} [$\Omega\cdot$m] even Al_2O_3 film is thicker than that of sample A. On the other hand, the sample C with B_2O_3 additive as 2.5 vol.% exhibited high ρ of 8×10^{9} [$\Omega\cdot$m] even though density was 98%. This observation implied that the cracks in the insulation films were almost repaired with B_2O_3 layers which prevent the direct contacts of metal particles.

Figure 4 shows the relationships between thermal conductivity (λ e) and density of the Al/Al_2O_3 materials. All these samples showed high ρ of over 10^4 [$\Omega\cdot$m]. Open circle and closed circle show the thermal conductivity of without and with B_2O_3 addition, respectively. It was shown that the λ e were strongly depend on the density.

Fig. 4 also shows the calculated thermal conductivity(λ e) based on the proposed model 1 and 2 as shown in figure 5. Model 1 shows that B_2O_3 was unevenly distributed such as in the triple point region. Model 2 shows that B_2O_3 was homogeneously distributed at the grain boundaries. Solid line and dashed lines show the calculated values of model 1 and model 2 respectively. Bruggeman's expression was applied for the calculation[7].

$$1-\mathbf{d} = (\lambda_e-\lambda_d)/(\lambda_c-\lambda_d)\times(\lambda_c/\lambda_d)^{1/3} \qquad (1)$$

where **d** was relative density, λ d and λ c were thermal conductivity of oxidized Al particles, and pore, respectively. For the calculation of λ d, we assumed that heat flow transmitted metal particle and insulation layer in series as shown by equation (2).

$$\lambda_d = \lambda_1\lambda_2/(V_1\lambda_2 + V_2\lambda_1) \qquad (2)$$

where λ_1, λ_2 were thermal conductivity of Al, insulation layer, and V_1, V_2 were volume fraction of them, respectively.

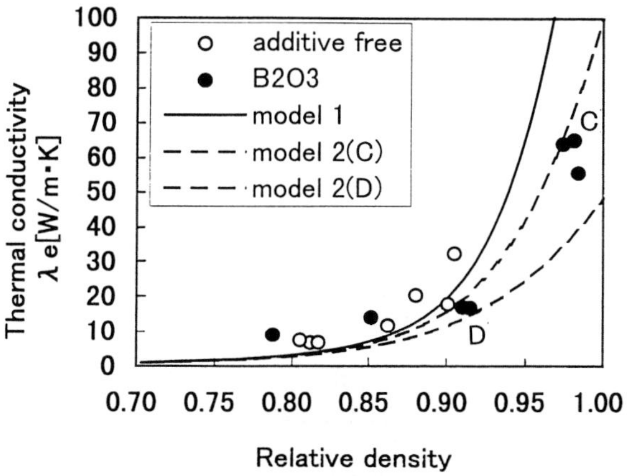

Figure 4. Density dependent thermal conductivity

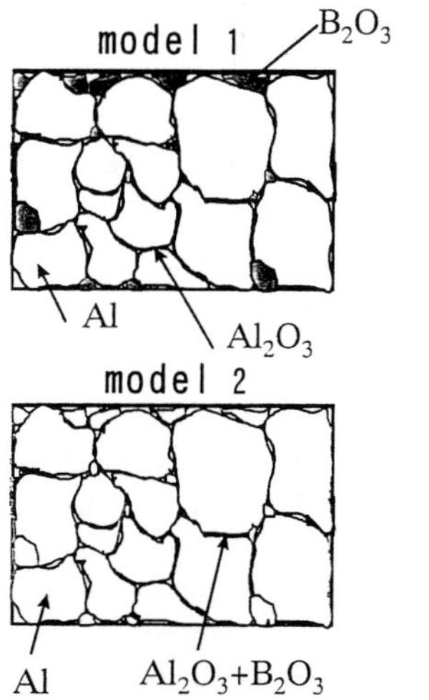

Figure 5. Model in distribution of B_2O_3

The λ_2 values were affected by the distribution of B_2O_3. In the case of model 1, we assumed that B_2O_3 behaved as pore for the calculation. On the other hand, in the case of model 2, we assumed that B_2O_3 formed insulation layer of oxidized Al grains. The measured λe value of sample C was in better agreement with calculated values of model 2 than that of model 1. This results supported that B_2O_3 were distributed homogeneously on the surface of oxidized Al grains.

Similar calculation based on the proposed model 1 and 2 were performed for apparent dielectric constant εr. When the metal particles were perfectly insulated by the oxide films, the εr values of Al/Al_2O_3 materials was expressed by the following equation (3)

$$\varepsilon r = \varepsilon_b \cdot s / \delta \tag{3}$$

where ε_b, s, and δ were relative dielectric constant of insulation layer, grain size and thickness of insulation layer respectively. Table 1 also shows the calculated εr values of sample C and D. The measured εr values were in better agreement with calculated values of model 2 than that of model 1.This observation also implied that B_2O_3 additives were distributed homogeneously such as model 2.

Consequently, the Al/Al_2O_3 materials with B_2O_3 additives exhibited high λe of 64[W/m·K], even though it showed high ρ of 8×10^9 [Ω·m]. This λe value was 3 times higher than conventional Al_2O_3 ceramics. The εr values of 83～2200 can be adjusted by the control of the insulation layer and the values are predicted from the theoretical values with consisting elements in Al/Al_2O_3 materials.

4. Conclusion

We have shown that some unique properties of Al/Al_2O_3 materials. They show the high thermal conductivity of metal (64 [W/m · K]) and high electrical resistivity of ceramics (8×10^9 [Ω·m]). The above thermal conductivity is 3 times higher than that of Al_2O_3 ceramics. The value of dielectric and thermal conductivity is good agreement with the calculated value predicted with consisting elements and the relative density.

Thus, the "Insulating Al/Al_2O_3 materials" can be applied to light substrate for heat sink or high capacity condenser.

References

1) S. Waku, A. Nishimura, T.Murakami, Rev. Elec. Commun. Lab., 19[5-6]665-80(1971)
2) Y.Sugaya, O.Inoue, K.Hirota and K.Kugimiya, Journal Ceramic Society of Japan, Int.ed.,vol.99[1], pp.72-76,1991.
3) Y.Sugaya, O.Inoue and K.Kugimiya, Ferrites, Proceeding of ICF 6, P.952(1992).
4) O.Inoue, Y.Sugaya, N.Matsutani and K.Kugimiya, Ceramic Transaction, Volume44: Materials Processing and Design II P.169-178
5) M.Usami et al: Proc. 1983 IEEE ICCD, pp272～275(1983)
6) G.A.Slack : J.Phys.Chem. Solids, Vol. 34, P321 (1973)
7) P.A.G. Bruggeman, Ann. Phys.,24, 636-664(1935).

Materials Science and Engineering Serving Society
S. Sōmiya, R.P.H. Chang, M. Doyama and R. Roy (Editors)
1998 Elsevier Science B.V.

EFFECT OF DOUBLE NOTCHED SPECIMEN CONFIGURATION ON THE SHEAR STRENGTH OF C/C COMPOSITES

Yohsuke ISHIGURO, Takashi AKATSU, Yasuhiro TANABE and Eiichi YASUDA
Materials and Structures Laboratory, Tokyo Institute of Technology
4259 Nagatsuta, Midori-ku, Yokohama 226, Japan

Kazuhiro MARUYAMA and Shigehiko YAMADA
Department of Materials, Faculty of Science and Engineering,
Teikyo University of Science and Technology
Yatsuzawa, Uenohara-machi, Kitatsuru-gun, Yamanashi 409-01, Japan

This research investigated the influence of geometry of a double notched specimen on evaluating interlaminar shear strengths of unidirectional C/C composites. Since the fracture behavior of the specimen varied with the depth and the interval of notches, variable shear strengths were evaluated. FEM analyses revealed that the stress state, shear stress distribution and existence of superimposed normal stress, in double notched specimens changed depending on the specimen's configuration.

1. INTRODUCTION

Carbon fiber reinforced carbon (C/C) composites are important materials classified into ultra high-temperature structural materials. In C/C composites as a structural component, the extremely high strength of reinforced fibers is often limited by other weak properties perpendicular to the fiber directions. It is, therefore, particularly important to understand shear strength and/or behavior, which is predominantly determined by inferior properties of matrix and/or interfaces. Generally, strength perpendicular to the layers is characterized as interlaminar shear strength. Several test methods mainly focusing on CFRP have been developed to measure shear strength[1]. However, standardized methods, e.g., short beam and Iosipescu, do not yield reliable strength for C/C composites, because the stress states realized in specimens are affected by superimposed normal stresses and because the actual shear stress increase/decrease with them[2]. On the other hand, compressive loading methods of double notched specimens are considered as one of the most applicable testing methods for shear strength of C/C composites. This method has merits: easy set-up and simple configuration of testing specimens.

The objective of this study is to evaluate the effect of double notched specimen geometry on shear strength of a unidirectional C/C composite, and to consider better methods for evaluating shear strength of C/C composites associated with stress states calculated by FEM analyses.

2. EXPERIMENTAL

The test material selected in this study was a unidirectional C/C composite (AC200-B10, provided by Across Co., Ltd.). All specimens were taken out of a 20mm thick plate, then machined to a short bar shape of 15mm width, 6mm thickness and 30mm length based on the ASTM standard[3]. Double notches, 0.5mm width, were introduced to the 15mm width x 30mm length side of the specimens. Depth and the interval of them were varied with 2.5mm to 4mm, 2mm to 10mm, respectively as shown Fig.1.

A special test fixture as shown Fig.2 is designed to suppress bending moment affecting test specimens. Compression test with 0.2 mm/min. cross head speed was carried out using a test machine (DSS-25T, Shimadzu Co., Ltd.). Five specimens were tested for each condition and shear strength was calculated using the maximum failure load.

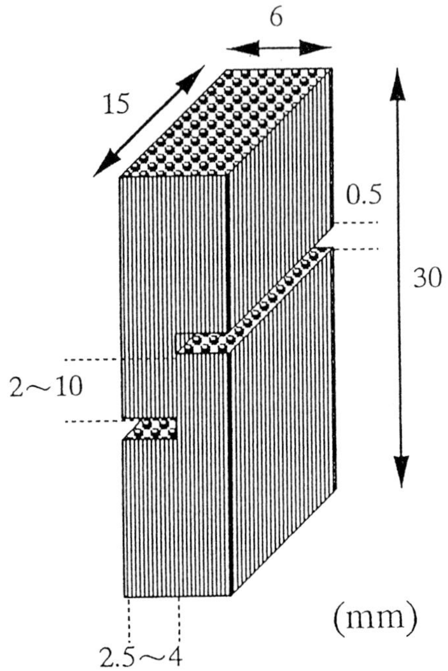

Fig.1 Schematic of double notched specimen.

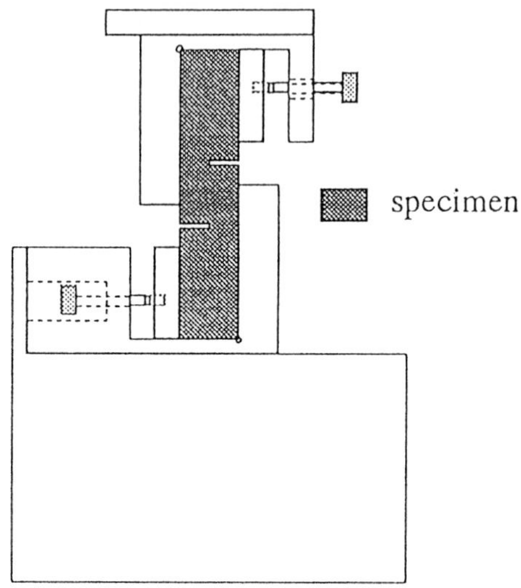

Fig.2 Experimental setup for double notch test.

3. STRESS ANALYSIS

The stress distributions in the specimens were calculated with an FEM-code of NASTRAN™. This program package has the capability of solving very general three dimensional problems of anisotropic elastostatics[4].

Material homogeneity and orthotropic properties were assumed in this calculation. Then internal stress concentration arising between fiber and matrix was ignored. The specimens were modeled consisting of twenty-node brick isoparametric elements. Since the stress between the two notches was of our main interest, smaller elements were concentrated in this area. The model parameters were follows; E_1 = 100GPa (Young's modulus in fiber direction), $E_2 = E_3$ = 10GPa, G_{12} = 5GPa , $\nu_{12} = 0.3$, $\nu_{23} = 0.25$, $\nu_{31} = 0.03$. The Young' moduli and the shear strength were measured whereas the Poisson's ratios were estimated.

A boundary condition for calculation was follows : a 0.1mm vertical displacement was applied to the specimen on the top face of the specimen, and the bottom and the both sides faced rigid walls.

4. RESULTS AND DISCUSSION

The stress distributions in a gage section line, imaginary line between both notch tips, in the ASTM type specimen of 3mm notch depth and 6mm notch interval is shown Fig.3. Shear stress and normal stress are normalized by averaged shear stress calculated from applied forces. The shear stress distribution is heterogeneous and the value at the vicinity of the notch tips is the highest, about 1.5 times higher than the averaged shear stress. Since tensile stress normal to the shear plane exists and the stress on the gage area is under I-II mixed mode, the shear strength obtained by this ASTM type specimen must be under-estimates. Therefore, the specimen configuration should be optimized by changing notch geometry.

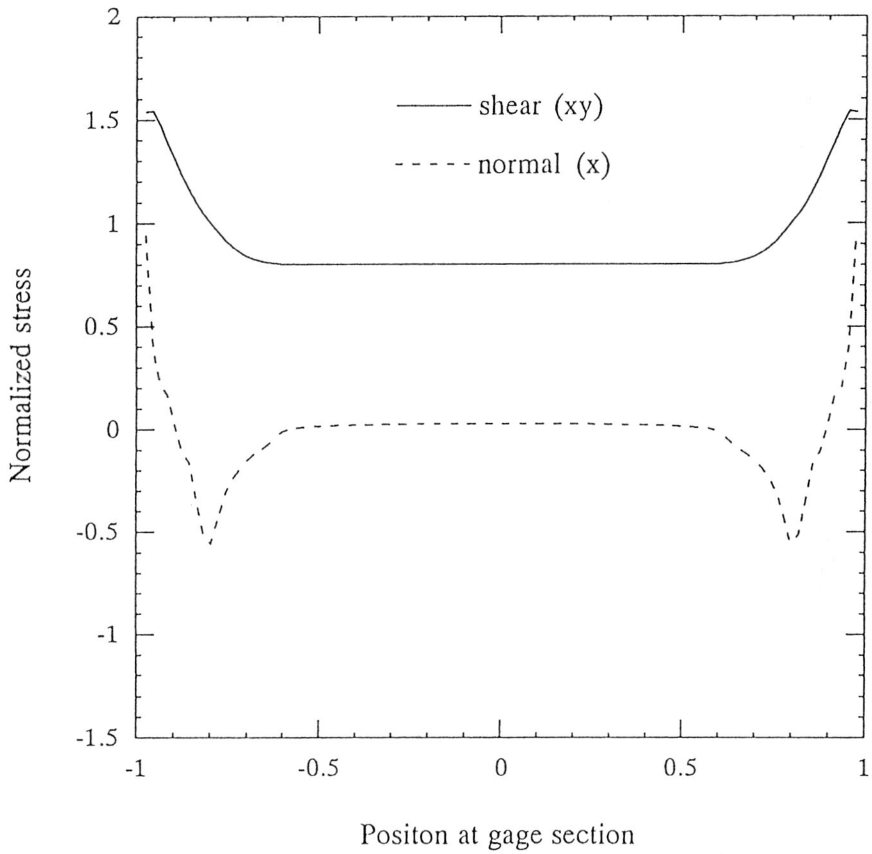

Fig.3 Stress profiles in a gage section line in double notched specimen.

Figure 4 represents the tested shear strength of the unidirectional C/C composite as a function of notch depth. When the depth was less than the half of the specimen thickness (under-cut), the shear strength was becoming higher drastically, and these specimens fracture with fiber breaking. These phenomena indicate that the shear strength obtained from under-cut specimens are not regarded as under-estimated values affected by the interlaminar properties of the specimen. In the case of depth = 3mm, on the contrary, some samples showed no complete separation into two pieces and measured strength were scattered. The measured shear strength were decreased gradually increasing with the depth, and the variation also decreased. Since the compliance of the specimen decreased with increasing notch depth, stress concentration at the notch tips increased, and the above reduced variation took place for depth = 3.25 and 3.50mm specimens as shown Fig.4. Tensile fracture at the side face was observed for d = 4.0mm specimen prior to shear fracture occurring. These tested results imply that over-cut notch configuration is profitable in this type specimen.

Figure 5 shows the shear strength of the unidirectional C/C composite as a function of notch interval. The measured shear strength were decreasing slightly with the interval. However their standard deviation was still high. This decreasing tendency was caused by the increment of the effective shear cross-section, i.e. effective volume/area affected the strength.

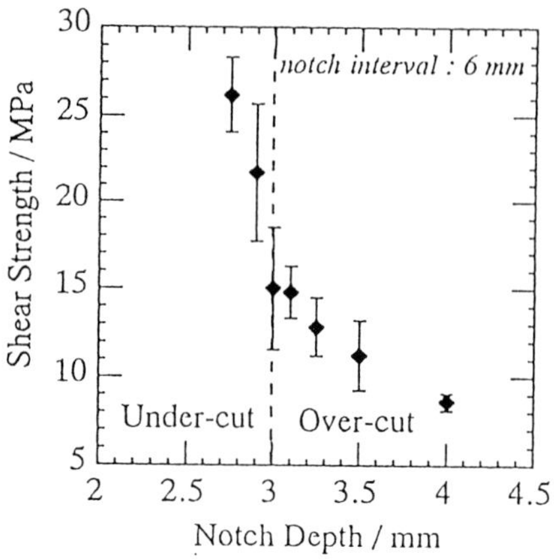

Fig.4 Notch depth dependence on shear strength.

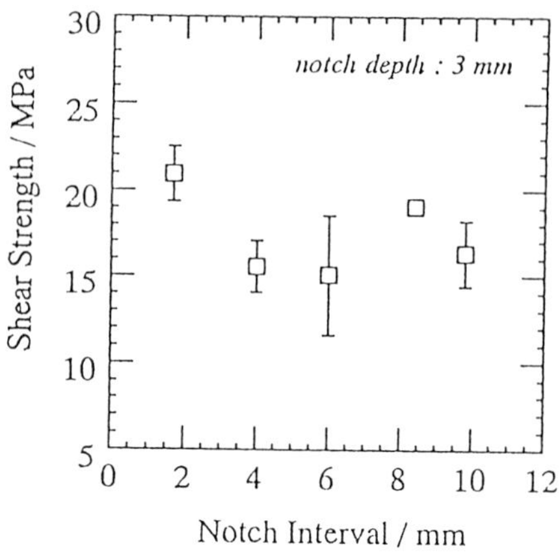

Fig.5 Notch interval dependence on shear strength.

5. CONCLUDING REMARKS

The authors studied the effect of double notched specimen configuration on evaluating shear strength of a C/C composite. Through experimental and numerical investigations, it was clear that notch geometry affected the fracture behaviors and the shear strength.

The followings are concluded :

1: FEM analyses revealed that shear strength obtained using ASTM type specimen should be under-estimated, which is caused by inhomoginity of shear stress distribution and the superimpose of tensile stress.

2 : The measured shear strength decreased gradually increasing with the depth because fracture behavior changed with the depth.

3 : The measured shear strength decreased slightly increasing with the interval, which due to the increase of effective area shear stress applied.

4 : A slight over- cut notch configuration is profitable in double notched specimen for evaluating shear strength of a C/C composite.

REFERENCES

1.The committee for carbon fibre composite research in Japan high polymer center, *Test report for carbon fibre composite research*, 2 (1986)

2. B. Thielicke, H.G.Maschke and U. Soltesz, *ECCM-7*, **1** (1996)

3. ASTM C 1292-95

4. *MSC/NASTRAN ver.68 static analysis handbook*, Japan MSC

Materials Science and Engineering Serving Society
S. Sōmiya, R.P.H. Chang, M. Doyama and R. Roy (Editors)

Reaction Synthesis of SiC-AlN Ceramic Alloys

Y. Kobayashi[a]*, M. Sugimori[a], J.-F. Li[b], A. Kawasaki[b], R. Watanabe[b]

[a]Graduate School, Tohoku University, Aramaki-aza-aoba Aobaku, Sendai, 980-77
[b]Department of Materials Processing, Faculty of Engineering, Tohoku University, Aramaki-aza-aoba Aoba-ku, Sendai, 980-77

In this paper, we report two processes for the reaction synthesis of SiC-AlN ceramic alloys. One process is suitable for the powder synthesis through the direct reactions between the constituent elements in the SiC-AlN system by reaction sintering of the mixed powders of Si, C and Al in N_2 atmosphere. As another process, dense SiC-AlN alloy compacts can be directly fabricated by in-situ HIP-sintering of premixture of Si, C and AlN powders. The mechanical properties including microhardness, strength and fracture toughness for the SiC-AlN alloys fabricated by both processes were evaluated.

1. INTRODUCTION

Both silicon carbide (SiC) and aluminum nitride (AlN) are important ceramics for structural and functional applications. Moreover, recent attention was paid on the SiC-AlN system because solid solutions form between SiC and AlN over a wide compositional range [1-5]. Up to date, two kind of processes have been adopted to synthesize the SiC-AlN solid solutions. Sintering of SiC and AlN powder mixtures is a direct process [3,4]. However, in such a process, solid solutions with a homogeneous composition distribution are quite difficult to obtain because of the low diffusivity characteristic of covalently-bonded compounds. The other processes utilize chemical reactions such as carbon reduction [2,5] and pyrolysis of organic precusor [6], which produce solid solutions with relatively uniform composition distributions.

In the present study, we have developed two new processes, in which the direct reactions between the constituent elements are used. The present processes are cost-effective because less-expensive Si and C powders are directly used to synthesize the SiC-AlN solid solutions or their composites. By reaction sintering of the mixed powders of Si, C and Al in N_2 atmosphere, porous SiC-AlN compacts hence SiC-AlN powders were obtained; dence SiC-AlN compacts were also directly prepared by HIP-reaction sintering. The mechanical properties of the synthesized dense HIP sintered bodies fabricated by the each processes were evaluated for hardness, fracture strength and fracture toughness.

2. PROCESSES

2.1. Synthesis by the reaction: Si+C+Al+N_2 → SiC-AlN

The SiC-AlN powders were synthesized from the mixed powders of Si, C and Al in N_2 according to the experimental flow chart shown in Fig.1(a). Fig.2(a) shows the X-ray diffraction patterns of the starting powder and of the reaction products obtained at various temperatures. From 973K, the AlN peaks appear, and the Al peaks disappear at 1073K, showing that AlN formation reaction occurs completely above 1073K. At 1673K, the Si peaks disappear completely with the formation of SiC and Si_3N_4; however, the formation of Si_3N_4 was not intended in this study. Further, at 2073K, the Si_3N_4 and C peaks disappear, and only SiC and AlN peaks remain.

*Research Fellow of the Japan Society for the Promotion of Science.

According to Gibbs free energy, Si_3N_4 forms prior to SiC at 1673K, but SiC is stabler than Si_3N_4 at higher temperatures. For this reason, the formed Si_3N_4 reacts further with C to produce SiC at 2073K [7]. To suppress the reaction between Si and N_2 at higher temperatures, the Si_3N_4 formation can be avoided by conducting two step reaction sintering which includes the primary nitriding reaction in N_2-10%H_2 at 1473K and the subsequent SiC formation reaction in Ar atmosphere at 1773K. The reaction products were so porous that they could be easily smashed with a mortar and a pestle, and fine powder with average particle size of about 0.6 μ m can be obtained by ball milling for 50h. The synthesized SiC-AlN powders were densified by HIP-sintering at 2123K and 150MPa to almost full density. The pressureless sintering is under investigation.

2.2. Synthesis by the reaction: Si+C+AlN → SiC-AlN

In the case of the starting powder of Si, C and AlN, a reaction occurs only between Si and C into SiC, so no diffraction peak of unfavorable reaction products is observed. As shown in Fig.2(b) from 1573K, the SiC formation reaction occurs with increasing intensity with increased temperature. The reaction proceeded rapidly at 1673K, and almost ended at 1773K.

This process consisting only of solid reaction enables the use of in-situ reaction HIP-sintering, so that reaction synthesis and densification occur simultaneously during HIPing. The details of HIPing procedures can be found elsewhere [8]. In each composition, a high density over 90% was obtained at 1973K, and the relative density reached almost 100% at 2123K. When compared with the SiC and AlN powder mixtures, it was confirmed that densification occurred at lower temperature in the mixed powders of Si, C and AlN.

3. MECHANICAL PROPERTIES

Table 1 shows the Vickers hardness, the fracture strength and the fracture toughness of HIP-sintered SiC-AlN compacts prepared by both processes including the reference data of the

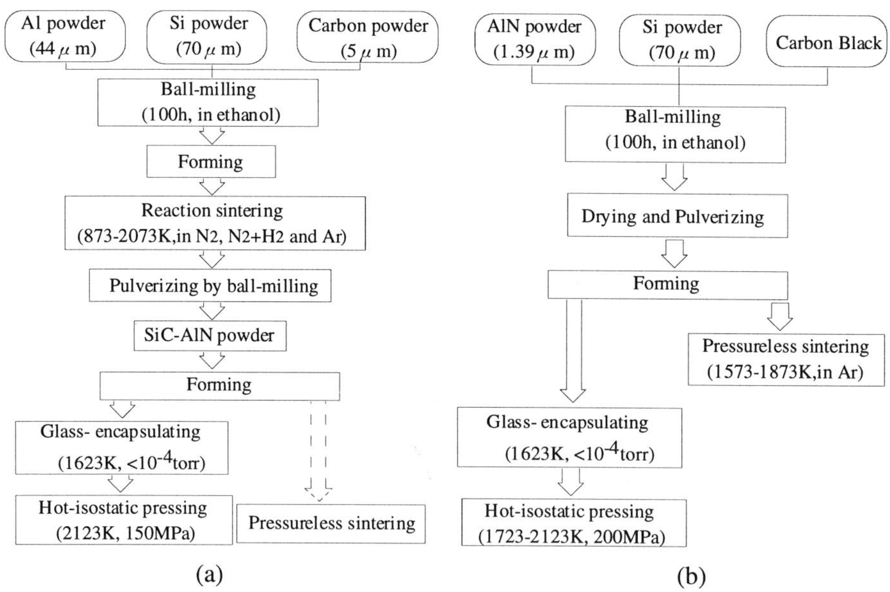

Figure 1. Flow charts of the present reaction processes for synthesizing SiC-AlN alloys from (a) Si, C and Al powders, (b) Si, C and AlN powders.

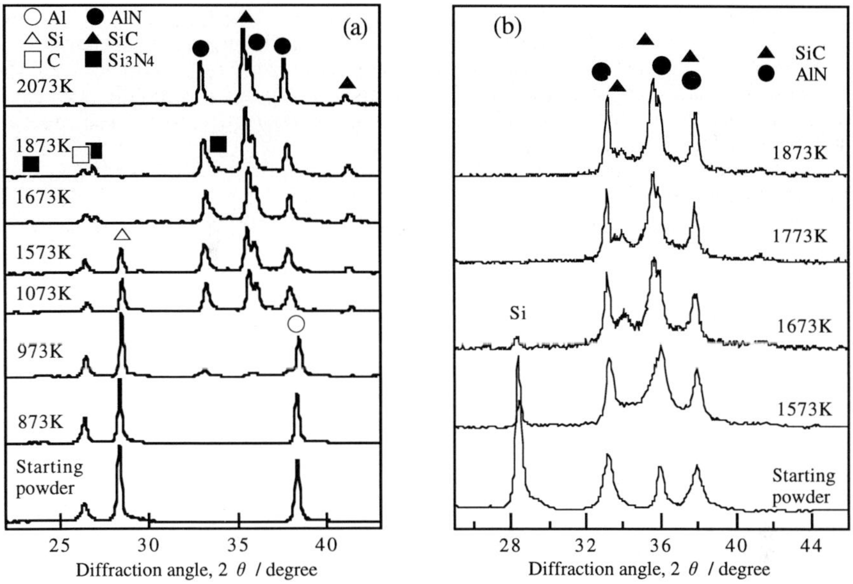

Figure 2. X-ray diffraction patterns of the reaction products obtained by sintering the mixture of (a) Si, C and Al powders in N_2 and (b) Si, C and AlN powders in Ar at the shown temperatures.

monolithic SiC (0.5mass% B_4C was added as sintering aids) and AlN (without additives), which were HIP-sintered at 2173K and 2073K (HIP pressure, 200MPa),respectively. The fracture strength was determined by a modified small punch test (MSP), which gives strength values comparable to that of 4-point bending tests [9]. The fracture toughness was measured by the indentation microfracture technique [10]. It was found that the values of all mechanical properties of the SiC-AlN alloys prepared by both processes are higher than the estimated values from the monolithic SiC and AlN by the rule of mixture. Particularly the fracture strengths are remarkably high; the SiC-AlN alloys even had strengths of 2.6 times that of the monolithic SiC. This strengthening of the SiC-AlN alloys may be attributed mainly to solid solutioning and the resulting

Table 1. Mechanical properties of HIP-sintered SiC-AlN alloys and the end-number materials. (average values with standard errors)

Starting powder	Material	Vickers hardness Hv/GPa	Fracture strength σ_{SP}/MPa	Fracture toughness Kc/MPa · $m^{1/2}$
SiC+0.5mass%B_4C	SiC	29.8±0.5	460±90	3.7±0.2
AlN	AlN	13.0±0.5	375±70	3.9±0.2
Si, C and Al	SiC-30%AlN	27.0±1.9	718±134	4.7±0.3
	SiC-50%AlN	23.7±0.8	1183±63	4.4±0.2
	SiC-70%AlN	21.4±0.9	984±176	6.0±0.4
Si, C and AlN	SiC-30%AlN	33.4±2.1	812±201	4.2±0.4
	SiC-50%AlN	28.4±1.6	573±146	3.9±0.2
	SiC-70%AlN	24.1±0.9	544±159	4.5±0.3

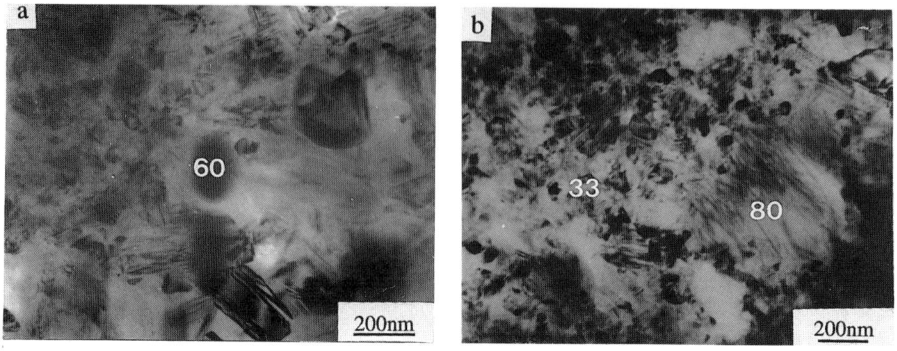

Figure 3. TEM micrographs of the HIPed SiC-AlN alloys with nominal composition of SiC-50%AlN obtained from (a) Si, C and Al (b) Si, C and AlN. The figures show the AlN content in local areas, which was determined by EDX.

ultrafine microstructure, as will be shown later. The average grain sizes of the monolithic SiC and AlN were 5-10 μ m, being at least 10 times larger than that of the obtained SiC-AlN alloys.

4.MICROSTRUCTURES

Fig.3 shows TEM micrographs of the HIP-sintered SiC-AlN compacts prepared by both processes. The obtained SiC-AlN alloys have similar ultrafine microstructures. The composition analysis by STEM-EDX revealed that fine grains like a matrix are rich in SiC while relatively large grains dispersed in the matrix are rich in AlN. A single-phase solid solution was not obtained because the single-phase area exists at temperatures above 2250K according to the phase diagram. Nevertheless, the present processes are effective in synthesizing SiC-AlN alloys with nearly equilibrium compositions in comparison with the synthesis of a mixture of SiC and AlN powders.

5. CONCLUSION

SiC-AlN alloys can be synthesized by the reaction between the constituent elements. The processes are cost-effective not only because the raw materials are less-expensive but also the process temperature is relatively low.

The obtained SiC-AlN alloys have better mechanical properties than the monolithic SiC and AlN, particularly the fracture strengths are outstandingly high because of their ultrafine microstructures.

REFERENCES

1. I.B. Cutler, P.D. Miller, W. Rafaniello, H.K. Park, D.P. Thompson and K.H. Jack, Nature, 275, 434 (1978).
2. W. Rafaniello, K.Cho and A.V. Virkar, J. Mater. Sci,.16, 3479 (1981).
3. R. Ruh and A. Zangvil, J. Am. Ceram. Soc.,65, 260 (1982).
4. R. Ruh and A. Zangvil, J. Am. Ceram. Soc.,71, 884 (1988).
5. J.-F. Li and R. Watababe, J. Mater. Sci.,26, 4813 (1991).
6. C.L. Czekaj, M.L. Hackney, W.J. Hurley, Jr.,L.V. Interrante and G.A. Sigel, J. Am. Ceram. Soc., 73, 352 (1990).
7. T. Hirai, K. Niihara and A. Ookubo, J. Jpn. Inst. Metals,41, 367 (1977).
8. Y. Kobayashi, J.-F. Li, R. Watanabe, J. Ceram. Soc. Jpn., Int. Ed., 104, 556 (1996).
9. J.-F.Li, A. Kawasaki and R. Watanabe, J. Jpn. Inst. Metals, 56, 1450 (1992).
10. K. Niihara, R. Morena and D. P. Hasselman, J. Mater.Sci. Lett.,1,13 (1982).

Materials Science and Engineering Serving Society
S. Sōmiya, R.P.H. Chang, M. Doyama and R. Roy (Editors)
1998 Elsevier Science B.V.

Microstructure and toughening of hot-pressed SiC-AlN solid solutions

Keun Ho Auh[a] and Chang-Sung Lim[b]

[a]Department of Ceramic Engineering, Hanyang University, Seoul 133-791, Korea

[b]Ceramic Processing Research Center(CPRC), Hanyang University, Seoul 133-791, Korea

ABSTRACT

The microstructure and toughening of hot-pressed SiC-AlN solid solutions were studied in the 1870 to 2030℃ temperature range. The reaction of AlN and β-SiC(3C) powder causing transformation to the 2H(wurtzite) structure appeared to depend on the temperature and the additives of α-SiC. The fracture toughness and hardness were examined by the indentation-fracture-test method.

1. INTRODUCTION

The 2H polymorph of SiC is isostructural with AlN and Al_2OC with a strong covalent bond[1]. The similarities between two structures and their properties suggest that alloying of one with other may provide the potential for property optimization. Cutler et al.[2-6] reported that a SiC-AlN solid solution in the range 2-100% AlN has been formed at 1600℃ by a vapor phase process using carbothermal reduction of amorphous silica and aluminum hydroxide in nitrogen. Recently, some studies of the phase relationship and microstructures of the SiC-AlN solid solutions have been reported by hot-pressing[7-13] and pressureless sintering with additives[14-16]. However, when all starting materials are in powder form, the covalent nature of the atomic bonding in SiC and AlN makes diffusion extremely slow, so that pressures of several hundred bars and temperatures of up to 2300℃ are required to fabricate the complete solid solutions. In this work, the fabrication of SiC-AlN solid solution is studied by the hot pressing under 2030℃ using additives of α-SiC to achieve the complete 2H solid solutions with an enhanced fracture toughness. It is aimed at attaining a basic understanding of the effect of the additives of α-SiC on the microstructure and toughening.

2. EXPERIMENTAL

Nominal compositions investigated ranged between 20% AlN/80% SiC and 90% AlN/10% SiC without additives and with the additives of 2, 5 and 8wt% α-SiC in Table 1. Compositions were prepared by wet milling the appropriate amounts of the powders for 24 h in isopropyl alcohol in a plastic jar with SiC balls. After milling, the powders were evaporated to dryness, broken with a mortar and passed through a 50-mesh sieve. Specimens with 3cm in diameter by 0.5cm thick were uniaxially hot-pressed in graphite dies lined with graphite washer. Hot-pressing was conducted under nitrogen at 1870℃ for 4 h and at 2030℃ for 1 h under 22.5 MPa(Table 1). Cooling was sufficiently rapid in this study, so that the high-temperature phase was quenched to room temperature. After the specimens were removed from the die, the surfaces were ground, and the density was determined by measuring weights and dimensions. Near theoretical densities were obtained in all cases.

The specimens were then polished using diamond pastes of 30, 15, 3, 1 ㎛. The polished samples were ultrasonically cleaned in ethanol, rinsed with distilled water and dried. All of the samples were examined using X-ray diffraction(XRD) with CuK_α. The polished samples were etched using Murakami's etch to reveal the microstructure. The etched sections were investigated using optical microscopy, scanning electron microscopy(SEM) and transmission electron microscopy(TEM). The indentation fracture test method was used with a load of

Table 1
Compositions and hot-pressing conditions of powder mixtures

Sample name	SiC/AlN mole ratio	Composition(wt%) SiC	AlN	α-SiC	Hot-press conditions Temp(℃)	Time(h)	Pressure(MPa)
A20	80/20	79.6	20.4	0	1870	4	22.5
					2030	1	22.5
A50	50/50	49.5	50.5	0	1870	4	22.5
					2030	1	22.5
A70	30/70	29.6	70.4	0	1870	4	22.5
					2030	1	22.5
A90	10/90	9.8	90.2	0	1870	4	22.5
					2030	1	22.5
2W50A	50/50	49.0	49.0	2	1870	4	22.5
					2030	1	22.5
5W50A	50/50	47.5	47.5	5	1870	4	22.5
					2030	1	22.5
8W50A	50/50	46.0	46.0	8	1870	4	22.5
					2030	1	22.5

10 Kgf, a loading speed of 0.1 mm/sec and a load time of 20 sec on the finished surfaces of the polished samples for the measurements of fracture toughness and Vickers hardness.

3. RESULTS AND DISCUSSION

As given in Table 2, the phases were listed in the order of the amount present. XRD indicated the coexistence of three or more phases in many of the samples. For the compositions of A20, A50 and A70 hot-pressed at 1870℃ for 4 h, the hexagonal 2H phase is the strongest with hexagonal 4H and 6H present and possibly some rhombohedral 15R and 3C, whereas for composition of A90 the 2H with 4H and 6H are the observed phases. For the composition of A20 hot-pressed at 2030℃ for 1 h, the 2H phase is the strongest with 4H and 6H present and possibly some rhombohedral 15R and 3C, while for the compositions of A50 and A70 the 3C phase was transformed to the hexagonal and rhombohedral phases. For the composition of A90 the hexagonal phases of 2H, 4H and 6H are present.

Table 2
Phases present in SiC-AlN solid solutions

Sample name	Hot-press conditions		Phases present
	Temp(℃)	Time(h)	
A20	1870	4	2H, 4H, 6H, 15R, 3C
A50	1870	4	2H, 4H, 6H, 15R, 3C
A70	1870	4	2H, 4H, 6H, 15R, 3C
A90	1870	4	2H, 4H, 6H
A20	2030	1	2H, 4H, 6H, 15R, 3C
A50	2030	1	2H, 4H, 6H, 15R
A70	2030	1	2H, 4H, 6H, 15R
A90	2030	1	2H, 4H, 6H
2W50A	1870	4	2H, 4H, 6H, 15R, 3C
5W50A	1870	4	2H, 4H, 6H, 15R, 3C
8W50A	1870	4	2H, 4H, 6H, 15R, 3C
2W50A	2030	1	2H
5W50A	2030	1	2H
8W50A	2030	1	2H, 4H

Results on the compositions of 2W50A, 5W50A and 8W50A as hot-pressed at 1870℃ for 4 h were similar to those obtained on the compositions of A50 hot-pressed at 1870℃ for 4 h. When these compositions were hot-pressed at 2030℃ for 1 h, single 2H solid solutions could be obtained for the compositions of 2W50A and 5W50A and with some 4H for the composition of 8W50A. These results indicate that the reaction of AlN and β-SiC powders causing the transformation to the 2H phase appeared to depend on the temperature and the amount of SiC/AlN with the additives present. Therefore, the phase transformation was influenced by the additives of α-SiC.

According to the XRD results, for the compositions of 49wt% SiC/49wt% AlN with an additive of 2wt% α-SiC and 47.5wt% SiC/47.5wt% AlN with an additive of 5wt% α-SiC, the complete solid solution with a single phase of 2H could be fabricated by hot pressing at 2030℃ for 1 h. The microstructures of the sample of 2W50A and 5W50A are relatively homogeneous with similar grain size distributions and compositional homogeneity of 2H phases. Sample 2W50A shows a somewhat smaller grains than that of 5W50A. However the sample of 8W50A exhibited an inhomogeneous size distribution, which suggests either untransformed or in the process of transformation to the 2H solid solutions. Fig.1 and Fig.2 show high magnification scanning electron micrographs of the samples of 2W50A and 8W50A. In Fig.1, small amount of columnar inter-grains was observed in a grain in the sample of 2W50A. The sample of 5W50A exhibits relatively uniform grains, which consist of a single phase of the solid solution. In contrast, a large proportion of heavy strained and faulted grains with exaggerated columnar inter-grains was observed in complicated grain mixtures in the sample of 8W50A in Fig.2. The columnar inter-grains were identified to be

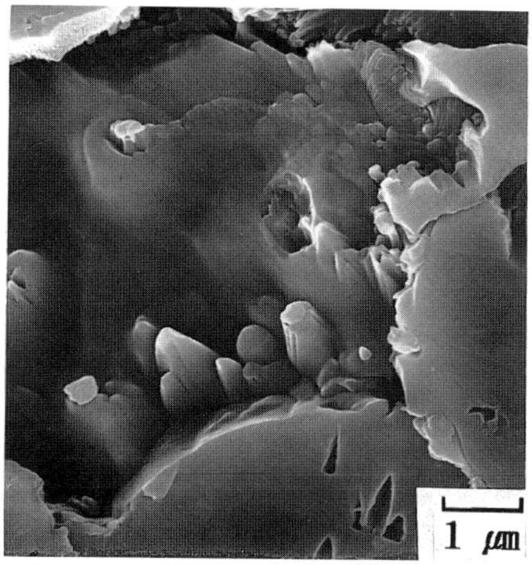

Figure 1. SEM micrograph of the SiC-AlN solids solutions as hot-pressed at 2030℃ for 1 h (2W50A).

Figure 2. SEM micrograph of the SiC-AlN solid solutions as hot-pressed at 2030℃ for 1 h (2W50A).

a 4H by TEM diffraction patterns. It is noted that a material with a high fracture toughness can be generated from the crack deflection of elongated grains such as the columnar inter-grains.

Table 3 shows the result of the fracture toughness of the 2W50A, 5W50A and 8W50A, which was calculated from the equation of Niihara, Charles & Evans and Lawn. The fracture toughness of the samples ranges between 6.4 and 6.9 MPa · $m^{1/2}$ from Niihara equation, between 5.0 and 5.4 MPa · $m^{1/2}$ from Charles & Evans equation and between 4.9 and 5.3 MPa · $m^{1/2}$ from Lawn equation. The average fracture toughness of the samples is much higher than that of the other work[16], which resulted in average fracture toughness values between 2.2 and 3.4 MPa · $m^{1/2}$ for various annealing temperatures of SiC-AlN solid solution ceramics. Among the samples, 5W50A indicates relatively higher value than that of 2W50A and of 8W50A. The fracture toughness could be enhanced by the crack deflection of the columnar inter-grains in the homogeneous microstructure with similar grain size distributions and compositional homogeneity of 2H phases, which was deduced by the additives of α-SiC.

Table 3
Fracture toughness and Vickers hardnessof the 2W50A, 5W50A and 8W50A

Sample name	Fracture toughness(MPa · $m^{1/2}$)			Vickers hardness (GPa)
	Niihara eq.	Charles & Evans eq.	Lawn eq.	
2W50A	6.709±1.21	5.288±1.23	5.150±1.16	17.085±0.15
5W50A	6.885±1.17	5.427±1.11	5.310±0.98	18.810±0.12
8W50A	6.410±1.35	5.052±1.54	4.940±1.04	16.683±1.34

The additive weight dependence for the improvement of the fracture toughness indicated the maximum value in the case of 5W50A. In the 8W50A, somewhat exaggerated columnar structures decreased the fracture toughness. The fracture toughness of the 2W50A was relatively lower than that of the 5W50A due to the inhomogeneous columnar structures.

Hardness measurements were made on the compositions of 2W50A, 5W50A and 8W50A. Experimental results are given in Table 3. The average hardness ranges between 16.7 and 18.8 GPa. There are no difference in hardness comparing with other work in literature[16]. Among the samples, 5W50A indicates relatively higher value than that of 2W50A and of 8W50A.

4. CONCLUSIONS

The reaction of AlN and β-SiC powders causing the transformation to the 2H phase appeared to depend on the temperature and the amount of SiC/AlN with the additives present. For the compositions of 49wt% SiC/49wt% AlN with an additive of 2wt% α-SiC and 47.5wt% SiC/47.5wt% AlN with an additive of 5wt% α-SiC, the complete solid solution with a single phase of 2H could be fabricated by hot pressing at 2030℃ for 1 h. The fracture toughness by the equation of Niihara ranges between 6.4 and 6.9 MPa · $m^{1/2}$. The fracture toughness could be enhanced by the crack deflection of the columnar inter-grains in the homogeneous microstructure with similar grain size distributions and compositional homogeneity of 2H phases, which was deduced by the additives of α-SiC. The additive weight dependence for the improvement of the fracture toughness indicated the maximum value in the case of 5W50A.

ACKNOWLEDGEMENT

This research was financially supported by the Korea Science and Engineering Foundation(KOSEF) through the research fund in 1996 under the contract number of 961-0802-014-2 and the Ceramic Processing Research Center(CPRC) at Hanyang University.

REFERENCES

[1] D. Bloor, R. J. Brook, M. C. Flemings, S. Mahajan and R. W. Cahn, The Encyclopedia of Advanced Materials, Vol. 4(1995) p. 2453.
[2] I. B. Cutler, P. D. Miller, W. Rafaniello, H. K. Park, D. P. Thomson and K. H. Jack, Nature 275(1987) 795.
[3] W. Rafaniello, K. Cho and A. V. Virkar, J. Mat. Sci. 16(1981) 3479.
[4] Y. Sugahara, K. I. Sugimoto, H. Takagi, K. Kuroda and C. Kato, J. Mat. Sci. Lett. 7(1988) 795.
[5] K. Tsukuma, M. Shimade and M. Koizumi, ibid. 1(1982) 9.
[6] Y. Sugahara, K. Kuroda and C. Kato, J. Am. Ceram. Soc. 69(1984) C247.
[7] A. Zangvil and R. Ruh, ibid. 71(1988) 884.
[8] S. Y. Kuo and A. V. Virkar, ibid. 73(1990) 2640.
[9] Y. Xu, A. Zangvil, M. Landon and F. Thevenet, ibid. 75(1992) 325.
[10] R. Sathyamoorthy, A. V. Virkar and R. A. Cutler, ibid. 75(1992) 1136.
[11] J. Chen, Q. Tian and A. V. Virkar, ibid. 75(1992) 809.
[12] I. Teusel and C. Rüssel, J. Mat. Sci. 25(1990) 3531.
[13] J. F. Li and R. Watanabe, ibid. 26(1991) 4813.
[14] R. R. Lee and W. C. Wei, Ceram. Eng. Sci. Proc. 11(1990) 1094.
[15] I. Teusel and C. Rüssel, J. Mat. Sci. Lett. 11(1992) 205.
[16] M. Miura, T. Yogo and S. I. Hirano, J. Mat. Sci. 28(1993) 3859.

Materials Science and Engineering Serving Society
S. Sōmiya, R.P.H. Chang, M. Doyama and R. Roy (Editors)

Formation of Intermetallic Compounds on Refractory Metals in Aluminum-Silicon Liquid

Makoto Nanko, Akio Takahashi, Takashi Ogura, Akihiko Kitahara, Katsuyuki Yanagihara and Toshio Maruyama
Department of Metallurgical Engineering, Tokyo Institute of Technology, 2-12-1 Ookayama, Meguro, Tokyo 152, Japan

The formation of $Mo(Si,Al)_2$ on Mo dipped into Al-Si liquid saturating Si were investigated, as well as $Nb(Si,Al)_2$ on Nb. The $Mo(Si, Al)_2$ layer was made of fine needle-like grains with submicrons in diameter. On the other hand, $Nb(Si, Al)_2$ consisted of submicron-order fine grains. Liquid phase consisting of Al-Si was observed at the grain boundary of the intermetallic layers and at the interface between the product and the refractory metals. This implies the formation of the intermetallics by solution-repricipitation process.

1. INTRODUCTION

For energy serving and ecological issue, an increase in operation temperature of gas turbine systems has been required to increase energy conversion efficiency. New high temperature materials have been required with higher applicable temperatures than that of Ni-based superalloys. Disilicides such as $MoSi_2$ has a promising material because of excellent resistance for high temperature oxidation. Since $MoSi_2$ has low creep strength and high temperature strength, the intermetallic compound is reasonable for application of coating to prohibit high temperature oxidation.

Dipping Mo into Al-Si liquid at 1000 K, $Mo(Si, Al)_2$ layer is appeared on the Mo surface[1]. Figure 1 shows the phase diagram of Mo-Si-Al at 1323 K[2]. The intermetallic compound coexists with liquid Al and solid Si. This phenomenon can be applied as a low-temperature coating technique for Mo and Mo alloys. This intermetallic compound has good resistance for high temperature oxidation, as same as $MoSi_2$ and better scale adhesion than $MoSi_2$[3]. The growth of $Mo(Si, Al)_2$ in Al-Si liquid is controlled by reaction between $Mo(Si,Al)_2$ and Mo[1]. Similar reaction was observed in the Nb dipping into Al-Si liquid[4].

The microstructure of the intermetallic layer is observed and discussed. Kinetics of the formation of the intermetallic layers on the refractory metals is discussed.

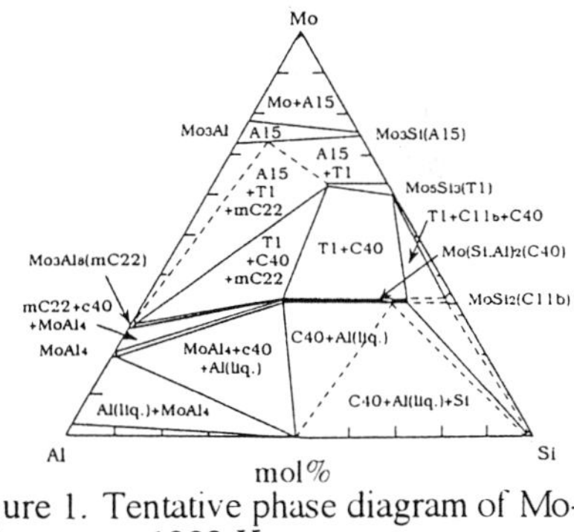

Figure 1. Tentative phase diagram of Mo-Si-Al system at 1323 K

2. EXPERIMENTAL PROCEDURE

Cold-rolled Mo or Nb plates were cut and were ultrasonically cleaned in alcohol. The plates were set with Al powder and Si grains into mullite containers in vacuum chambers. The vacuum chambers were heated at dipping temperatures from 940 K to 1200 K. Dipping of Mo was also carried out in air. There were not remarkable differences in the formation of intermetallic layers by dipping.

Dipped samples were evaluated by X-ray diffraction (XRD), optical microscopy, scanning electron microscopy (SEM) and electron probe microanalysis (EPMA). Growth rate of the product layer was obtained from a consumption thickness of Mo or Nb during dipping.

3. RESULTS

Figure 2 shows the cross section of Mo dipped into Al-Si liquid at 973 K for 350 ks. Figure 3 shows Nb dipped at 1025 K for 85 ks. Dense product layers appear on Mo and Nb. Figure 4 shows chemical composition in the product layers by the Mo dipping. Figure 5 shows one in the Nb dipping. Numerical compositions were $Mo(Si_{0.83}, Al_{0.17})_{2.3}$ in the Mo dipping and $Nb(Si_{0.88}, Al_{0.12})_{2.1}$ in the Nb case. The values of the compositions of the intermetallic layer were independent on the dipping conditions.

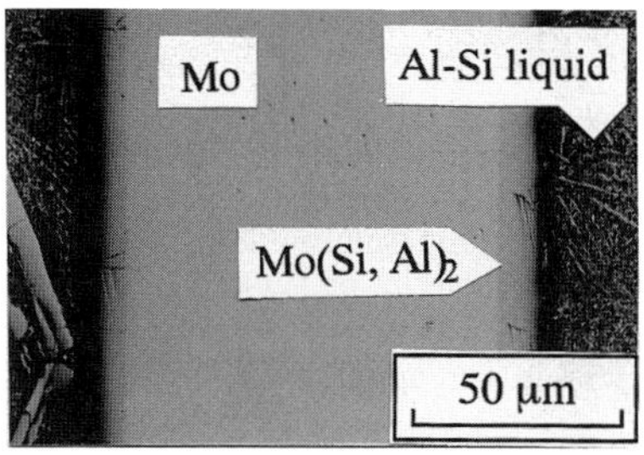

Figure 2. Cross-section of Mo dipped into Al-Si liquid at 973 K for 350 ks.

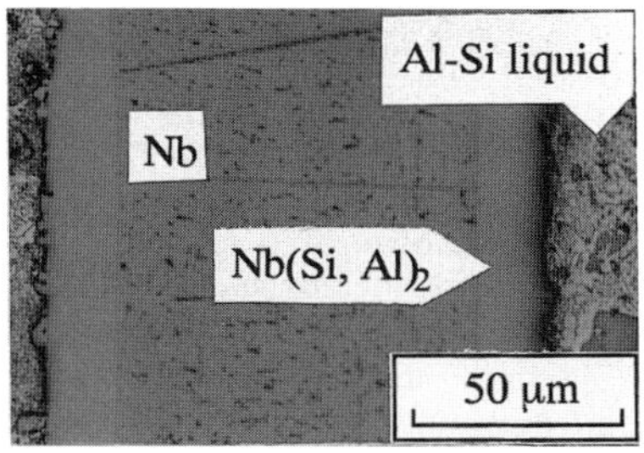

Figure 3. Cross-section of Nb dipped into Al-Si liquid at 1025 K for 85 ks.

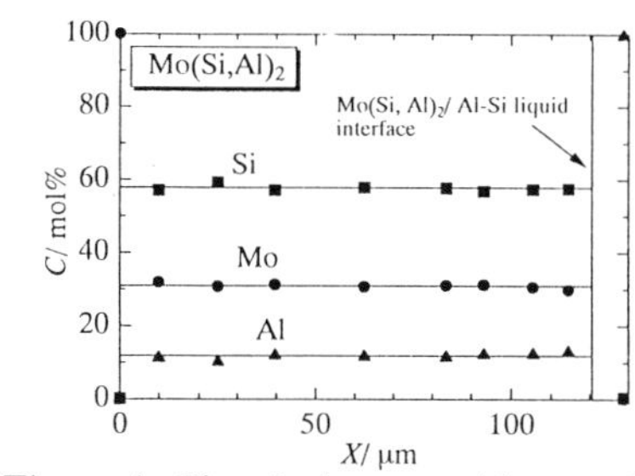

Figure 4. Chemical composition profile of the product layer on Mo by the dipping at 973 K for 350 ks. The interface between Mo and the product layer is X =0.

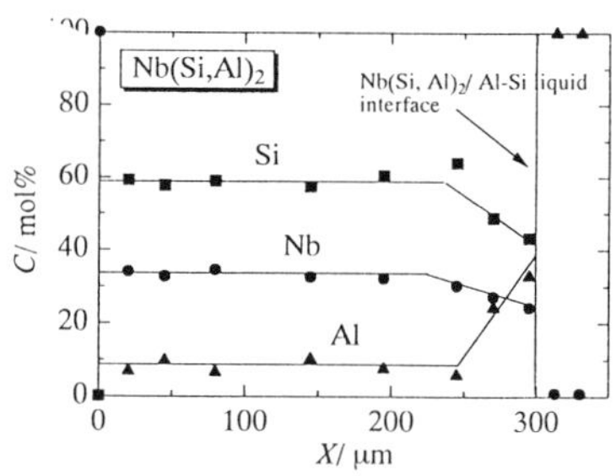

Figure 5. Chemical composition profile of the product layer on Nb by the dipping at 940 K for 847 ks. The interface between Nb and the product layer is X =0.

Figure 6 shows the fractured surface of the $Mo(Si, Al)_2$ layer produced by dipping at K for ks. The $Mo(Si, Al)_2$ layer consists of needle-like grains, which grows perpendicular to Mo surface. The grain size was sub-micron order in diameter. The detail of the fractured surface of the $Mo(Si, Al)_2$ layer are given in Fig. 6. Roundish grains and bridges by liquid phase exists at the grain boundary. On the other hand, the $Nb(Si,Al)_2$ layer consists of fine granular grains, as shown in Fig. 7. Grain size of $Nb(Si, Al)_2$ is less than 0.1 μm near the Nb/ Nb(Si, $Al)_2$ interface. The grain size near the liquid/ $Nb(Si, A)_2$ interface is larger ($\approx$3 μm) than one at the interface between Nb/ $Nb(Si, Al)_2$ interface. The secondary phase can be clearly observed between grain boundary, as shown in Fig. 7. Taking account of the analysis of chemical composition, the secondary phase in the intermetallic layer is probably Al-Si liquid.

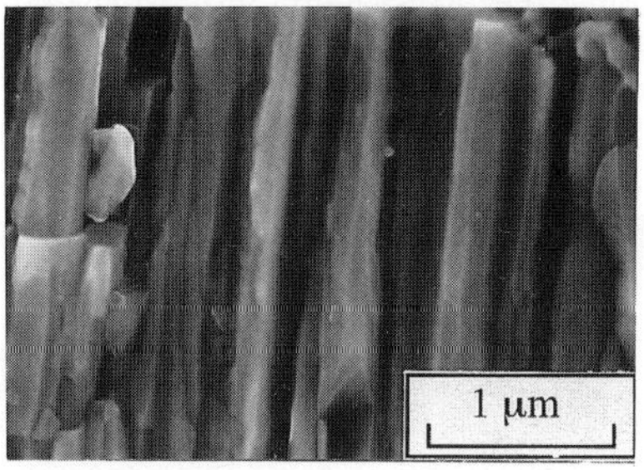

Figure 6. Fractured surface of $Mo(Si, Al)_2$ made by the dipping at 973 K for 85 ks.

Figure 7. Fractured surface of $Nb(Si, Al)_2$ made by the dipping at 1140 K for 14 ks.

4. Discussion

According to the previous works[1, 4], the formation of the product layer is controlled by the reaction at the refractory metal/ the intermetallics interface. This means that the supply of Al and Si to the interface is faster than consumption of Al and Si by the reaction. In the present study, the product layer was dense up to 200 μm in thickness. Figure 8 shows temperature dependence of k_p on the formations of different disilicides. The values of k_p on the Mo dipping and Nb dipping were calculated from the linear growth rate, k_l, by

$$x_{\text{product}} = k_l t \geq \frac{2k_p}{x_{\text{product}}} t \qquad (1)$$

where thickness of the product layer, x_{product}, is equal to be 200 μm which is the maximum value in the present study. The values of k_p of the Mo dipping and Nb dipping were much smaller those ones in the other disilicides. The value of k_p depends on diffusion coefficient and the Gibbs energy change. Since the Gibbs energy changes in formation of disilicides probably have no significant difference, the difference of k_p is caused by the diffusion coefficient. However, It is not reasonable that the diffusion of Si and Al in $Mo(Si, Al)_2$ and $Nb(Si, Al)_2$ is remarkably faster than the other disilicides, in particular VSi_2 with C40 structure. Considering the detail observation and the analysis of chemical composition on the product layers, Al and Si are supplied to the interface through the Al-Si liquid at grain boundary in a product layer. An assumption that Al-Si liquid passes in 10 nm-diameter capillaries at the grain boundary of the Mo(Si, $Al)_2$ by the capillary force of 0.1 MPa allows

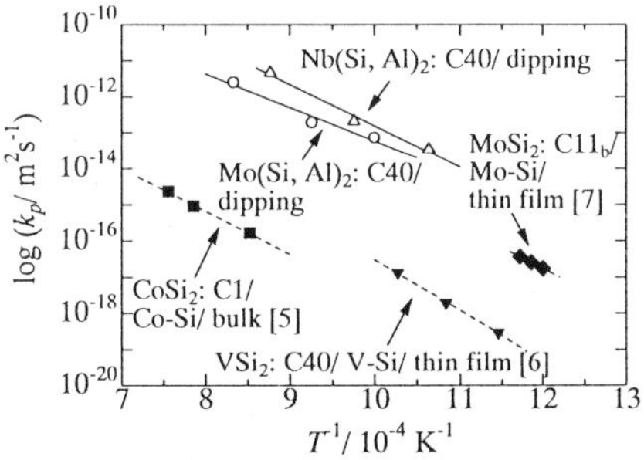

Figure 8. temperature dependence of parabolic rate constant, k_p, on the growth of disilicide.

to supply an enough amount of the liquid for the reaction at the Mo/ $Mo(Si, Al)_2$ interface.

Figure 9 shows the SEM images on Mo surface stripped the product layer. The surface consists of smooth area, rough area and nodules containing Al and Si. As shown in Fig. 9 (d), the nodules are Al-Si grains solidified Al-Si liquid because the shape is totally different with the needle-like grains of $Mo(Si, Al)_2$ formed by the dipping. Almost of the interface seems to be not connected between Mo and $Mo(Si, Al)_2$. This implies that Al-Si liquid is used for the formation during cooling. The intermetallic product, therefore, is probably formed by the solution-repricipitation process through Al-Si liquid.

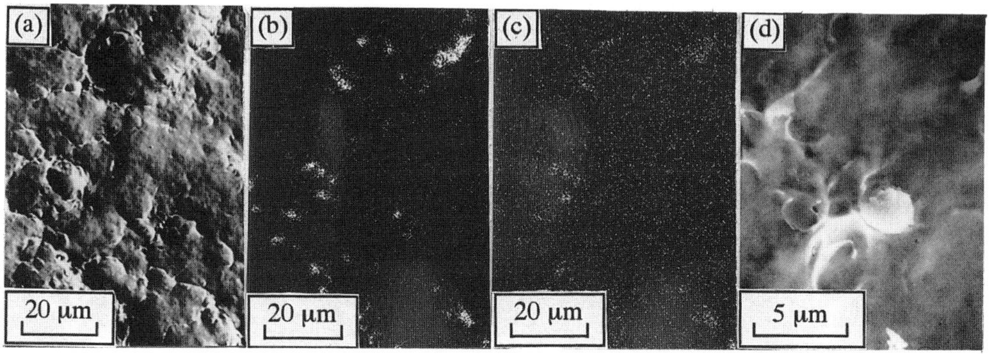

Figure 9. Surface of Mo stripped from $Mo(Si, Al)_2$ layer dipping at 973 K for 86 ks. (a) SEM image, (b) Si map, (c) Al map and (d) a nodule containing Al and Si.

5. CONCLUSIONS

Dipping the refractory metals, Mo and Nb, into molten Al liquid saturating Si, alumino-disilicides were formed on the surface of the metals. The existence of Al and Si at the grain boundary of the intermetallic layer and the interface between the refractory metals and the intermetallic layer was revealed from the results of detail observation and EPMA. The supply of Al and Si is carried out by passing the Al-Si liquid at the grain boundary. The reaction for the formation of alumino-disilicide occurs at the Al-Si liquid at the interface between the refractory metals and the intermetallics.

REFERENCES

1. K. Yanagihara, T. Maruyama and K. Nagata, Irons and Steels, 80, 90-94 (1994).
2. T. Maruyama, X. F. Bi, and K. Nagata, High Temperature Corrosion of Advanced Materials and Protective Coating, Edited by Y. Saito, B. Onay and T. Maruyama, Elsevier, 1992, pp291-299.
3. K. Yanagihara, T. Maruyama and K. Nagata, Mater. Trans. JIM, 34, 1200-1206 (1993).
4. A. Takahashi, K. Yanagihara, T. Maruyama and M. Nanko, Abst. 118th Jpn. Inst. Metals, Japan Institute of Metals, Sendai, 1996, p90.
5. C. H. Jan, C. P. Chen and Y. A. Chang, J. Appl. Phys., 73, 1168-1179 (1993).
6. K. N. Tu, J. F. Ziegler and C. J. Kircher, Appl. Phys. Lett., 23, 493-495 (1973).
7. J. Y. Cheng. H.C. Cheng and L. J. Chen, J. Appl. Phys., 61, 2218-2223 (1987).

Materials Science and Engineering Serving Society
S. Sōmiya, R.P.H. Chang, M. Doyama and R. Roy (Editors)

Optimal Control of Steel Strip Temperature in Continuous Annealing Processes

Naoharu Yoshitani [a]

[a] Dept. of Materials Science and Engineering, Teikyo University, Utsunomiya, 320 Japan. E-mail: n-yoshi@koala.mse.teikyo-u.ac.jp .

This paper presents a mathematical model and an optimal temperature control of steel strip based on the model at the heating furnace in continuous annealing processes. The model was derived from first-principles for use in design of on-line control system. The control system consists of "Optimal preview control" and "Temperature tracking control." The former optimizes the line speed and strip temperature trajectory, while the latter controls strip temperature to its optimal trajectory (target trajectory), by employing the self-tuning adaptive control, one of advanced control technologies.

The control has been successfully working in several real plants with the improvement on strip quality, production rate and production cost.

1. INTRODUCTION

A continuous annealing process is a highly efficient heat treatment process after cold rolling in steel works. It produces steel strips of high tensile strength and high formability. The process consists of the heating, the soaking, the cooling and the overaging furnaces, as shown in Fig. 1. The heat pattern (reference temperature) for continuous annealing is shown in Fig. 2.

The accuracy of strip temperature is decisive in obtaining desired property of the strip, Strip temperature is measured and controlled at the outlet of each furnace. Among these furnaces, the heating furnace (HF) influences large effects on production rate, strip quality, stability of operation, etc.

On the other hand, the strip temperature at the outlet of the heating furnace shows some complicated characteristics because of slow dynamics, dead time, thermal interactions between the strip and hearth rolls which support the strip, and setup changes. These make it difficult for a conventional control system with such as PID controllers to achieve satisfactory control performances.

To overcome this problem, we have developed a new optimal control system [1–3]. The control is based on a new mathematical model describing the static and the dynamic characteristics of strip temperature in the heating furnace. This model is called "the plant model" hereafter.

This paper presents the model and the control system.

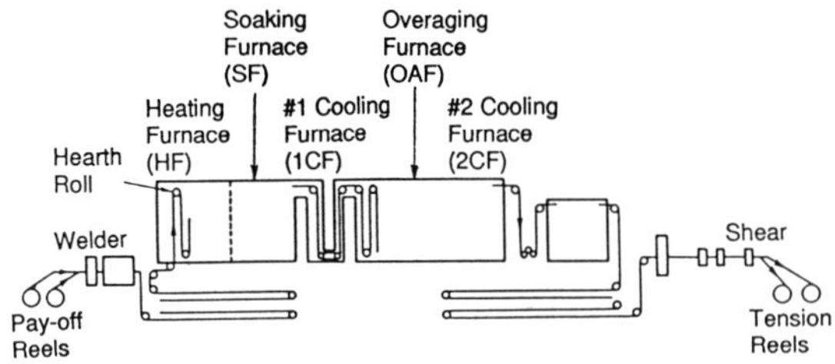

Fig. 1 Continuous annealing process

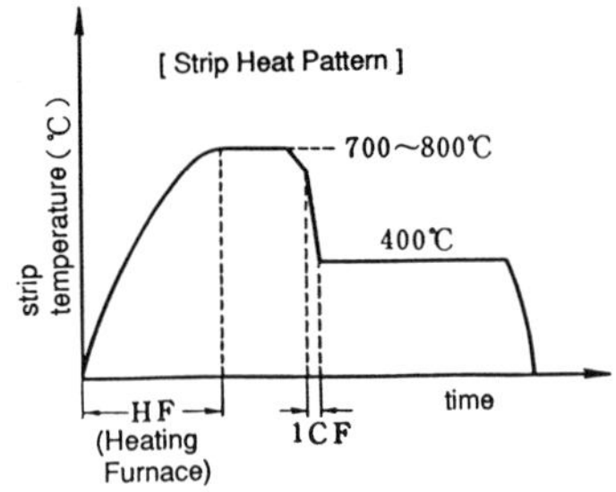

Fig. 2 Strip heat pattern

2. THE PLANT AND ITS CHARACTERISTICS

Fig. 1 shows a typical continuous annealing process called C.A.P.L. (Continuous Annealing and Processing Line). Here, the material for annealing is a cold-rolled strip coil, which is put on a pay-off reel on the entry side of the line. The head end of the coil is then pulled out and welded with the tail end of the preceding coil for continuous annealing and other processing. On the delivery side, the strip is cut into a product length by a shear machine and coiled again by a tension reel.

The heat pattern of a strip, as shown in Fig. 2, is determined according to the composition and the product grade of the strip. The actual strip temperature must be within the defined ranges from the heat pattern to achieve desired property of the strip. The value of the heat pattern at the outlet of the heating furnace is the reference temperature for the control. In most cases, the strip in the heating furnace is heated indirectly with gas-fired radiant tubes.

The heating furnace is 400 to 500 m in strip length and is split into several zones. The furnace temperature and fuel flow rate are measured at each zone, while the strip temperature is measured only at the outlet of the furnace with a radiation pyrometer. It takes a few minutes for each part of the strip to go through the furnace.

Fig. 3 shows observed characteristics of the strip temperature at the outlet of the heating furnace (expressed as "out-strip temperature", or simply "strip temperature" hereafter) in a real plant. Fig. 3(a) shows static characteristics, while Fig. 3(b) $\sim$ (d) show dynamic ones. As seen from Fig. 3(b), the response of out-strip temperature against the change of fuel flow rate has a dead time and a large time constant. Next, Fig. 3(c) and (d) show the response against the change of strip thickness and the line speed, respectively. The first-order lag response in both figures is due to the heat transfer between the strip and hearth rolls. The ramp-like change of the temperature in Fig. 3(d) is caused by the ramp-wise change of the strip heating time due to the stepwise change of the line speed.

3. PLANT MODEL

The plant model was developed so that it represents all the characteristics described in Section 2. The model consists of the dynamic model and the static model of strip

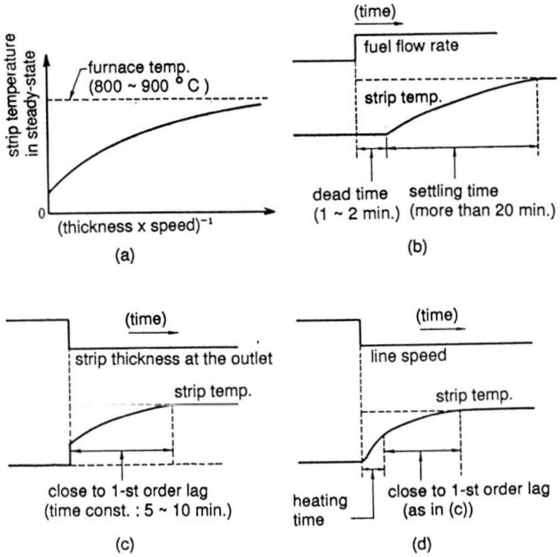

Fig. 3 Characteristics of strip temperature

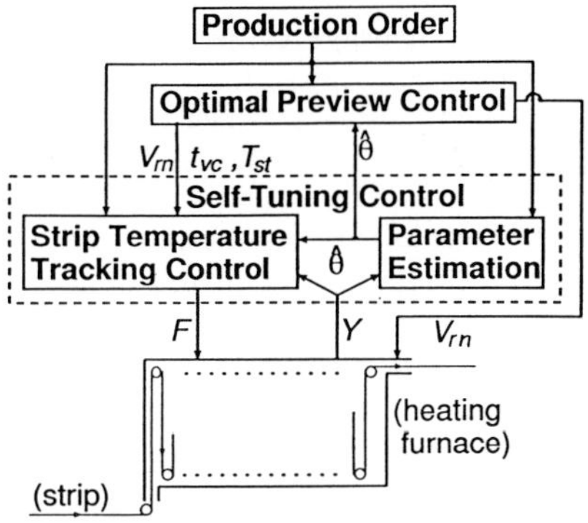

Fig. 4 Strip temperature control system

temperature.

The dynamic model describes the dynamic relationship between the out-strip temperature and the manipulated variable (fuel flow rate in the heating furnace), with additional disturbance terms which include the strip size and the line speed.

On the other hand, the static model describes characteristics of out-strip temperature in steady-state where the above mentioned disturbances, the strip size and the line speed, remain constant.

3.1. Model Equation

The dynamic model equation is written as follows:

$$y(t+d) = \alpha_0 y(t) + \sum_{i=0}^{m} \beta_i u(t-i) + \sum_{i=1}^{4} d_i v_i(t+d), \tag{1}$$

$$\text{where} \quad y(t) = T_s(t) - T_s^*, \quad u(t) = D_{sf}(t)[F(t) - F^*], \tag{2}$$

t : time [min] (one sampling period = 1 min.),

d : dead time [min],

F : total fuel flow rate in the heating furnace [10^4Nm^3/hr] (Nm^3 (normal m^3) denotes m^3 at 0°C),

T_s : out-strip temperature of the heating furnace [100°C],

m : non-negative integer (to be determined experimentally),

v_i $(i = 1, \cdots, 4)$: disturbance terms,

D_{sf} : nonlinear function of strip thickness and the line speed derived from the static model,

F^*, T_s^* : average values of F, T_s respectively in normal operations;

$\alpha_0, \beta_0, \cdots, \beta_m, d_1, \cdots, d_4,$: unknown model parameters (to be estimated on-line).

As seen in the above equations, the dynamic model is chosen to have a predictive form for the prediction of future strip temperature. In the model, v_i $(i = 1, \cdots, 4)$ and D_{sf} are nonlinear functions of disturbances, strip size and the line speed. D_{sf} is derived from the static model, which, not shown here, is represented by Fig. 3(a). D_{sf} is used as a nonlinear gain for fuel flow rate and strip-mass flow rate. The future values of disturbance terms (v_i $(i = 1, \cdots, 4)$) can be predicted in advance because processing schedule is known beforehand. The accuracy of the dynamic models is improved by including such nonlinear functions.

The dynamic model was derived from both the dynamic heat balance equation in the heating furnace and actual observation of the dynamic characteristics of the temperature shown in Fig. 3(c) and (d).

Thus, the plant model is derived from both physical theories and actual plant behaviors to establish the reliability of the model. The model accuracy is confirmed satisfactory with actual plant data under various operating conditions.

4. CONTROL SYSTEM

4.1. Outline of the control

Fig. 4 schematically illustrates the control system. The system is two-layered, consisting of "Optimal preview control" at the higher layer and "Temperature tracking control" at the lower layer, and is installed in a process control computer.

The higher layer performs preset control. It calculates speed reference V_{rn} after the approaching setup change, the optimal timing of speed change t_{vc} and the target trajectory of strip temperature T_{st}. It uses pre-determined production order which is sent from a business computer for production management. The production order includes size, grade and reference temperature T_{sr} of strips.

Next, the lower layer performs feedback control and parameter estimation. The controller calculates a manipulated variable, which is the command value of the total fuel flow rate F in the heating furnace (denoted by "fuel flow rate" hereafter), at a certain control interval (typically 1 min.). The controller uses measured values of plant variables Y and employs T_{st} as the control target. Here, Y includes F, T_s, T_f and V; where T_s, T_f and V denote the out-strip temperature, the mean value of the furnace temperature over all zones of the furnace (expressed simply as "furnace temperature" hereafter), and the line speed, respectively. V is not chosen as a manipulated variable of temperature feedback control, because frequent speed changes sometimes cause fluctuations in controls at other furnaces. Normally V is changed, if necessary, only at setup changes. Estimated values $\hat{\theta}$ of unknown parameters of the plant model are also calculated on-line at this level. The plant model is used at both layers.

4.2. Strip temperature tracking control and recursive parameter estimation

The following conditions are required for temperature tracking control:

C1 : To minimize the control error.

C2 : To avoid large fluctuations of fuel flow rate.

C3 : To adapt the controller to any possible changes of plant characteristics.

C1 and *C2* are common to most control systems. *C3* is necessary to cope with the changes, especially those caused by aging, maintenance, or alteration of the plant. Considering these conditions, we applied the self-tuning control (STC), [1–4], one of advanced control technologies which adjusts control parameters on-line according to the changes of plant characteristics in order to keep desired control performances.

Unknown parameters of both the dynamic and the static models are recursively estimated by using actual plant data. The estimation for the dynamic model is performed at every sampling instance (typically every 1 min.) with REVVF (Recursive Parameter Estimation with a Vector-type Variable Forgetting Factor) [5]. REVVF assigns a different value of forgetting factor to each unknown parameter and thus makes it possible to adjust the adaptation rate for each estimate independently.

Fig. 5 shows on-line parameter estimation with actual operation data. After about 100 observations (100 minutes), parameter estimates roughly converged to reasonable values, which illustrates the validity of the model and the estimation algorithm. Next, Fig. 6 shows an example of actual control. Here, the target trajectory of strip temperature has a ramp-wise change before and after a setup change and actual strip temperature follows its target with good accuracy. Thus, the figure illustrates the effectiveness of the control.

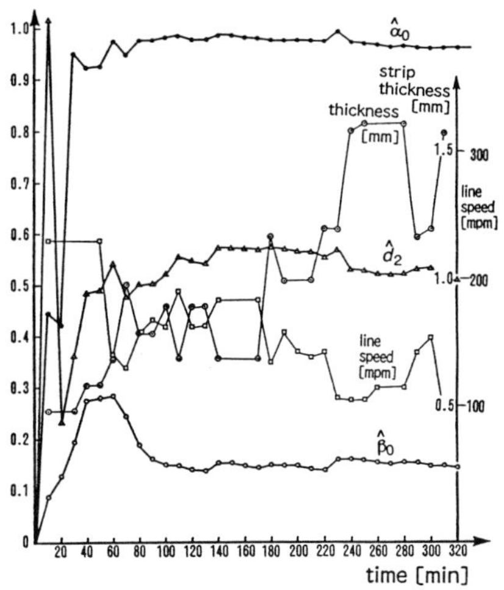

Fig. 5 On-line parameter estimation

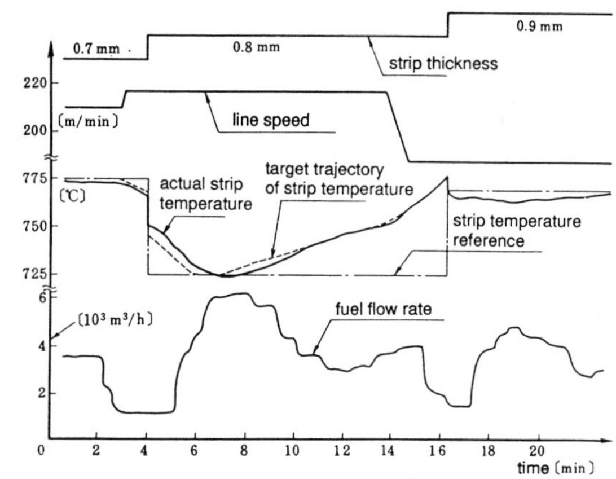

Fig. 6 Actual control

5. CONCLUSION

The strip temperature control at the heating furnace in continuous annealing processes has been developed by employing the self-tuning control technology based on a newly-developed plant model.

The control has been implemented in several real plants and is planned to be applied to some more ones, in Nippon Steel Corporation, Japan, and other steel-manufacturing companies. In real plants, the control has been working quite successfully, contributing to production rate increase, energy conservation, quality improvement, et al.

Acknowledgments

The author would like to express thankfulness to all the people who contributed to the development of the control in Nippon Steel Corporation, Japan, and to Prof. D.Q. Mayne and Mr. M. Rafiq for their suggestions and encouragements in the research on REVVF at Imperial College, U.K.

REFERENCES

1. N. Yoshitani, Proc. the 5-th IFAC MMM Symposium, Tokyo, Japan, pp. 380-385, 1986.
2. N. Yoshitani, Proc. IECON'93, Hawaii, pp. 469-474, 1993.
3. A. Hasegawa, Proc. IEEE CCA'94, 1994.
4. A.Y. Allidina and F.M. Hughes, IEE Proc. (Part D), vol. 127, no. 1, pp. 13-18, 1980.
5. N.Yoshitani, Transactions of the Society of Instrument and Control Engineers, Japan, vol. 25, no. 5, pp. 71-77, 1989 (in Japanese).

Materials Science and Engineering Serving Society
S. Sōmiya, R.P.H. Chang, M. Doyama and R. Roy (Editors)
1998 Elsevier Science B.V.

Mechanical Properties and Oxidation Behavior of TiB2 Particle Dispersion Strengthened TiAl Intermetallic Compounds

Shunji KIKUHARA[a] and Shouichi OCHIAI[b]
[a] Graduate School, [b] Department of Materials,
Teikyo University of Science & Technology, Uenohara-machi, Kitatsuru-gun, Yamanashi-ken 409-01, Japan

ABSTRACT TiAl -TiB_2 composites with 0, 10, 15 and 20 vol%TiB_2 were produced by mechanical alloying (MA) process and subsequent hot isostatic press (HIP) method using pure Ti, Al and TiB_2 powders as starting materials. Microstructures, mechanical properties and oxidation behavior of the composites were investigated.

1. INTRODUCTION

TiAl(γ) intermetallic compound possesses attractive properties like high melting temperature, low density, high modulus, and good oxidation resistance. These properties are desirable for the use as high temperature structural material [1]. However, its poor ductility at low temperatures and poor workability at high temperatures are limiting the variety of applications. One promising method for the improvement of workability is the refinement of microstructures [2].

Mechanical alloying (MA) process is known to be effective on refining microstructure of intemetallics. Furthermore, the powder metallurgical process enables the homogeneous distribution of ceramic dispersoids in intermetallic matrix [3]. Combining intermetallics with hard and thermally stable ceramic particles hold promise for enhancing the elevated temperature strength, creep strength and the wear resistance. However, reports on the TiAl base composite material have been limited up to now. Incidentally, since TiB_2 ceramic particles show good compatibility with titanium alloy, they appear to be proper candidates for reinforcements for TiAl [4].

In this study, TiAl -TiB_2 composites with 0, 10, 15 and 20 vol%TiB_2 were produced by mechanical alloying (MA) process and subsequent hot isostatic press (HIP) method. Microstructures, mechanical properties and oxidation behavior of the composites were investigated.

2. EXPERIMENTAL

As starting powders, pure Ti powder (purity : >99.1%, mean particle size: about 45μm), pure Al powder (purity : 99.5%, mean particle size: about 100 μm) and Ti B_2 powder (mean particle size: about 1.6 μm) were used. The starting powders were so blended as to make composites with 10, 15 and 20 vol%TiB_2 as well as the matrix alloy with the composition of Ti-48mol%Al. These four kinds of mixed powders were subjected to mechanical alloying treatment using a planetary type ball mill up to 36ks. The green compacts of MAed powders were subjected to the HIP treatment under the condition of 1473K x 2000 atm x 7.2ks. Constituent phases of the sintered alloys were identified by the x-ray diffraction (XRD).

In order to examine mechanical properties, compression test was undertaken for consolidated rectangular specimens with 3 x 3 x 6 mm^3 at temperatures from room temperature to 1373K at the initial strain rate of 1.4×10^{-3} s^{-1}. Hereafter, oxidation behavior at 1273K for 86.4ks in an air was examined using specimens with the size of 0.5 x 3 x 6 mm^3. Mass gain during the oxidation reaction was measured for each sample. The section of scales were observed and analyzed using EPMA.

3. RESULTS AND DISCUSSION

3.1. Microstructures

Matrix sintered alloy exhibited equiaxed grains with mean diameter of under 4 μm. There is the tendency that the microstructure becomes finer with the increase of TiB_2 vol%. Therefore, it can be stated conclusively that MA process is very effective for refining microstructures of titanium aluminide.

According to the XRD study, matrix alloy was found to be composed of TiAl and Ti_3Al. In composites, existence of TiB_2 phase was clearly confirmed besides TiAl and Ti_3Al phases. This indicates that TiB_2 particle is thermally stable in the TiAl matrix.

3.2. Mechanical properties

The result of Vickers hardness measurement is presented in Fig.1, showing that the hardness has the tendency to increase with increasing the volume fraction of dispersoids. The temperature dependence of 0.2% flow stress is portrayed in Fig.2. As can be seen, not only flow stress decreased gradually with increasing temperature, but also became larger when the vol% of TiB_2 increases. In particular 20vol%TiB_2 alloy exhibited larger stress by 500MPa than the matrix alloy at 873K.

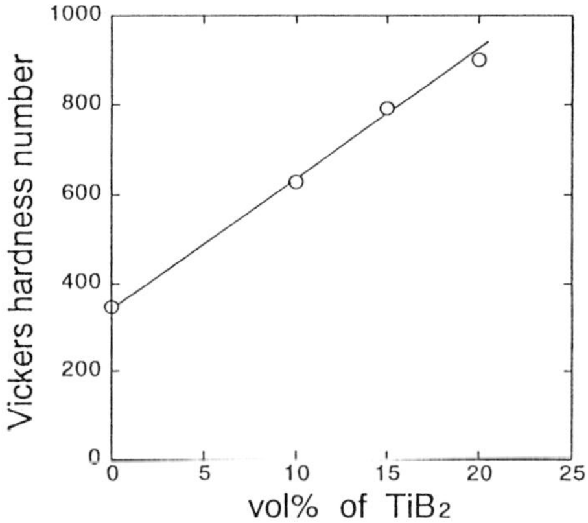

Fig.1 Relation between the Vickers hardness and the vol% of TiB_2.

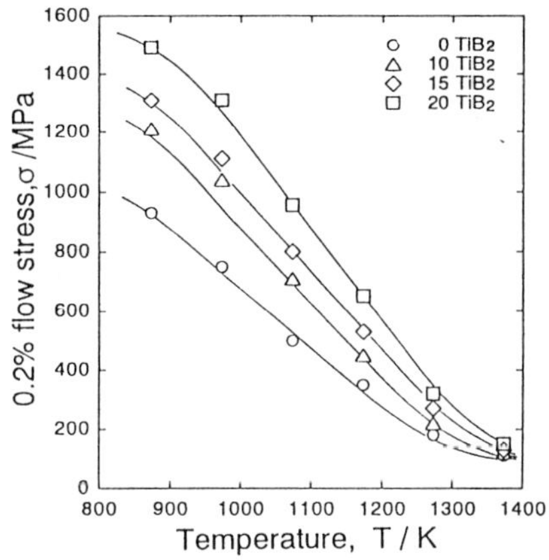

Fig.2 Temperature dependence of the 0.2% flow stress.

But, this stress difference tends to decrease with increasing temperature, so each flow stress was reduced to under 150MPa at 1373K.

Compressive strain rate change tests were conducted at high temperatures in order to elucidate the deformation mechanism. Backofen et al. regarded the strain rate sensitivity exponent (m value) of above 0.3 as one of the measures for the development of superplasticity [5]. In this condition, it is suggested that the homogeneous elongation occurs because of the restraint of the necking. Fig.3 presents the temperature dependence of m value. There was the tendency that m values increased with increase of temperature. Although m value tends to decrease by the addition of TiB_2, m values for any specimen are higher than 0.3 at 1373K.

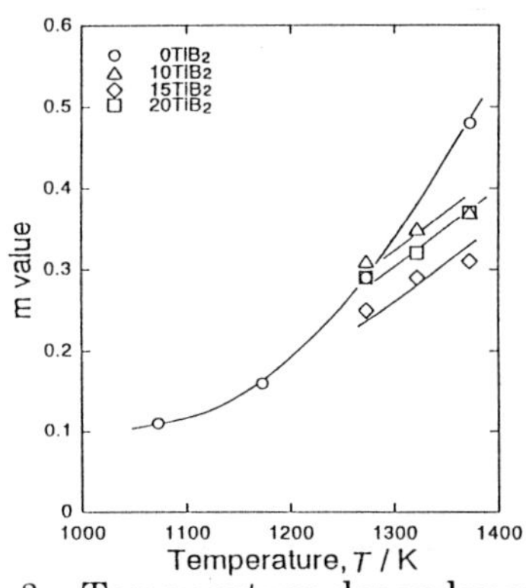

Fig.3 Temperature dependence of the m values.

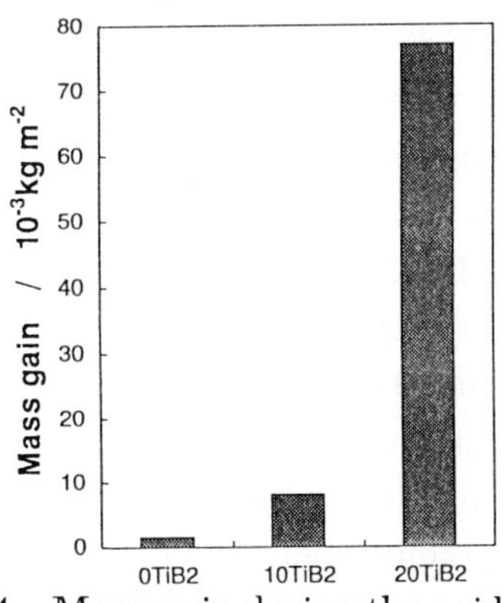

Fig.4 Mass gain during the oxidation test at 1273K for 86.4ks.

At this temperature, the occurrence of cracks was not observed irrespective of existence of TiB_2. These facts are indicating the development of superplastic deformation.

3.3. Oxidation behavior

The oxidation behavior of TiAl-TiB_2 composites was examined in an isothermal oxidation test at 1273K for 86.4ks. Obtained mass gain is illustrated in Fig.4. Matrix alloy showed the least mass gain of 2 g/m^2. On the other hand, mass gain tends to increase with increase of vol% of TiB_2 and one of 80g/m^2 was measured for 20vol%TiB_2 alloy.

In the matrix alloy, only the alumina layer is existing. However, not only alumina layer of 10μm in width was found on the bulk surface, but also the TiO_2 layer was seen to form on the outer side of alumina layer. Therefore it is clear that the restrain of the growth of TiO_2 layer is necessary for the improvement of oxidation resistance for the TiAl-TiB_2 composites.

4. SUMMARY

Compacts with the addition of TiB_2 particles were consisting of TiAl, Ti_3Al and TiB_2 phases. The strength of compacts increased with increase of vol% of TiB_2. This fact indicates the obvious dispersion strengthening effect of TiB_2 for TiAl intermetallic alloy. The strain rate sensitivity exponents (m values) were found to be larger than 0.3 at 1373K for any specimen, suggesting the development of superplasticity. Although mass gains during oxidation test of the matrix and 10TiB_2 alloys were very small of less than 10 g/m^2, large one of 80g/m^2 was measured for 20TiB_2 alloy. It was observed that the oxidation scale near the surface consisted of alumina layer and the volume of TiO_2 scale increased with increasing TiB_2 volume.

REFERENCES

1. H.A.Lipsitt, D.Shechtman and R.E.Schafrik, Metall. Trans., 6A (1975) 1991.
2. E.M.Schulson and D.R.Barker: Scripta Metall., 17 (1983) 519.
3. J.M.Panchal and T.Vela, Advances in Powder Metallurgy, Metal Powder Industries Federation, 3 (1989) 359.
4. S.Ochiai, M.Inoue and K.Kokubo, Powder and Powder Metallurgy, 42 (1995) 391.
5. W.A.Backofen, I.R.Turner and D.H.Avery, Trans. ASM, 57 (1964) 980.

Materials Science and Engineering Serving Society
S. Sōmiya, R.P.H. Chang, M. Doyama and R. Roy (Editors)

Molecular Dynamics Simulation of Fine Particles

H. Masuyama and Y. Kogure

Teikyo University of Science & Technology
Uenohara, Yamanashi 409-01, Japan

Molecular Dynamics Simulation of fine particles of a noble metal(Cu) has been made. The configulation and the motion of atoms are monitored by radial distribution functions. It is seen that the radial distribution functions depends on the number of atoms in particles.

1 INTRODUCTION

Fine particles or clusters of metals are shown disinguished properties from bulk materials in morphologic,thermal,electrnic and magnetic properties. Sometimes the atomic structures are different from bulk crystal. The melting temperature is lower and depend on size. The low temperature specific heat is smaller due to the enhanced discreatness of quantum energy levels of electrons and phnons. The fluctuation effects is large in superconducters. These properties are due to limited size and large surface effects of particles.

Molecular dynamics simulation of fine particles of copper has been made in the present study.

2 METHOD OF SIMULATION

The embedded atom method potential developed by Doyama and Kogure [1] is used in the simulation. The potential is expressed by simple functions. The potential is truncated between third and forth nearest neighbors. The number of atoms in a particle are between several hundreds to several thousands, and the time intervals for molecular dynamics simulations are chosen to be 5×10^{-15} sec. As an initial condition the atoms are arranged in a fcc crystal structure.

Under these initial conditoins, molecular dynamics simulation of structure and collision has been simulated.

2.1 Slow cooling of particles

First, each of two particles(1099 atoms and 666 atoms) is quenched at temperature 0K. These result is an equilibrium strucutre of each particles. Next, a velocity with random distribution is given to each atom is an equilibrium strucutre, then the system attains a liquid state. Lastly, the molten particle is cooled slowly and an equilibrium configuration is obtained. Slow cooling is made by decreasing kinetic energy in every 500 MD Steps(cooling 1) or 2000 MD Steps(cooling 2). Configuration of atoms in the particle is examined by means of radial distribution function(RDF) and morphology.

2.2 Collision of two particles

The interaction of two particles has been simulated. Two particles with similar sizes are formed (1099 atoms and 172 atoms systems,172 atoms and 172 atoms systems) and a particle velocity is given to one or both of the particles.

3 RESULTS

3.1 Slow cooling of particles

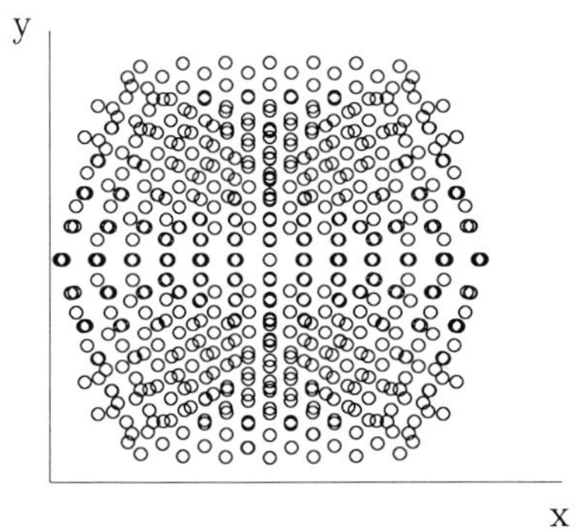

(a) Configuration

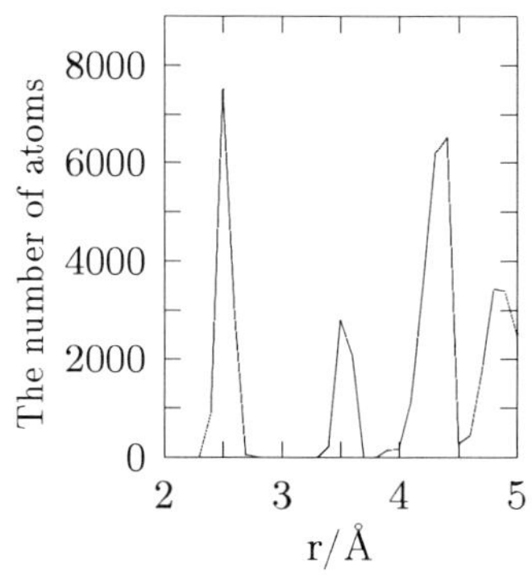

(b) 0K-RDF

Figure 1. Atomic configuration and RDF at 0K of a particle with 1099 atoms

RDF of a particle with 1099 atoms simulated at 0K(10000 MD Steps) is shown in Figure 1-(b) and RDF of simulated at 1600K(10000 MD Steps) is shown by a thin line in Figure 3-(b). RDF of cooling 1 is shown by a thick line in Figure 3-(b), and RDF of cooling 2 is shown in Figure 2-(b). RDF of liquid of 1600K shows a broad-peak nearby a lattice constant, 3.6Å, of copper. And, the peak in RDF of cooling 1 is as low as the

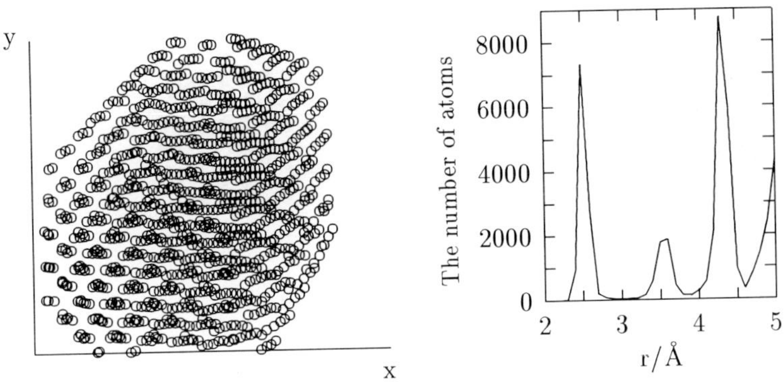

(a) cooling 2-XYZ (b) cooling 2-RDF

Figure 2. Atomic configuration and RDF of cooling 2 of a particle with 1099 atoms

peak of 1600K. On the contrary, the peak nearby 3.6Å of RDF of cooling 2 is as high as the peak of RDF of 0K. For comparison, these RDF are summarized in Figure 3-(a)(b). It is seen that RDF of 0K and cooling 2 in Figure 3-(a) have same shape. By the slow cooling, arrangement of atoms became more and more crystal like structure. It can be seen from the configuration of cooling 2(Figure 2-(a)).

RDF(0K and cooling 2) of particle with 666 atoms simillary caluculated is shown in Figure 3-(c). In this case, the correspondence of RDF is not so clear as 0K and cooling 2 of perticle of 1099 atoms. These difference of RDF may come from the difference of particle size and the numbers of surface atoms in the particle.

The systematic investigation is continuously in progress.

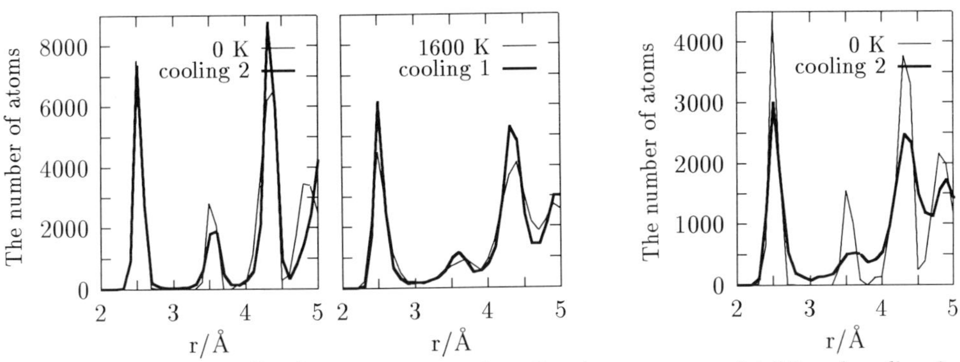

(a) 0K and cooling 2 (1099 atoms) (b) 1600K and cooling 1 (1099 atoms) (c) 0K and cooling 2 (666 atoms)

Figure 3. RDF of particles after quenched or cooling

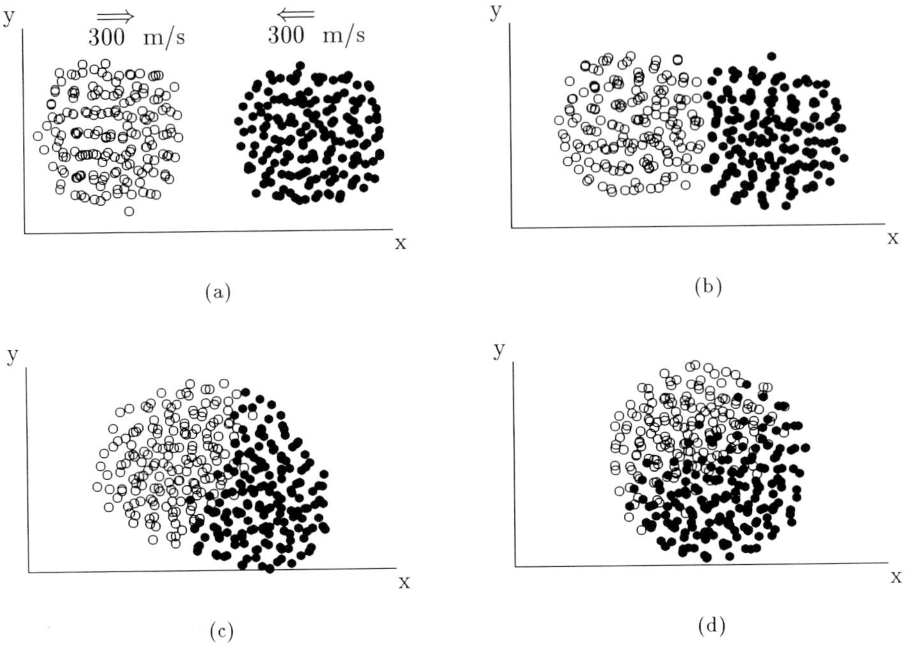

Figure 4. Collision of two particles

3.2 Collision of two particles

Collision of two particles consisting of 172 atoms is shown in Figure 4. A particle velocity of 300 m/s is given to both of the particles and head-on collision has been made. Immediately after the collision the particle is seen to be distorted(Figure 4-(c)). But as time passed, morphology of particles became spherical(Figure 4-(d)). This changes of the shape of particles may come from the large surface effects.

REFERENCES

[1] M.Doyama and Y.Kogure Radiation Effects and Defects in Solids ,**141**,565(1997)

[2] Yoshiaki Kogure Proc. of Intern. Conf. Microstructures and Functions of Materials, 317(1996)

Materials Science and Engineering Serving Society
S. Sōmiya, R.P.H. Chang, M. Doyama and R. Roy (Editors)

Characterization of Silk Produced in 12th Century Japan with the Aid of ^{13}C NMR

Riichirô Chûjô

Department of Materials Engineering, Teikyo University of Science and Technology, 2525 Yatsusawa, Uenohara-machi, Yananashi 409-0193, Japan

1. INTRODUCTION

Silk from the cocoon of silkworm has been appreciated for a long time as the most beautiful fiber. For instance, a plain-weave silk was excavated from the Arita remains close to Fukuoka, Japan. It was attached to a thin-type bronze *ko.*[1] It is estimated to the production of the early Yayoi Perid (2nd century of B.C.). At the present stage, it is the oldest silk fabric excavated from Japan. The existence of silk such in an old age of Japan has been supported by ancient documents of surrounding countries. One example appeared in the official chronicle of Wei dynasty which governed northern China in the 3rd century of A.D. In this chronicle there is a description on the existence of silk in Japan when it was edited. The most famous collection is in Shosoin Treasure-House at Nara in which silk fabrics are still reserved (not excavated) since 7th century. Silk is, therefore, the best materials for the archaeological study through polymeric materials. Actually, Nunome[2] has carried out elaborate scientific study of archaeological silk mainly with the aid of optical microscope.

The aim of this paper is the characterization of archaeological silks with the aid of Nuclear Magnetic Resonance (NMR). NMR has been widely accepted as an excellent chaaracterization method in physics, chemistry, polymer science, and biology as well as medical diagnosis. In principle this method is based on the measurement of the magnetic moment of atomic nucleus. We must, therefore, select the most appropriate atomic nucleus for the measurement. In this paper the nucleus is confined to ^{13}C. This is a stable isotope of carbon. Carbon is the most basic element in organic compounds. Among stable carbon isotopes ^{12}C is the most abundant, however it has no magnetic moment. The natural abundance of ^{13}C is only 1.108%. Nevertheless, this isotope is frequently used in NMR due to its non-zero magnetic moment.

In Hiraizumi, a city in the north-eastern Japan there was a semi-independent kingdom in 12th century. In the precient of Chusonji temple at Hiraizumi which was founded by Kiyohira Fujiwara, the first chieftain of the kingdom there is (still exists) Konjikido, the Gold Colored Hall. Mummies of three chieftains have been reserved under the floor of this hall. Three chieftains are:

Kiyohira Fujiwara (died in 1128)
Motohira Fujiwara (died in 1157)
Hidehira Fujiwara (died in 1187).

The fourth chieftain, Yasuhira Fujiwara was killed in other area in 1189 by the Kamakura Shogunate. By his death the Hiraizumi kingdom became to be ruined. It means the mummies of all chieftains are still reserved. In the coffins of these mummies many silk materials were used; they are clothes, winding fabrics, and pillows. In this paper NMR is applied to these silk materials.

2. PRIMARY STRUCTURE OF SILK AND ANCIENT CLIMATE

In this chapter solid-state high resolution ^{13}C NMR is applied to the silk samples from the coffins of three chieftains. These samples are generous gifts from Drs. W. Kawanobe and C. Sano, Tokyo National Research Instituteof Cultural Properties. All samples were used for measurements without any pretreatment in order to avoid the further damage of samples. In high resolution NMR there are solid- and solution-state ones. For the static measurement the latter is excellent due to its higher resolution. However, the samples are insoluble to any solvent after degumming. The former is, therefore, used for measurements. The measurements were done with JNM-A 500 spectrometer (125 MHz for ^{13}C nuclei) with CP (Cross Polarization)/MAS (Magic Angle Spinning) mode (spinning frquency of 5.8 - 6.1 kHz). Spectral deconvolution was done with the aid of the deconvoluiton software, NM-ACFT installed in the spectrometer. These harewares and software were equipped by JEOL.

Figure 1 shows the solid-state ^{13}C high-resolution CP/MAS NMR spectra of fablics winding the mummies of Kiyohira (top), Motohira (middle), and Hidehira (bottom).[3] The peaks appearing in 170 -174 ppm have been assigned to $\underline{C}$=O of the peptide bonds in fibroin molecules. Splittings appearing in this region are due to the difference of constituent amino acid residues (alanyl, glycyl, *etc.*) as well as those of secondary structures (silk I and silk II). The former affects more significantly than the latter.[4] Chemical shift values are 177.0, 172.3, 170.7, and 169.5 ppm frpmthe peak of tetramethyl silane for alanine in silk I, in silk II, glycine in silk I, and in silk II, respecttively for fibroin from *Bombix mori*. The silk I is close to random-coil and the silk II is essentially anti-parallel β-sheet.[5] As a first approximation, observed $\underline{C}$=O peaks were deconvoluted into two parts, namely, the lower and the higher peaks. "First approximation" means the neglect of the existence of amino acid reisdues other than glycyl and alanyl residues. Figure 2 shows the mole fraction of alanyl residues determined from the spectra in Figure 1. The symbols, HK, HM, and HH appearing in the abscissa are the abbreviations of Kiyohira, Motohira, and Hidehira of Hiraizumi, respectively. JA1, I, and T denote modern Japanese, Italian, and Thai silks, respectively. The latter three are used for the convenience of comparison. The fraction is not constant, but changes with generation. The best way promoting the reliability of trend concluded from any figure is the increase of data points corresponding to the abscissa. This is, however, impossible in this study, because of the limitation of the numberof mummies. Alternative strategy was adopted; namely, we observed NMR spectra for plural number of samples from the same coffin. Plural number of points in HK, HM, and HH are for samples in the same coffin. The deviation of points between intergeneration is larger than the scattering in intrageneration. This finding means the above changes with generation is not artefact but realistic.

T.Mitsutani[6] of Nara National Research Institute of Cultural Properties is compiling tremendous data on dendrochronology. According to his analysis, data from southwest Japan are correlated with atomospheric temperature as well as rain fall, while those from northeast Japan are exclusively done with the temperature. Comparing the above NMR data with the dendrochronological data of the dead years of three chieftains, we can imagine the mole fractions of amino acid residues in silk are correlated with climate, namely, atmospheric temperature.

3. PRIMARY STRUCTURE OF MODERN SILKGLAND REARED IN SEVERAL TEMPERATURES

In order to confirm the above imagination, we are doing several experiments. The comparison between inter- and intrageneration deviations of the mole fractions of amino acid residues is one example. The comparison between the mole fractions of modern cocoon and the record of climate is the second example. Besides these two, the primary structure is analyzed in this chapter for modern silkgland reared in several temperatures. There are two candidates affected by atmospheric temperatures; they are silkworm itself and mulberry leaves fed by silkworms. As a working hypothesis the former is adopted in this chapter. The hybrid between Shunrei No.1 and Shogetsu No.1 was reared in variable temperature cultivator, whose temperatures were controlled to the following six levels; 20, 22, 24, 26, 28, and 30℃. At the middle stage of fifth instar the silkgland was picked up from these silkworms after anatomy and dissolved into LiBr aqueous solution (9M). For the hemolymph the same procedure was done to obtain the samples. These sample preparations were done after separating to male and female. Due to direct leading of the silkgland and the hemolymph to solvent, the concentration was unable to determine. In this case solution-state ^{13}C NMR is applicable and good resolution is obtained.

Figure 3 shows an example (Gland-28℃-Female) of solution-state ^{13}C NMR spectrum.[7] As indicated in the figure five peaks from $C\alpha$'s are observ edwith sufficient intensities; they are Ala, Gly, Ser, Tyr, and Val. The normalized mole fractions obtained from NMR spectra are plotted against temperature for the silkgland of male in Figure 4. Qualitatively speaking, the fractions of major two amino acids (Ala and Gly) tend to the values at 25℃ where is the optimum temperature for rearing. The non-linear least-square fitting with quadratic function is much better than that with linear function. The fractions of Ala is sensitively depending on temperature in male, while not so sensitive in female. Interestingly, the behavior of the temperature dependence of hemolymph is comprehensive with that of silkgland. In other words, consumption of one amino acid species brings the decrease of the same species in hemolymph. This seems to support the plausibility of the working hypothesis. Suzuki of Kanagawa Prefectural Agricultural Research Institute has found that the body temperature of silkworms changes with rearing temperature. In the rearing temperature lower than 25℃ the body temperature becomes higher than the rearing temperature, and *vice cersa*. Even in insects so-called homeostasy seems to be partly realized. This homeostasy may be controlled by the above-mentioned change of amino acid fraction in hemolymph. The temperature dependence found in this chapter supports the plausibility of the imagination in the previous chapter.

4. ESTIMATION OF PHYSICAL PROPERTIES OF DEGRADED ARCHAEOLOGICAL SILK FROM ^{13}C NMR SPIN-LATTICE RELAXATION TIME WITH THE AID OF FLUCTUATION DISSIPATION THEOREM

For modern silks the physical quantites such as initial compliance and initial tensile strength are used for the estimation of physical properties. We would, therefore, like to observe the same quantities for the archaeological silks. It is, however, impossible, due to stress centralization to degraded parts. The measurements of so-called fluctuation is possible. Fortunately, the equivalence is guaranteed between fluctuation and dissipation such as initial compliance and others by the fluctuation dissipation theorem. Both fluctuation and dissipation are measured for fresh samples (such as JA1 in Figure 2), reduction formula is obtained from these two measurements, and the measured value of fluctuation for aged samples are converted to dissipation with the aid of the formula; they are our strategy. Details are described elsewhere.[8)]

5. CONTRIBUTION TO ARCHAEOLOGY

There is an important problem in the arrangement way of three mummies in Chusonji Temple. In Buddhist temples there are two definitions on the left and right; one is the definitions from Buddha image side and the other is from the prayers side. Before the academic interdisciplenary survey of 1950, three mummied were believed to arrange in the order of Motohira (left), Kiyohira (center), and Hidehira (right) from the Buddha image side. This belief was due to the combination between a document in the temple and the circumstance of Buddhist temples. The name tags used in this paper were written in this definition. Based on the anthropological information, majority of academism changed the definition from the prayers side in 1950.[9)] The finding in chapters 2 and 3 is consistent with the definition before 1950. The author would like to introduce the existence the opinion insisting the definition before 1950.[10)]

REFERENCES

1. J. Nunome, Me de Miru Sen-i no Kokogaku (Archaeology of Fiber seen by the Reader's Eyes), Senshoku to Seikatsu Sha, Kyoto, 1992
2. J. Nunome, Kinu to Nuno no Kokogaku (Archaeology of Silk and Fabrics), Yuzankaku, Tokyo, 1988
3. R. Chûjô, A. Shimaoka, K. Nagaoka, A. Kurata, and M. Inoue, Polymer, 37 (1996) 3693
4. M. Ishida, T. Asakura, M. Yokoi, and H, Saitô, Macromolecules, 23 (1990) 88
5. R. D. B. Fraser and T. P. MacRae, Conformations in Fibrous Proteins, Academic Press, New York, 1973
6. T. Mitsutani, private communication
7. R. Chûjô, H. Fukuda, T. Kusaka, M. Suzuki, and I. Okumura, Rep. Prog. Polym. Phys. Jpn., in press
8. R. Chûjô, Polym. Prepr., 38 (1997) No.1, 896
9. N. Mori, Chusonji Goitai Gakujutsu Chosa Saishu Hokoku (Final Report on the Academic Survey for the Mummies in Chusonji Temple), ed. by Chusonji, Chusonji, Hiraizumi, 1994
10. Y. Haniwara, *ibid.*

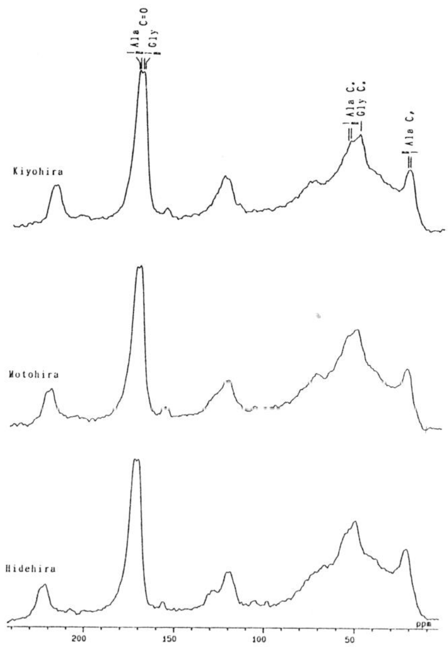

Figure 1 Solid-state ^{13}C high-resolution CP/MAS NMR spectra of fabrics winding the mummies of Kiyohira (top), Motohira (middle), and Hidehira (bottom).

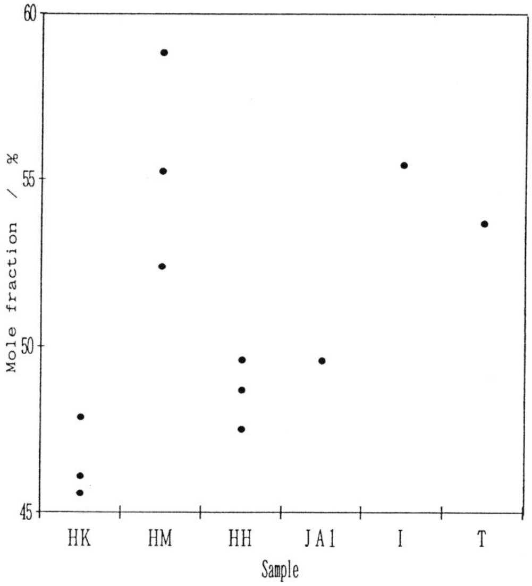

Figure 2 Mole fraction of alanyl residues determined from the spectra in Figure 1. HK: Kiyohira, HM: Motohira, HH: Hidehira, JA1: modern Japanese silk, I: modern Italian silk, and T: modern Thai silk

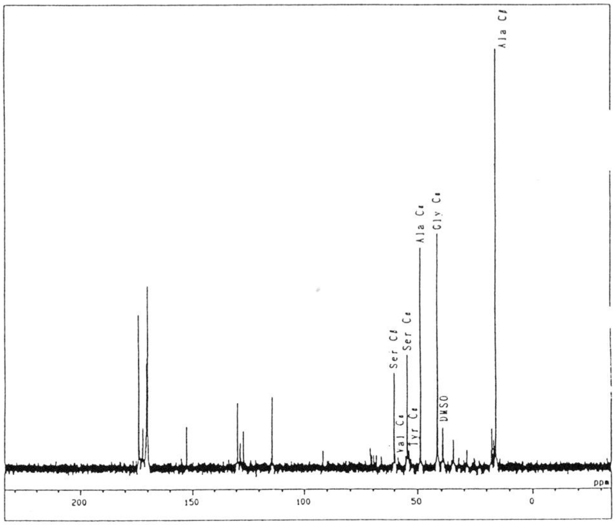

Figure 3 An example (Gland-28℃-Female) of solution-state ^{13}C NMR spectrum

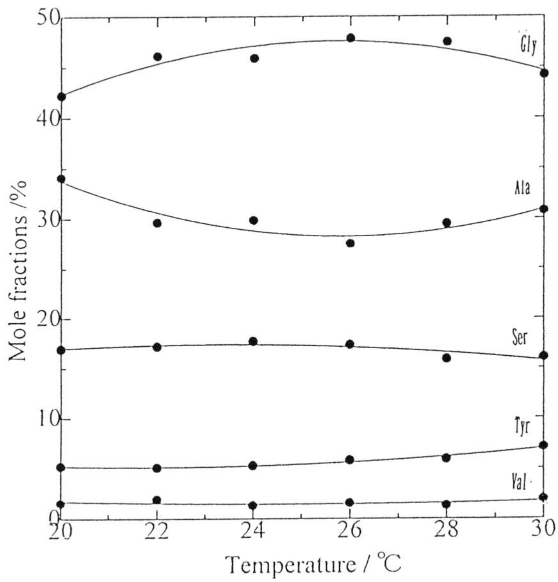

Figure 4 Normalized mole fractions obtained from NMR spectra for the silkgland of male

Materials Science and Engineering Serving Society
S. Sōmiya, R.P.H. Chang, M. Doyama and R. Roy (Editors)

Optical Properties of the Natural Quantum-Well System $(C_6H_5C_2H_4NH_3)_2(CH_3NH_3)_{m-1}Pb_mX_{3m+1}$ (X; halogen)

Nobuaki KITAZAWA and Shigeki KASHIWAMURA
Department of Materials Science and Engineering, National Defense Academy,
1-10-20 Hashirimizu, Yokosuka, Kanagawa 239, Japan

Thin films of microcrystalline $(RNH_3)_2(CH_3NH_3)_{m-1}Pb_mX_{3m+1}$ (m=1 and 2) were successfully fabricated on glass substrate by the spin-coating technique. In these films, strong exciton band was observed even at room temperature. With increasing the number of inorganic layers, m, red shift in the exciton band as well as the band gap was observed.

1. INTRODUCTION

The optoelectronic properties of excitons in low-dimensional structures have been growing interest due to their interesting quantum functions. The self-organized layered perovskite compounds, with the general formula $(RNH_3)_2(CH_3NH_3)_{m-1}Pb_mX_{3m+1}$ (R: alkyl group, X: halogen, m=1-3), are typical natural quantum-well systems consisting of metal halide semiconductor layers sandwiched between organic insulator layers.[1] Owing to their large exciton binding energy induced by the dielectronic and quantum confinement effects, these compounds show pronounced excitonic effects such as photoluminescence (PL) and large third-order optical nonlinearity.[2-4] Further, the electronic band structure of these materials can be controlled by (1) replacement of halogen and (2) addition of CH_3NH_3X.[5-7] In this study, optical properties of microcrystalline $(RNH_3)_2(CH_3NH_3)_{m-1}Pb_mX_{3m+1}$ and their mixed crystals were investigated.

2. EXPERIMENTAL PROCEDURE

Thin films (~20nm) of microcrystalline $(RNH_3)_2(CH_3NH_3)_{m-1}Pb_mX_{3m+1}$ (R; $C_6H_5C_2H_4$, X; halogen and m=1 and 2) were prepared on SiO_2 glass substrates using the spin-coating method as reported previously.[5] These films were characterized by X-ray diffraction (XRD) using CuKα radiation and optical density

(OD) measurements at room temperature. Note that the incident light beam is kept perpendicular to the substrate surface for the OD measurement.

3. RESULTS

It should be noted that the films fabricated on glass substrate are highly oriented, with the c-axis perpendicular to the substrate surface.

Figure 1 shows the optical density spectra of (a) $(RNH_3)_2PbX_4$ and (b) $(RNH_3)_2(CH_3NH_3)Pb_2X_7$ (X: Cl, Br and I) films measured at room temperature. In the $(RNH_3)_2PbX_4$ films, strong absorption peaks with narrow bandwidth were observed in the Visible to UV regions. The $(RNH_3)_2(CH_3NH_3)Pb_2X_7$ films also showed clear absorption bands. These absorption bands have been attributed to the excitons which were formed in the $[Pb_mX_{3m+1}]^{4-}$ layers by the transition of Pb^{2+}(6s) to Pb^{2+}(6p).[8-10]

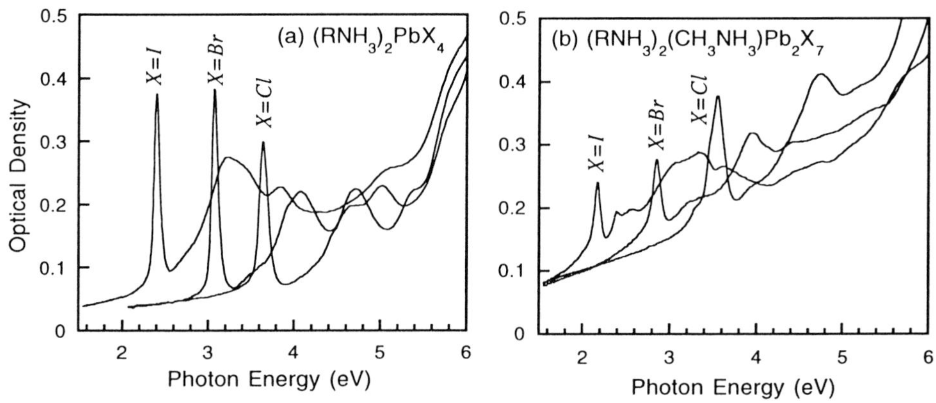

Fig. 1. Optical absorption spectra of (a) $(RNH_3)_2PbX_4$ and (b) $(RNH_3)_2(CH_3NH_3)Pb_2X_7$ (X: Cl, Br and I) films measured at room temperature.

Peak positions of the exciton absorption bands of $(RNH_3)_2(CH_3NH_3)_{m-1}Pb_mX_{3m+1}$ films are shown in Fig. 2. The positions of the $(C_6H_5C_2H_4NH_3)_2Pb(Cl, Br, I)_4$ mixed crystals are also plotted in the figure.[7] In the $(RNH_3)_2PbX_4$ films, the peak position of the exciton absorption bands shifted towards high photon energy by the replacement of halogen species. With increasing the number of inorganic layers, *m*, the red shift in the exciton band was occurred.

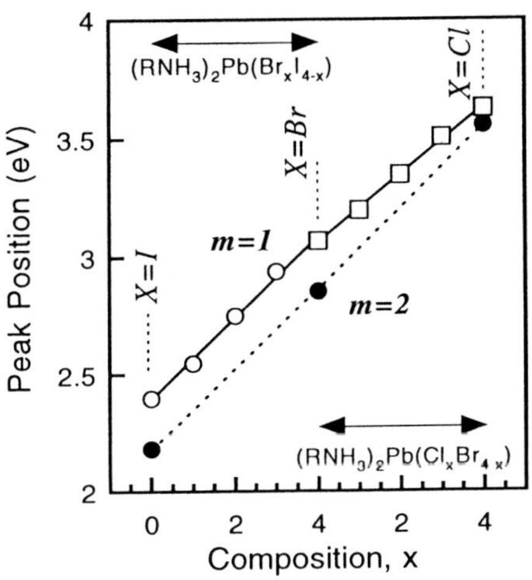

Fig. 2. Peak position of the exciton absorption band of $(RNH_3)_2PbX_4$ and $(RNH_3)_2(CH_3NH_3)Pb_2X_7$ (X: Cl, Br,I) films.

4. DISCUSSIONS

In this study, preparation optical properties of microcrystalline $(RNH_3)_2(CH_3NH_3)_{m-1}Pb_mX_{3m+1}$ films were were investigated. In the $(RNH_3)_2PbX_4$ films, the peak position of the exciton absorption bands shifted towards high photon energy depending on the halogen species. It is well known that the fundamental transition in $(RNH_3)_2PbI_4$ is attributed to the 6s-6p transition of Pb^{2+} and is similar to that in PbI_2.[11] In PbI_2, the top of the valence band is composed of the Pb(6s) orbital which is hybridized with the I(5p) orbital, and the bottom of the conduction band mainly has Pb(6p) character. In PbX_2 (X=Br, Cl and F), the energy level of the valence band decreases because the energy level of np orbital (n=2~4) of these halogens are relatively low compared with that of the I(5p) orbital. Therefore, the blue shift in the exciton band observed for the $(RNH_3)_2PbX_4$ films was attributed to the hybridization effect of X(np) orbital into the Pb(6s) orbital. In the $(RNH_3)_2(CH_3NH_3)Pb_2X_7$ films, in contrast, the red shift in the exciton band was occurred with increasing the number of inorganic layers. In these films, both the thickness and dielectric constant of the organic insulator (ε_b=2.34) layers remained unchanged.[11] In the inorganic semiconductor layer, on the contrary, the well layer thickness increases two times compared with $(RNH_3)_2PbX_4$ films. Regarding the dielectric constant of the well layers, ε_w, the value of ε_w are 6.05 for $(RNH_3)_2PbI_4$ and 6.48 for $(RNH_3)_2(CH_3NH_3)Pb_2I_7$ crystals.[12] Therefore, the dielectric confinement effect of $(RNH_3)_2(CH_3NH_3)Pb_2X_7$ crystals is larger than that of $(RNH_3)_2PbX_4$ crystals. However, the quantum confinement effect of $(RNH_3)_2(CH_3NH_3)Pb_2X_7$ crystals is equal to half that of $(RNH_3)_2PbX_4$ crystals because of the well layer thickness. Consequently, the weaker quantum confinement effect of $(RNH_3)_2(CH_3NH_3)Pb_2X_7$ films than $(RNH_3)_2PbX_4$ films leads to shift in the exciton band towards low photon energy.

5. SUMMARY

Thin films of microcrystalline $(RNH_3)_2(CH_3NH_3)_{m-1}Pb_mX_{3m+1}$ (m=1 and 2) were successfully fabricated on glass substrate by the spin-coating technique. The exciton properties of these materials were strongly influenced by the halogen species and number of the inorganic layers. The blue shift in the exciton band observed for the $(RNH_3)_2PbX_4$ films was attributed to the hybridization effect of X(np) orbital into the Pb(6s) orbital. With increasing the number of inorganic layers, m, red shift in the exciton band as well as band gap were observed.

ACKNOWLEDGEMENT

This work was supported in part by Kanagawa Academy of Science and Technology.

[1] J. Calabrese, N. L. Jones, R. L. Harlow, N. Herron, D. L. Thorn and Y. Wang, *J. Am. Chem. Soc.* **113** (1991) 2328.
[2] X. Hong, T. Ishihara and A. V. Nurmikko, *Phys. Rev.* **B45** (1992) 6961.
[3] M. Era, S. Morimoto, T. Tsutsui and S. Saito, *Appl. Phys. Lett.*, **65** (1994) 676.
[4] X. Hong, T. Ishihara and A. V. Nurmikko, *Solid State Commun.*, **84** (1992) 657.
[5] N. Kitazawa, *Jpn. J. Appl. Phys.*, **35** (1996) 6202.
[6] N. Kitazawa, *ibid*, **36** (1997) 2272.
[7] N. Kitazawa, *Mater. Sci. Eng. B* in print.
[8] T. Ishihara, J. Takahashi and T. Goto, *Phys. Rev.* **B42** (1990) 11099.
[9] E. Doni, G. Grosso, G. Harbeke, E. Meier and E. Tosatti, *Phys. Status Solidi.* **B68** (1975) 569.
[10] M. Fujita, H. Nakagawa, K. Fukui, H. Matsumoto, T. Miyanaga and M. Watanabe, *J. Phys. Soc. Jpn.* **60** (1991) 4393.
[11] I. Ch. Schluter and M. Schluter, *Phys. Rev.* **B9** (1972) 1652.
[12] E. A. Muljarov, S. G. Tikhodeev, N. A. Gippius and T. Ishihara, *Phys. Rev.* **B51** (1995) 14370.

Materials Science and Engineering Serving Society
S. Sōmiya, R.P.H. Chang, M. Doyama and R. Roy (Editors)

What I Learned from My Forty-years' R&D Experience on Ceramics and Refractories

Kiyoshi Sugita

R&D, Nippon Steel Corporation
3-35-1 Ida, Nakaharaku, Kawasaki, 211 Japan

Abstract

History is a great mentor. This saying can be applied to the history of career persons, which could be effective to the education or self-development of young scientists and engineers including college students.

The author, Dr. K.Sugita, Executive Advisor, R&D, Nippon Steel Corp., has been working mainly on ceramics and refractories since he joined the company in 1954. He realizes that he has been learning really a lot during the past long years, and believes that some of the lessons he took can be utilized even today as guidelines or suggestions for the younger generation especially who want to work in materials R&D sectors in industry.

The major lessons the author has picked out by looking back over his years are :

(1) Form a clear-cut and encouraging target from your work.
(2) Make your job enjoyable by getting interested in it.
(3) Expose yourself to different cultures and environments.
(4) Have a historical view over your field.
(5) Solidify the basics or fundamentals in your area.
(6) Diversify or broaden your specialty.
(7) Develop yourself balancedly as a company member, a professional, and a citizen.

In the paper, the backgrounds for those lessons are described together with author's experiences.

How did I start my career ?

My major in college was Applied Chemistry. In the 1950s, organic chemistry related to petrochemicals or synthetic polymers was most popular among the students. Inorganic chemistry did not look attractive to us. And I wanted to do research for my diploma paper on some subject based on organic chemistry. Unfortunately, however, too many students followed the trend, applying to organic-chemistry laboratories. We had to decide our diploma laboratories by drawing a lottery, which eventually allocated me to a laboratory for inorganic materials such as refractories and cements. My graduation thesis was on "Property Measurements of Chemically-Bonded Chrome-Magnesia Bricks Imported from the U.S."

I joined Nippon Steel (then Yawata Steel) in April 1954, still hoping to get an organic chemistry-based job like the coal chemicals processing. After two-month introductory training, I was interviewed to finalize the job assignment for a newly-employed college graduate. The Personnel Section Manager said, "You studied refractories in college. We've been waiting for you. As you know, we're going to introduce a new steelmaking process, and the refractories will make a key to our success in the new technology. Good luck with your job at the Refractories Department!" That was the very moment when I started off my career as a refractories engineer.

Lesson 1 : Form a clear-cut and encouraging target from your work

My first task in my career was surprisingly a big one — to develop the manufacturing process of the tar-bonded dolomite brick for the LD converter. The technologies needed for the new steel- making were entirely transferred from Europe, except for the refractories. So we had to start from scratch. And the deadline for commercializing the brick production was not far away in the future.

Eventually we succeeded in it. Even now I feel very proud that I, just a young researcher, was able to play a key role in such a big project. At that time I had to carry out a lot of different work concurrently. Of a number of factors having led to the success in addition to some luck, the most contributing one was that the target or goal had been set very clearly, timewise and technologywise. It also provided us with a strong sense of mission, which heavily pressed on us and highly motivated us as well.

This project gave me an opportunity to work outside my company. I was sent to Kurosaki Refractories Co., working there for about a year to help build and start up a commercial dolomite brickplant. It was a valuable experience.

Lesson 2 : Make your job enjoyable by getting interested in it

In the course of the R&D activities on the tar-bonded dolomite brick, I had been gradually attracted to refractories technology. When I began working back at Yawata Steel's Research Lab. in 1958, I got really crazy about doing research on refractory materials.

I think I enthusiastically did a lot of highly original researches, especially on the high-temperature behaviors of refractories : the wettability of refractories with molten steels, the wear mechanism of dolomite bricks in LD converters, an EPMA study on direct-bond structures in chrome-magnesia bricks, to mention a few.

In the early 1960s, my interest and job extended to the surface treatment of steel with ceramics : flame-sprayed ceramic coatings, porcelain enamels, and the like. I enjoyed doing research on different things. I am curious about many things even now, let alone when I was much younger.

If you enjoy your work, I can assure you that you will come up with a lot of good ideas. You should try to make your job enjoyable.

Lesson 3 : Expose yourself to different cultures and environments

In general, every Japanese company has its own culture —— management philosophy, people's mentality, workplace practice, and so forth. I learned a great deal during my working at Kurosaki Refractories, and I had a good chance to objectively look at Nippon Steel from outside and to copy Kurosaki's good points.

So far I have visited foreign countries more than 30 times, meeting with a great variety of characters and enjoying different cultural and historical attractions overseas. Learning foreign languages, I believe, has a similar effect of stimulating and activating yourself in different cultures.

I have long been interested in a historical fact that many of great inventors and discoverers, including the Novel Prize winners, had been strongly affected by multicultural backgrounds. Creativity, which is one of the most important or requisite capabilities for R&D activities, seems to have a close relation with the diversity of your surroundings.

Lesson 4 : Have a historical view over your field

In 1959 I happened to get curious about old literatures on ceramics and refractories. First I chose the Journal of the American Ceramic Society. Kicking off with the Journals published in the 1950s, I read on, only some papers carefully and others very roughly, sequentially back to those in the 1920s. I finished it with the Journals for 1921, the oldest ones available then at the library of Yawata Steel's Research Lab.. I did the same with other major journals. Such reading helped me effectively form a historical view about ceramics and refractories.

Some senior researchers I knew were quite against reading old literature because it would restrict researchers' flexible thinking, saying "You can be more knowledgeable, but less creative." But I am of the opinion that you will be able to find even the future in the history and develop your creativity or intuition based on your historical view. Recently I published a book titled "Refractories for Iron and Steel-making : A History of Battles over High-Temperatures" (Chijin Shokan Co., in Japanese, 1995.) Writing the book has made me more confident in my opinion.

Lesson 5 : Solidify the basics or fundamentals in your area

In addition to the dolomite brick, I was assigned to the development of the fireclay brick for blast furnaces in 1955. The raw materials for the brick were mainly imported from the U.S. and China. I was very interested in the property difference between domestic and overseas fireclays. I keenly studied the basics of clay minerals including their crystallographic characteristics, formation mechanisms and hydrothermal reactions. Since then, I have always been trying to learn the basics behind my subject whenever I got involved in a new theme.

One of the most noticeable current trends is that your professional knowledge is acceleratedly depreciating due to the recent rapid change or advance in science, technology and industry. According to some studies, the half-value period of the knowledge in my specialty is estimated about 10 years when I joined Nippon Steel, compared with 5 years for today.

I am sure that enhancing the basics or fundamentals for your specialty areas will be one of the most reliable and effective measures for you to survive such a fast-changing world. Basics always works well and pays off, even in the distant future and in new other areas.

Lesson 6 : Diversify or broaden your specialty

History teaches us that many of inventions or new technologies were made by combining known principles or techniques, and that promising research areas very often lie in the boundaries between the existing specialized areas. In order to promote interdisciplinary approaches, each individual researcher should diversify his specialty areas in addition to more enhanced command of his core specialty. This attitude is also needed for better collaborative activities with people from other areas.

The flame-spray gunning repair method can be, I believe, one of the big technologies developed by Nippon Steel. About 30 years ago, I happened to come up with the basic concept for the method by linking the flame-spray ceramic coating method for steel sheets to the traditional gunning repair process for steelmaking furnaces.

I know a collage professor who majored in Biology for his B.S. and Electronics for his M.S., now doing a lot of creative R&D in computer technology. In the U.S., much more students seem to take such cross-major courses as above. In Japanese companies, job rotations and organizational reformations can often offer you good chances to diversify or broaden specialty.

Lesson 7 : Develop yourself balancedly as a company member, a professional, and a citizen.

When I recollect how I have been following my career, I really feel obliged for the encounter with so many attractive personalities who achieved great success in their R&D both inside and outside Nippon Steel. One of the most influential lessons I took from them is that they are all honorable citizens, always making efforts to improve their entire personality. I am convinced of the close relation between their success and personality.

Several years ago, I wrote a book, titled "Towards A Career in Industrial R&D : Some Guidelines for Young Scientists and Engineers" (Japan Technical Information Service, in Japanese, 1991). Its Chinese version, translated by Prof. Na Bao Kui and his son, was also published in Beijing in 1995. In the book, I placed a special emphasis on the target of your self-development, suggesting that you should not confine your target to some specific aspects of your ability or personality as a specialist.

Ethics for scientists and engineers is drawing public attention these days throughout the world. I would like to conclude the paper by stressing that you should make a better member of society in addition to following a great career in the R&D world.

Materials Science and Engineering Serving Society
S. Sōmiya, R.P.H. Chang, M. Doyama and R. Roy (Editors)
1998 Elsevier Science B.V.

CF_4 Plasma Etching of Ta-Al Alloy Thin Films

Seung Ho Shin[a], Yong-Sun Chung[b] and Keun Ho Auh[a]

[a]Department of Ceramic Engineering, Hanyang University, Seoul, 133-791, Korea

[b]Ceramic Processing Research Center(CPRC), Hanyang University, Seoul, 133-791, Korea

Reactive ion etching(RIE) of Ta-Al alloy thin film with the thickness of 1000 Å was studied. It was confirmed that CF_4 gas could be effectively used to etch the Ta-Al alloy thin film. The etching rate of the thin film which the molar ratio is 1:1 is about 67 Å/min. The etching rate of the SiO_2 layer was 12 times faster than that of the Ta-Al alloy thin film. In addition, it was observed that photoresist of AZ5214 was more useful than Shiepley 1400-27 in RIE with the CF_4 gas.

1. INTRODUCTION

Ta-Al metal alloy is used as the important material for imaging system. Ta-Al metal alloy, which the molar ratio is 1:1, has a strong chemical stability. Also, the alloy keeps a constant resistance while Al-Si(1 wt%) metal alloy shows a linear variation with a temperature. It is simply considered because of the interaction between the refractoriness of Ta and the high conductivity of Al.

When Ta-Al metal alloy is applied to a display device, the alloy film typically has to be etched to micro-dimensions. There are two types in etching process; one is the wet etching and the other is the dry etching. If the wet etching is applied to etch Ta-Al alloy, it is very difficult to obtain an uniform etching because the etching rates between Ta and Al are different and the etching solution of each is also different. Therfore, the dry etching is used as a method to form a micro-pattern on the Ta-Al alloy film[1].

There are many methods in the plasma etching - for example, sputtering, reactive ion etching, ion beam etching(IBE), etc. Especially, the reactive ion etching(RIE), which has a high etching rate and a good selectivity (compared with sputtering and IBE), is a effective method. However, a selection of gas for etching is very critical. In case of alloy etching, the characteristics of the alloy elements are very different so that the selection of a etching gas is very strictly limited[2].

In this study, CF_4 gas, which is used for an etching gas of refractory metal, is chosen as a etchant and its etching characteristics on the Ta-Al alloy thin films is observed.

2. EXPERIMENTAL PROCEDURE

All silicon wafers discussed in this study were Boron-doped (100) with the thickness of 300 ㎛ and the diameter of 10.16 cm. Ta-Al alloy thin film (thickness = 1000 Å) was deposited on SiO_2 layer (1.6 ㎛), which was grown on semi-insulating Si substrates by a thermal chemical vapor deposition method, by a RF sputtering method. Photoresists, Shiepley 1400-27 and AZ5214, were spin-coated at the speed of 4000 rpm for 30 sec. The coating thickness of Shiepley 1400-27 was Wafers were lithographically patterned with Shiepley 1400-27 and AZ5214 photoresist to produce features with dimensions between 5 ~ 40 ㎛. The etching was performed in a NE500c system and etching conditions were listed in Table 1.

The etching rates were measured by an α-step after the removal of the resist by ALEG 355 PR stripper. Scanning electron microscopy (SEM) and α-step were used to examine the surface morphology and the etching depth of the samples.

Table 1. The Plasma etching condition of Ta-Al alloy thin film

Gas flow rate (CF_4)	60 sccm
Pressure	30 mTorr
In. Power	600 watts
Re. Power	0.7 watts
Bias Voltage	400 volts

3. RESULTS and DISCUSSION

Figure 1 shows the SEM micrographs of the tilted cross-sectional features etched with CF_4 at 30 mTorr. Fig. 2(a), 2(b) and 2(c) shows that the etching is not completed while Fig. 2(e) shows overetching. Fig. 2(d) shows that the etching depth is 1000 Å. The etching rate of Ta-Al alloy thin film is about 67 Å/min. The overetching time of SiO_2 layer is 5 min and the measured etching depth is 4000 Å. The etching rate of SiO_2 layer is 800 Å/min and 12 times faster than that of Ta-Al alloy thin film.

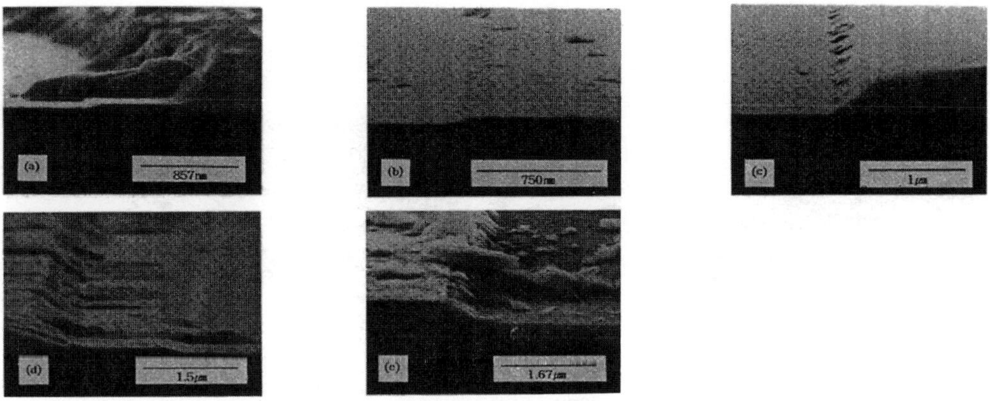

Figure 1. SEM micrographs of cross-sectional features etched using CF_4 60 sccm at pressure 30 mTorr and Chuck bias voltage 400 volts ; (a) 5 min, (b) 10 min, (c) 13 min, (d) 15 min and (e) 20 min.

Fig. 2 shows the variation of etch rate as a function of time. The etching rate is decreasing as time increases. The etching rate, however, is increasing after 13 min. It is considered that the chemical etching begins to act mainly after 13 min. In RIE process, there are both the physical etching and the chemical etching. In the beginning of the etching, the physical etching is dominant as the main mechanism. However, after a critical point, the chemical etching becomes the main mechanism[2].

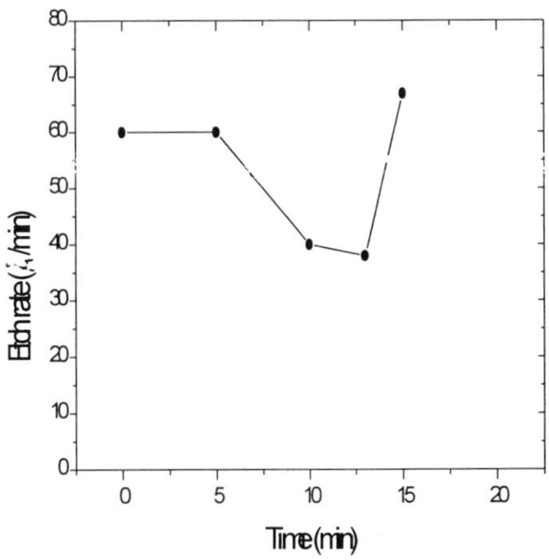

Figure 2. Etching rate of Ta-Al metal alloy as a function of time in CF_4 60 sccm at pressure 30 mTorr and chuck bias voltage 400 volts

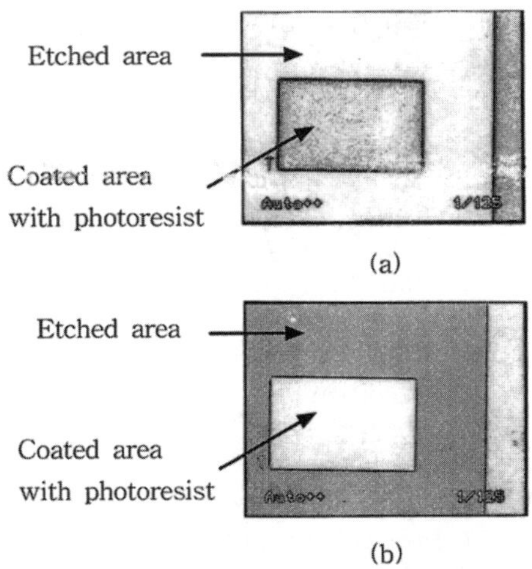

Figure 3. Micrographs of features after the photoresist stripping using ALEG 355 PR stripper ; (a) Shiepley 1400-27 and (b) AZ5214.

Fig. 3 shows the micrographs after the photoresist is removed. ALEG 355 PR stripper is used without rubbing or ultrasonic vibration. Shiepley 1400-27 is not stripped while AZ5214 is stripped completely. The coating thickness of Shiepley 1400-27 is 1.3 μm and that of AZ5214 was 1.2 μm. The stripping is performed in same conditions but the results were very different. According to this result, it was considered that the photoresist of AZ5214 was more useful than Shiepley 1400-27 in RIE with the CF_4 gas.

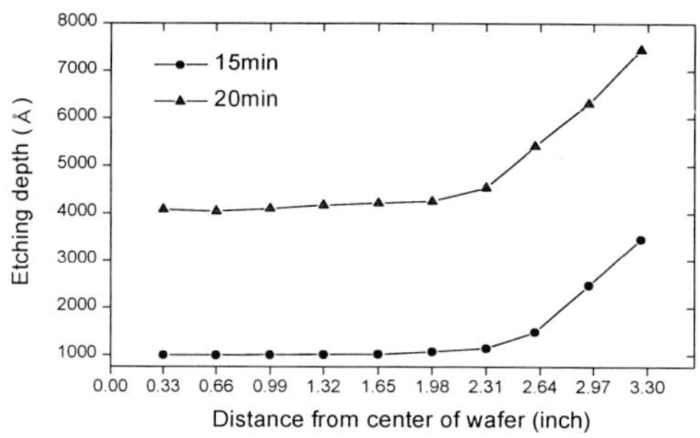

Figure 4. Distribution of the etching depth in wafer.

Figure 4 shows the uniformity on a wafer by α-step scan. In case of RIE for 15 min, the etching depth is constant while the etching depth for 20 min is not uniform. The etching uniformity is affected by the deposition conditions of a thin film and a gas flow rate. It is considered that the non-uniformity is originated from the large gas flow rate(60 sccm)[3].

4. CONCLUSIONS

RIE using CF_4 gas is effective for Ta Al alloy thin film and the etch rate is about 67 Å/min. It is achieved at conditions of CF_4 60 sccm, 30 mTorr. The etching rate of SiO_2 layer was 800 Å/min, 12 faster than that of Ta-Al alloy thin film. It is also observed that the photoresist AZ5214 was more useful than Shiepley 1400-27 in RIE with the CF_4 gas. The uniform etching on a wafer is achieved when the etching is performed at the conditions of CF_4 60 sccm, 30 mTorr, 15 min.

REFERENCES

1. S. Wolf and R. N. Tauber, "Silicon Processing for the VLSI Era Vol. 1-Process Technology", Lattice Press, P542, 1986
2. T. J. Bisschops and F. J. deHoog, "On the Plasmaphysics of Plasma Etching," *Pure and Appl. Chem.*, 57, 1311 (1985)
3. D. A. Danner and D. W. Hess, "The Effect of T and Flow Rate on Al Etch Rates in RF Plasmas," *J. Electrochem. Soc.*, 133, 151 (1986)

Subject Index